SACRED WORLDS

This book, the first in the field for two decades, looks at the relationships between geography and religion. It represents a synthesis of research by geographers of many countries, mainly since the 1960s.

No previous book has tackled this emerging field from such a broad, inter-disciplinary perspective, and never before have such a variety of detailed case studies been pulled together in so comparative or illuminating a way. Examples and case studies have been drawn from all the major world religions and from all continents. Many historical examples complement the contemporary ones in this wide-ranging review.

Major themes covered in the book include the distribution of religion and the processes by which religion and religious ideas spread through space and time. Some of the important links between religion and population are also explored. A great deal of attention is focused on the visible manifestations of religion on the cultural landscape, including landscapes of worship and of death, and the whole field of sacred space and religious pilgrimage.

Chris C. Park is Senior Lecturer in Geography and Principal of the Graduate College of Lancaster University.

SACRED WORLDS

An introduction to geography and religion

Chris C. Park

London and New York

First published 1994
by Routledge
11 New Fetter Lane, London EC4P 4EE

Transferred to Digital Printing 2003

Simultaneously published in the USA and Canada
by Routledge
29 West 35th Street, New York, NY 1001

Typeset in Garamond by
Mathematical Composition Setters Ltd
Printed and bound by Antony Rowe Ltd, Eastbourne

British Library Cataloguing in Publication Data
A catalogue record for this book is available from the British Library

Library of Congress Cataloging in Publication Data
Park, Chris C.
Sacred worlds : an introduction to geography and religion/Chris Park.
p.   cm.
Includes bibliographical references and index.
1. Sacred space. 2. Religion and geography. 3. Human geography.
4. Geographical perception. I. Title.
BL580.P37 1994
291.3'5--dc20   93-36709
CIP
ISBN 0-415-09012-1   0-415-09013-X (Pbk)

This book is dedicated to
Elizabeth Ann Park
with love

CONTENTS

FIGURES

ix

TABLES

TABLES

PREFACE

Not for the first time have I tried to find a book which offered a broad introduction to an area of geographical inquiry, and not for the first time have I set about the task of writing it myself when the search failed! My aim has been simply to pull together and package in a sensible framework a wealth of geographical research and writing on the theme of religion which has appeared particularly over the last decade.

The central theme of this book is geography *and* religion, not the geography *of* religion. The distinction is important, because whilst there are many fascinating questions to be answered concerning spatial patterns of religions, these represent only a part of the field. To focus only on the geography of religion would be to overlook many important geographical themes, and it would be a travesty to the many geographers whose work explores religion in other contexts.

The book neither assumes nor requires any particular religious belief or sympathy of its readers. Neither does it seek to endorse any particular religion and criticise others. I have tried hard to remain objective throughout, and treat all of the themes with an even hand.

Doubtless most readers will find some sections over-played and others underplayed, and many may be surprised at the inclusion of some themes and the exclusion of others. Inevitably in a book of this sort the selection partly reflects the author's interests, but the main constraint has been the availability of geographical work on different themes. I have tried to capture the diversity of work within the field, and I hope that the book might offer a map of interesting and important themes which readers will find useful.

A note on terminology. In referring to dates, I have adopted the recent convention of referring to dates more than 2,000 years ago as BCE (Before Christian Era) rather than BC (Before Christ) and dates less than 2,000 years ago as CE (Christian Era) rather than AD (Anno Domini). As Hinnells (1984a: 13) suggests, 'these make no assumptions about a person's religious position' and are less antagonistic to people of non-Christian religions than BC and AD.

ACKNOWLEDGEMENTS

Writing this book has been a great joy, and much of pleasure has come from working with and being surrounded by a special bunch of people. Many work at Lancaster University. I owe a great debt to the staff of the Inter-Library Loan section of the University Library for their invaluable assistance in getting hold of much of the material on which the book is based. My colleagues in the Department of Geography have once again tolerated my indulgence of exploring new territory, and were sensible enough not to enquire too often how the writing was coming along. Matthew Ball drew most of the figures, with flourish and enthusiasm, and some figures were drawn by Claire Jarvis who is now teaching's gain and cartography's loss. I have been working on the book for over 18 months, but benefited greatly from a term's study leave in which to complete the writing.

I am grateful to Tristan Palmer at Routledge for his encouragement, his commitment to this book, and his belief throughout that I could write it and he could defend it! I owe special thanks to David Livingstone (School of Geosciences, The Queen's University, Belfast), David Ley (Department of Geography, University of British Columbia) and Joan Taylor (Department of Religious Studies, University of Waikato) for their many helpful suggestions on how to improve the manuscript, and accept full responsibilities for any errors or inaccuracies which remain.

As ever, my immediate family have met most of the personal costs of writing a book and they have done so with customary generosity. My son Sam and daughter Elizabeth think nothing of their father being glued to the computer all day long, and must often wonder how other kids' dads spend their days. Angela, the powerhouse of the Park family machine, has long since come to understand and accept how preoccupied I am while writing, and is getting very skilful at juggling often irreconcilable demands on her time, space and energy. She is the true hero of the piece, without whom it would have been inconceivable for me to even think of tackling this venture. I hope she is pleased with the outcome!

Chris Park
Lancaster

1

GEOGRAPHY AND RELIGION
Context and content

In matters of religion and matrimony I never give any advice; because I will not have anybody's torments in this world or the next laid to my charge.

Earl of Chesterfield, Letter to A.C. Stanhope, 1765

INTRODUCTION

At first glance, geography and religion seem to be curious bedfellows. However, even a brief reflection reveals a myriad of ways in which the two interact – religion affects people and their behaviour in many different ways, and geographers have traditionally been concerned with the spatial patterns, distributions and manifestations of people and environment. Glacken notes how,

> in ancient and modern times alike, theology and geography have often been closely related studies because they meet at crucial points of human curiosity. If we seek after the nature of God, we must consider the nature of man and the earth, and if we look at the earth, questions of divine purpose in its creation and of the role of mankind inevitably arise.
>
> (Glacken 1967: 35)

None the less, the study of geography and religion remains peripheral to modern academic geography. The Earl of Chesterfield was doubtless correct in insisting that 'religion is by no means a proper subject of conversation in a mixed company', but the real reason for this marginality lies more in the assumed rationality of post-Enlightenment science, which dismisses as irrational (thus undeserving of academic study) such fundamental human qualities as mystery, awe and spirituality – reflections of the very essence of humanness.

Yet a geography that ignores what we might call 'the supernatural' neglects some of the most deeply rooted triggers of human behaviour and attitudes, is blind to some critical dimensions of humanity and overlooks some profoundly significant implications of geographical patterns of human activity and behaviour. It is not the intention of this book to advance any particular religion or belief system, nor to propose that geographers should become religious to advance their subject. However, the book *does* have a missionary objective – and that is to bring the study of geography and religion back on to the geographical agenda, by raising awareness of the richness and diversity of work in the field and highlighting emerging themes and approaches.

A clarion call of some leading geographers recently has been to 'reclaim the high ground' for geography, by turning away from empirical reductionism and arid theoretical debate, to the 'big questions' that trouble society (Stoddart 1987). Surely questions concerning ultimate meanings and purpose, humanity's very reason for existence, human suffering and inequality deserve a place on the geographical agenda.

Before we look at how the field of geography and religion has evolved, it is perhaps useful to provide a broad context for the field by reflecting on some of the many ways in which religion affects people and environment.

INTERACTION

This review is not intended to be exhaustive, and many of the themes will be explored in greater depth in subsequent chapters. It does, however, serve to illustrate the manifest variety of links between geography and religion, and perhaps to beg the question 'Why is the field not more prominent within contemporary geography?' Examples are drawn from a variety of sources (including Broek and Webb 1973; Morrill and Dormitzer 1979; Wynne-Hammond 1979; de Blij and Muller 1986; Jordan and Rowntree 1990).

Geographical distribution of religions

One of the most obvious areas of interaction is the geographical distribution of religions. This is the focus of Chapter 3, but we can note in passing that each of the major world religions tends to have its own geographical range and territory. For example, Christianity is most common in Europe, America and other regions of European settlement, and Islam is dominant in the Middle East, northern Africa and western Asia. Buddhism is concentrated in central Asia, and India is predominantly Hindu. Even at this coarsest of spatial scales, it is evident that religion exerts powerful influences on human activities and patterns.

Religious imprint on the cultural landscape

Religion is often strongly imprinted on the cultural landscape, through distinctive styles of architecture (see Chapter 7). The most obvious imprints are centres of religious worship (such as mosques, temples, churches and cathedrals) and other religious symbols (shrines, statues) and structures (cemeteries), which

2

often give a distinctive identity and character to an area. Many settlements, such as the cathedral or abbey towns of Europe, were founded and have evolved for religious reasons.

Many religions recognise sacred space and sacred places such as caves, groves, lakes, mountains (such as Mount Fuji and Mount Sinai) and rivers (such as the Ganges and Jordan). These reflect the formative influence of environment in the evolution of different religions and in turn encourages the preservation of sacred landscapes (see Chapter 8, pp. 249–57). Landscape also bears the imprint of religion through the duplication of facilities by different religions or denominations. For example, many countries (such as Britain and the United States) have separate school systems for Roman Catholics and Protestants, and in Israel there are separate schools for Jews and Arabs.

Impacts of religion on lifestyle and commerce

The list of interactions between religion and geography goes much further than just visible landscape elements. Spatial variations in religious belief influence and are influenced by social, economic, demographic and political patterns in many different ways. Religion prohibits certain activities, restricts others, and encourages others. So as well as shaping people's philosophy of life, religion also exerts powerful influences on their behaviour and patterns of activity.

Visible manifestations of religion on lifestyle include the adoption of codes of dress (such as the wearing of veils by many Muslim women, and turbans by Sikh men), and personal habits (such as the beards worn by Sikh and Jewish men). Differences in religion within a country might be associated with differences in language, ethnic identity, educational achievement and opportunity, economic security, occupational distribution.

Religion can exert strong influences over commerce. In medieval Europe, for example, the Christian Church was strongly opposed to money-lending at interest (usury), and because Jews were not bound by these religious rules they took on the role of money-lenders. Until quite recently banking institutions have not developed among Muslims because the Prophet prohibited acceptance of interest from borrowers. On the other side of the coin, literally, are the vast sums of money exchanged by religious pilgrims to holy sites (see Chapter 8, pp. 258–85). Pilgrimage plays a significant role in the economy of religious centres such as Mecca in Saudi Arabia, Lourdes in France, and Banaras in India. Religion can also strongly influence what type of employment a person has, particularly in Hindu society where caste prescribes certain duties and occupations by birthright rather than suitability.

Religious taboos on food and wildlife

Religious beliefs, particularly amongst the primitive religions, often include taboos on the use of certain plants and animals. Some plants, for example, are

3

cult symbols of specific religions – examples include the lotus and pipal (bo) tree in Buddhism, the conifer in Shinto, and the oak and spruce in ancient Germanic ritual (hence the Christmas trees widely used today).

As well as encouraging the preservation of specific species of plants and animals, such taboos can determine which crops and livestock are raised by farmers (Simoons 1961). Pig-rearing is uncommon in Muslim countries and Jewish areas, for example, because Muslims and Jews regard the pig as an unclean animal that cannot be eaten. Similarly Buddhism prohibits stock-raising for meat and wool, thus creating a distinctive farming landscape in Japan and dominantly Buddhist parts of India, China, and Sri Lanka. Hindus do not generally eat fish, eggs or meat, and this also translates into visible differences in the farming landscape.

The distribution of wine-growing areas has also been associated with religious factors. Stanislawski (1975) suggests that the diffusion of vines across southern Europe almost 3,000 years ago was closely associated with the spread of the Great Mother cult (with which Dionysus – later to be the Greek god of wine – was affiliated) in pre-Christian Europe. Religious prohibition prevents Muslims from drinking wine, and so eastern parts of the Mediterranean – which have favourable climates but are predominantly Muslim – have no well-developed wine industry. The heavy dependence of post-colonial Algeria on its traditional wine industry for valuable foreign exports has created serious dilemmas for this Islamic country, and encouraged the conversion of many vineyards to other land uses (Sutton 1990). The dispersion of the citron from Palestine by the Romans has also been associated with religious diffusion because the citrus fruit was used in some Jewish celebrations (Isaac 1959a, 1959b).

Religion and demography

Some of the strongest imprints of religion on human geography arise through demographic constraints and practices (see Chapter 6, pp. 169–76). Most aspects of population dynamics, from cradle to grave, can be heavily influenced by religious belief. Many religious groups (most prominently the Roman Catholic Church) oppose contraception and encourage large families to safeguard continuity of belief. Islam does not officially oppose contraception, but it is custom in most Islamic countries to have large families. Consequently birth rates (and, in turn, population growth rates) are higher and family sizes are larger in countries where these religions dominate (such as Catholic Italy, Southern Ireland and Latin American countries) than they are elsewhere. Patterns of marriage in many countries are also influenced by religion. Islam, for example, permits polygamy for men whereas Christianity insists on monogamy and most high Hindu castes forbid remarriage of a widow. Inter-marriage between religious groups is also tightly controlled in some religions. Some Hindus and Muslims in central India still practise *purdah* (the seclusion

4

of women from society, particularly from males outside the family; marriages are often still arranged by parents of the same religion and caste).

Religion, politics and conflict

Some of the most widely talked about impacts of religion arise from the political conflict created by the coexistence of religious groups and minorities (see Chapter 6, pp. 180–4). This is reflected in geographical patterns at various scales, from the local (such as residential segregation of Catholics and Protestants in Belfast, and the persistence of Jewish enclaves in many Western cities), through the regional (such as the separation of Muslim and Christian areas in Lebanon, the former to the east and south and the latter to the north), to national conflicts (such as the partition of India in 1947 and the establishment of the state of Israel in 1948). The importance of religious differences in shaping the political climate within and between countries must not be underestimated. Brook (1979: 513) maintains that 'it would not be possible to understand many political events in the post-war world without taking into consideration the religious situation in the country in question'.

Religion and culture

Geographers have traditionally been interested in the culture of different places, and there is no escaping the fact that religion is a major determinant of culture. De Blij and Muller define religion as one of the

> foundations of culture [and] vital strands in the fabric of societies. . . . In many societies less dominated by technology than [the United States], religion is the great binding force, the dominant ruler of daily life. From eating habits to dress codes, religion sets the standards for the community.
>
> (de Blij and Muller 1986: 181)

For Jordan and Rowntree (1990: 190) 'religion is part of culture . . . [it] is an essential hue in the human mosaic' which geographers would be foolish to ignore. Religion has often been the basis for identifying cultural regions, such as The Mormon West and The South in the United States (Shortridge 1982: 177).

In many parts of the world (particularly in Africa and Asia, and throughout the Islamic world) religion is a vital and dynamic part of culture, which exerts powerful control over behaviour and attitudes. This is true not just for the people who profess and practise the religion in question, because a dominant religion can set norms for acceptable behaviour that even non-believers are expected to comply with. For example, the ban on drinking alcohol in Muslim countries applies to everyone, whether they are residents or visitors, Muslims or 'infidels'.

The powerful influence of religion on culture is perhaps nowhere more apparent than in India, where 'loyalty to one's religion has coloured the entire political, economic, social and moral life' (Mamoria 1956: 189). In India,

> being a Hindu . . . Muslim, Sikh, Jain, Parsee, or Christian conditions or even determines one's diet, dress, calendar, holidays, and, even more important, education, social status, and economic activities; in short, religion is the main constraint on one's code and mode of life.
>
> (Broek and Webb 1973: 183)

In contrast, religion is clearly becoming less of a cohesive force in industrialised, urbanised Western societies. Zelinsky (1973: 95) notes that 'unlike most pre-modern communities of the Old World where daily existence and religion are mixed inseparably, [in the United States] the religious factor does not enter into the layout of town, field, or house, or the shaping of economic activities'. Carl Sauer (1963) had earlier described the community on the Middle Border, where the country church played a key role in social communication (especially among Catholics and Lutherans), parochial schools extended social connections, church festivals were frequent, and Sunday observance was the norm.

Religion and environmental attitudes

Many of the interactions between religion and geography outlined above reflect an indirect influence, via attitudes and through behaviour. Many types of behaviour exhibit some imprint of religious influence if not control.

Environmental theology (Doughty 1981, Livingstone 1983) has already attracted geographical attention as an important ingredient in trying to understand the value-systems that underpin the contemporary environmental crisis. A recent report by the Church of Scotland (Pullinger 1989), for example, views the greenhouse effect as ultimately a religious matter that is the legitimate focus of interest by scientists, policy-makers and theologians. A key ingredient in the environmental debate is the role played by values and world views in determining both attitudes and behaviour towards environment. The question of ethics cannot be avoided, and there is a large and growing literature on environmental ethics (see, for example, LaFreniere 1985, Hargrove 1986, Moss 1985).

American historian of science Lynn White (1967) argued that 'human ecology is deeply conditioned by beliefs about our nature and destiny – that is, by religion'. He concluded that the historical roots of our modern environmental crisis lie in medieval Christianity, which promoted aggressive and exploitative attitudes towards nature. A number of other writers, including Toynbee (1972) and Brett-Crowther (1985), have eagerly seized upon the White thesis as a plausible explanation for many environmental attitudes and practices which are anthropocentric and destructive. Others, such as Tuan (1968b), disagree and point to major discrepancies between the stated ideals of most religions and what happens in reality. Buddhism is sometimes held up as the most nature-centred religion, for example, yet large areas of forest have been felled in Buddhist areas to produce wood for cremation and for building religious structures. Native religions, such as those of the American Indians, also appear on the surface to be more environmentally sensitive than Christianity (Vecsey 1980, Cornell 1985) but in practice they have been associated with widespread environmental damage. Doughty (1981) suggests that the White thesis is too simplistic and insists that 'Western Christian thought is too rich and complex to be characterised as hostile toward nature'.

Inevitably, Christian geographers – including Aay (1972) and Moss (1973, 1975, 1976, 1978) – have responded to the challenge of the White thesis, and pointed out that he bases his argument on a partial and particular interpretation of key passages in Genesis that speak of man's dominion over nature. Houston (1978) and Kay (1988, 1989) try to redress the balance, by setting Old Testament views on nature into their proper context. My own book *Caring for Creation* (Park 1992) provides a detailed critique of the White debate.

EMERGENCE AND EVOLUTION

It is useful to set recent interest in geography and religion into its proper historical context because this allows us to trace the evolution of ideas. A historical perspective reveals that the field is long-established, and that its character at any one point in time is strongly shaped by the prevailing intellectual climate.

Classical roots and branches

The historical roots of geographical interest in religion can be traced back more than a thousand years. Kong (1990: 356) notes how 'concerns linking geography and cosmology in the mind of the religious person lay at the heart of early geography, and in that sense a geography that incorporated religious ideas was evident from the earliest times'. Greek geographers reflected a world-view heavily influenced by religion, and they saw spatial order as a manifestation of a religious principle, as is clearly evident in their recurrent concern with cosmological models and world diagrams and maps (Isaac 1965: 2–5, Gay 1971: 1).

Islam's intense interest in geography is well represented by the writings of early Muslim geographers (Kish 1978). Muslims traded extensively, travelled widely, and explored incessantly; they wrote some classic early geography books that project an overtly Islamic view of the known world and its peoples. One particularly prominent writer and traveller was Al-Muqaddasi, a well-educated man born in what is now Jerusalem in 946 CE (Christian Era). His book *The Best Divisions for the Classification of Regions* – written in 985 CE and, according to Siddiqi (1987: 2), 'one of the most valuable treatises in Arabic literature' – was one of the very first regional geography texts. It

> described the Muslim Empire complete with a description of the deserts and the seas, the lakes and the rivers, its famous cities and noted towns, the sources of spices and drugs, the hills, plains and mountains, the soils, the cold and hot regions, and the places of industry.
>
> (Conde 1980: 92)

A third early source of material on geography and religion is the early Celtic monastic schools in Ireland, which were major seats of scholarship and learning from the sixth to the eleventh centuries, attracting monks and students from throughout Europe. Many teachers in the monastic schools composed educational poems that summarised the main body of facts about their subject and could be memorised by students. Fahy (1974) describes several such poems written by Airbheartacht MacCosse prior to 991 CE, which described practically all that was known about the universe and about the geography of the world as they knew it (Europe, Africa and Asia).

Golden era

Little is known in detail about developments in geography and religion during the Middle Ages (between about 1000 and 1500 CE), although the growing interest in magic and cosmology was doubtless a potent force. Sack notes how, although they were never fully accepted by the Church,

> in the late Middle Ages and Renaissance, the magical principles, and especially their use of the spatial properties of things to establish similitudes, were elaborated and abstracted in a tangled magical scheme which mirrored the world view of the time. This age viewed the objects of nature as reflections of sympathies, antipathies, and chains of influences, based on the resemblances of things to one another.
>
> (Sack 1976: 312)

The relationship between geography and traditional sources of knowledge during the period of the Scientific Revolution of the sixteenth and seventeenth centuries has been examined in some detail by Livingstone (1988, 1990). He concludes that even though it confirmed traditional beliefs derived from magic and astrology, sixteenth-century geography challenged dogma by its anti-authoritarian emphasis on experience. The geography of the day had much to say about magic, cosmography, astrological meteorology and discovery, and it played an active part in debates about human origins and the origins of modern science.

We pick up the main thread of our intellectual history again during Britain's Tudor and Stuart periods when much English scholarship was under the strong influence of theology (Taylor 1930, 1934). During this time 'the entire subject of geography (and all other sciences) at Lutheran universities beginning with Wittenberg, underwent that change from a philosophy conducted in the spirit of the Greeks to a theologically guided discipline' (Buttner 1980: 89). Four key developments during the sixteenth and seventeenth centuries were the emergence of theologically orientated Christian geographies, the study of other religions, biblical geography, and natural theology.

Theologically oriented Christian geographies

Buttner (1974: 164) argues that the history of the geography of religion really begins in the sixteenth century with the Reformation 'when geography as a whole (as also the other sciences) changed from a philosophy pursued in the Greek spirit to a theologically oriented discipline'. Many sixteenth-century geography books were the work of theologians, and reflections on God's handiwork in creating the earth and its inhabitants permeated much contemporary writing. Geographers at this time were particularly motivated to describe the spread of Christianity around the world.

Much of the writing on geography and religion during the sixteenth and seventeenth centuries is what Isaac (1965: 10) terms '*ecclesiastical geography*'. This work placed great emphasis on mapping the advance of Christianity around the world, was mission-orientated, and was strongly assisted by the Christian Church (Kong 1990).

Buttner's (1974, 1979, 1980) studies of developments in Reformation Germany provide a rich source of information on its development in this formative period. One particularly important figure was Philipp Melanchthon (1497–1560), a leading German Protestant reformer, friend of Luther and founder of the school and university system in Lutheran Europe. Melanchthon insisted that

> only subjects which can be placed at the service of the evangelical doctrine ... should be taught at our schools and universities. Subjects which do not fulfil this requirement are either to be struck from the canon of subjects or recast in such a way as to correspond to the set standards.
>
> (Buttner 1980: 89)

Most of the important geographical work at this time came from the German school. Good examples include books by Caspar Peucer (1556, *De dimensione terrae*) on geography as the science of the visible side of the Divine revelation, and by Neander (1583, *Orbis terrae partium succincta explicatio*) which describes Europe mainly in terms of religion and religious places and ecclesiastical history (Buttner 1980).

The movement was not confined to Lutheran Germany. Gilbert (1962) discusses the work of two Oxford divines who proved that geography was not incompatible with divinity. George Abbot (1562–1633), Vice Chancellor of

9

the University of Oxford and later Archbishop of Canterbury, wrote a geographical book entitled *A Briefe Description of the Whole Worlde* that was first published in 1599. Peter Heylyn (1599–c.1662), former Master of Magdalen College, published *Microcosmus: Or a little description of the great world* in 1621, and a more important book *Cosmographie* first published in 1652. *Cosmographie* includes interesting reflections on the possible location of seven regions (including Utopia, New Atlantis, Fairy Land, the Lands of Chivalry and the New World in the Moon) which Heylyn thought must lie within the (then) unexplored Southern Continent if anywhere. Another leading early post-Reformation English geographer was Nathanael Carpenter (1589–c.1628), whose book *Geography Delineated Forth* (1625) similarly reflects the importance of theology as a context for geography (Baker 1928, Livingstone 1984b).

It is widely thought that the term 'geography of religion' was first used by Gottlieb Kasche in 1795, in his German book *Ideas about Religious Geography* (Buttner 1987). Kasche's concluding remarks capture the spirit of the mission-orientated ecclesiastical geography. He wrote:

> The geography of religion convincingly teaches the advantages of Christianity as compared to any other positive religion. In the same way as the geographer of religion is compelled to describe the doctrines of each faith in an unbiased way, it is the duty of the Christian observer to delight in the brighter light which illuminates the way for him ahead of so many of his fellow-men. All this can only be achieved if a geography of religion presents the existing types of faith in their perfection and their coarseness side by side, thus facilitating comparisons for the philosopher.
>
> (Quoted by Buttner 1987: 223)

The study of other religions

From its essentially Christian origins, Buttner (1980: 92) claims that ecclesiastical geography broadened its remit and perspective during the mid-seventeenth century, when work describing the spheres of influence of other religions began to appear. Varenius's (1649) *Descriptio Regni Iaponia* was the first major work about non-Christian religions. This diversification, Buttner (1980) argued, liberated geography from a purely Christian influence and gave it a more neutral position. The real aim of the change was less neutral, however, because it was intended to determine which religions Christian missionaries found in different parts of the world, and how mission was progressing among them.

Despite the partisan motives, this diversification was highly significant, because 'for the first time an attempt was made systematically to subdivide the various religions as they differ from Christianity and to describe their areas of distribution … thereby making a beginning of what we might today call

10

"description of the spatial distribution of religions"' (Buttner 1974: 165). To this time, therefore, we can trace the beginnings of modern interest in geography and religions.

Although Kasche was probably the first writer to use the term 'geography of religions' in 1795, Buttner (1987) classifies Kasche's book as systematic theology rather than geography. He regards Immanuel Kant as the founder of the subject. Kant (1724–1804) published *Religion within the Boundaries of Pure Reason* a generation before Kasche. It describes different religions in the context of country, society, environment and culture and is welcomed by Buttner as a geographical book.

Biblical geography

In parallel with the evolving ecclesiastical geography of the sixteenth and seventeenth centuries, an emerging interest in biblical or scriptural geography is clearly evident. This focus of geographical inquiry, which Isaac (1965: 8) describes as the 'historical geography of biblical times', involved 'attempts to identify places and names in the bible and to determine their locations' (Kong 1990: 357). It was an inevitable product of a geography of religion which in the sixteenth century was pursued almost exclusively as the geography of Christianity (Buttner 1980: 104).

Delano Smith's (1987) study of maps in Bibles in the sixteenth century throws useful light on some early forms of biblical geography. Martin Luther and John Calvin (leading figures in the Reformation) apparently deliberately used maps as vehicles for the propagation of their religious views, and this left a legacy of early Bible maps that reveal as much about their theological provenance as they do about contemporary cartographic knowledge. The earliest printed Bible known to contain a map (a heavily stylised map of the Holy Land) is dated 1525. Eight types of map appeared regularly from the 1520s onwards – dealing with the Exodus of the Old Testament, Canaan, the Holy Land, the Eastern Mediterranean, the Land of Promise (Coverdale), Daniel's Dream, the Garden of Eden, and Paradise. Delano Smith (1987: 11) notes how 'geographical 'realism' in bible maps was increased as the [sixteenth] century progressed and as the urgency of their doctrinal messages was either accommodated to or was overcome by the impact of cartographic advances outside the bible world'.

Biblical geography was also prominent in the nineteenth century, when many Westerners – including geographers of the calibre of Ritter and Huntington – visited the Holy Land and documented its historical geography. As Ben-Arieh (1989: 70) notes, much of the nineteenth-century literature 'was written by Western people who travelled in Palestine or resided there; it includes itineraries, letters, memoirs, reports to missionaries, physicians, consuls, etc.' and most 'was written either by scientists engaged in biblical research, or by travellers who had been attracted to this land by their deep interest in the Bible and other holy scriptures'. One prominent writer was George Adam Smith whose influential book *The Historical Geography of the Holy Land* was first published in 1894. Smith was 'a prominent Old Testament scholar and theologian . . . [with] a special talent for narrative and description' (Butlin 1988: 381). His book included detailed descriptions of biblical places and sites, and emphasised the historico-geographical uniqueness of the Holy Land.

Like the two trends outlined above, biblical geography has survived through to the present day, albeit in much reduced form (e.g. Baly 1957). Exploration and excavation of archaeological sites associated with events in the Bible continue throughout the Holy Land. Production of reliable and informative guide books for the large number of visitors to the Holy Land has maintained an outlet for writing of this genre.

One particularly interesting manifestation of an enduring interest in the geography of the Bible is studies that seek to recreate biblical events on the basis of present understanding. A graphic example is the reconstruction of weather conditions which shipwrecked Saint Paul on Malta during his voyage from Caesarea to Rome during the winter of 61–2 CE, as recorded in Acts chapters 27 and 28 (Figure 1.1). The biblical description of the violent storm with gale-force winds as Paul's ship passed Crete – called Euroclydon in the King James's version of the Bible and translated 'North-eastern' in other Bible translations – appears to be quite consistent with the known climatic character of the area at that time of the year (Hayward 1982, Coones 1986). This evidence, it is argued, lends authenticity to the recorded story.

Natural theology

Another strand in the linkage between religion and geography was natural theology, which emerged in the late seventeenth century and was particularly strong in the eighteenth and nineteenth centuries. Glacken (1967) provides a good description of this new theology in *Traces on the Rhodian Shore*. In natural theology 'nature was seen as a divinely created order for the well-being of all life. Scholars adopting the physicotheological stance ardently defended the idea that in living nature and on all the Earth, evidence could be found of the wisdom of God' (Kong 1990: 357). It was quite natural, therefore, for theological explanations of all geographical phenomena to be regarded as the only acceptable ones at a time when the task of geography as an academic discipline (particularly in Lutheran Germany) was seen as the description of God's created world (Buttner 1976). Schemes were devised which offered theological explanations for the distribution of climates, plants and animals, landforms and other environmental features.

A fascinating illustration of this influential line of thinking is offered in Tuan's (1968a) study of how the concept of the hydrological cycle was understood and explained between about 1700 and 1850. The cycle was first

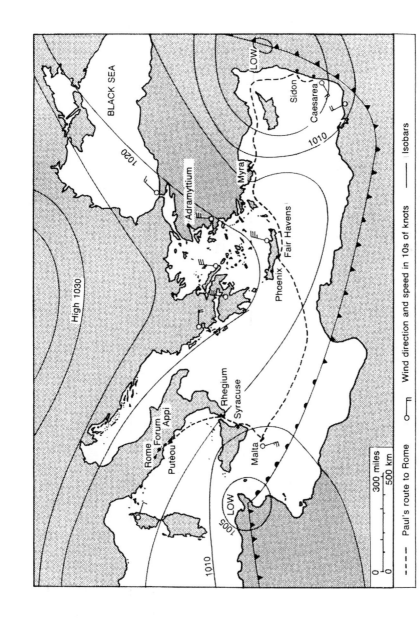

Figure 1.1 Weather conditions in the western Mediterranean typical of St Paul's voyage from Caesarea to Rome in 61–2 CE

Note: The map is a composite of weather conditions in the area on 18–19 and 23 January 1950.
Source: After Hayward (1982).

described in a modern form by English naturalist John Ray, in his popular book *The Wisdom of God Manifested in the Works of the Creation* (first published in 1691, and in its twelfth edition by 1759). Tuan argues that

> the eighteenth century was a period of faith in reason . . . [and] until the concept of the hydrological cycle was introduced and elaborated, it was difficult to argue convincingly for rationality in the pattern of land and sea, in the existence of mountains, in the occurrence of floods, etc.
>
> (Tuan 1968a: 5–6)

There is no doubt that the sixteenth and seventeenth centuries represent the golden age of studies of geography and religion. This was unquestionably *the* formative period for the field, to which we can trace the roots of most subsequent developments. The vitality of geographical studies in this era has never since been recaptured or rekindled.

Nineteenth century

Rationality and reason replaced faith and passion in the Enlightenment times that followed, and which heralded the arrival of a new intellectual climate and the creation of new arenas for debate about the relevance of religion to the modern mind. Much of the innocence of the golden age was lost forever.

Some themes from the golden age persisted into the nineteenth century. Certainly there was enduring enthusiasm to project Christianity as the only sensible religion. This is clearly evident in early (pre-Civil War) geography texts published in New York and New England, many of which were written by devoted Christians (Vining 1982). Books such as *Geography Made Easy* by Jedidiah Morse (a Congregationalist minister) which first appeared in 1784, *Elements of Geography* by the same author (1801), and *An Introduction to Ancient and Modern Geography* written by J.A. Cummings (1817) gave much more information about Christianity than about other major religions, included pictures of strange religious activities (Figure 1.2) and implied that Christianity is preferred to all other religions. Vining (1982: 24) concludes that 'because virtually all pre-college students studied geography, . . . and at almost every grade level, it is reasonable to assume that the writings of the early geographers contributed to religious intolerance, suspicion, and perhaps even hatred'.

Without doubt the most significant, persistent and wide-ranging intellectual change of the nineteenth century was the radical reappraisal of the place of people within the natural order of things. Natural theology was to be eclipsed by New Science which through observation and experimentation was discovering new laws of nature. There was little need for God in an emerging world-view that saw the earth and its features as products of immutable scientific laws. So began 'the move away from the older theological paradigm to the immanent design of natural law, aided it would seem by the advent of

14

Figure 1.2 An early geographical illustration of Islamic worship: pilgrims at the temple in Mecca

Note: The woodcut first appeared in William Woodbridge's (1825) *Rudiments of Geography*.
Source: After Vining (1982).

Neo-Lamarckism' (Livingstone 1984b: 22–3). Jean Lamarck (1744–1829), a French naturalist, first proposed a theory of organic evolution in *Philosophie Zoologique* (1809), a full half-century before Charles Darwin's more infamous *Origin of Species* (1859). Lamarck proposed that characteristics of an organism that are modified during its lifetime are inheritable, thus opening the prospect of progressive adaptation through successive generations. Neo-Lamarckism, which emphasises evolutionary change driven by environment, was a powerful catalyst for much broader change within natural and social science during the nineteenth century (Livingstone 1984b), along with its sibling Neo-Darwinism (Stoddart 1966).

The Enlightenment brought a dramatic shift in emphasis within the geography of religion, which directly reflected the growing influence of Neo-Lamarckism. Buttner captures the spirit of this new heroic age, in which

> all at once it became important, with complete disregard of what had once been the task of the geographer of religion, to assess how far religion was determined by its environment, particularly by climate. Attempts were even made (in the sweep of the rising materialistic currents) to explain the essential nature of various religions, especially that of Christianity, in terms of the respective geographical environments.
>
> (Buttner 1974: 166)

This new era of environmental determinism, where the primary focus of study became the influence of environment on religions, continued well into the twentieth century. Semple (1911), for example, described how different societies' notions of hell were strongly influenced by their own environments (to Eskimos, hell was a place of darkness, storm and intense cold, whereas to Jews it was a place of eternal fire). Huntingdon (1945) adopted a similar deterministic way of explaining why different cultures have different types of deity (in India, for example, the rain god is important because rain is unreliable there).

Twentieth century

An important turning point for academic studies in the impact of religion came during the 1920s, according to Buttner (1974), when it was clear that materialism had run its course and the time was ripe for a fundamental change of direction. Max Weber's pioneering work on the influence of religion on social and economic structures through the 1920s was instrumental in opening up potentially significant new areas of study, but geographers were not to adopt Weber's perspectives for at least another three decades. The stranglehold with which determinism dominated geographical studies in the early decades of the century would take a long time to break down!

After the Second World War Troll

> succeeded in giving the geography of religion a new orientation. In his opinion (in which he was not alone) the materialistic orientation of the geography of religion was an erroneous development in that the geographers who had conducted research in this field led it into an area for which they were not and are not expert. They were lacking the training in ... theology and philosophy necessary for investigating the environment's incorporation into the notional fabric of religion in a form that would somehow be tenable and scientifically acceptable.
>
> (Buttner 1980: 96)

Troll urged his friend Fickeler 'to describe in a fundamental essay in *Erdkunde* what a geography of religion based on purely geographical principles should look like' (Buttner 1980: 96). This Fickeler duly did, and the paper – originally published in German in 1947, but subsequently translated into English (Fickeler 1962) – raised fundamental questions in the geography of religions. The review was wide-ranging, and it identified as suitable for geographical study such themes as sanctification and ceremonial expression, religious tolerance, cult symbolism and the relevance of sacred colours, ceremonial sounds, sacred directions and positions, sacred motion, time and numbers, impacts of religion on landscapes, sacred plants and animals, and protected landscapes.

Fickeler (1962) emphasised that relationships between religion and environment are mutual, and suggested that the science of religion should tackle the

16

question of how environment affects religion, and that the geography of religion should tackle the question of how religion affects people and landscape. After Fickeler made this proposition, Buttner (1980: 96) argues, the geography of religion split into two streams – one dealing with the geographical aspects, the other with the science of religion. Sopher's (1967) book *Geography of Religions* brought the two strands back together again and tried to establish a new direction (Buttner 1980: 98).

Whilst the German school was progressive and innovative, a French school – though numerically stronger, and active through the 1940s and 1950s – continued the largely descriptive and deterministic style imported from the nineteenth century. Reviews of religion and geography were written by Delaruelle (1943) and Le Bras (1945), but they were published in French-language journals and failed to spark much lasting interest among the English-speaking community. The most influential French geographer of the genre was Pierre Deffontaines (1953a, 1953b), whose *Geographie et religions* (1948) offered a wide-ranging though entirely descriptive treatment of relevant themes (see Table 1.1). Although Eliade (1959) was not a geographer, he has had a lasting impact on the development of geographical studies of religion, particularly in distinguishing the sacred from the profane (secular).

Across the Atlantic, a productive and influential school of American geographers of religion has evolved particularly since the early 1960s. This school leads the way today, and many of the studies described throughout this book come from this particular stable. Zelinsky's (1961) innovative study of the religious geography of the United States based on patterns of church membership in 1952 remains a classic of its type, and it was very instrumental in inspiring the American school that has since diversified and consolidated. Other pioneering American geographers of religion include Isaac who championed the study of relationships between religion and landscape (see Chapter 7), Sopher who championed studies of the geography of pilgrimage (see Chapter 8), and Shortridge and Stump who championed studies of

Table 1.1 Summary of the contents of *Géographie et religions* by Deffontaines (1948)

Part I Religion and the geography of residence (includes chapters on the influence of religion on human residence, dwelling place of the dead and dwelling places on gods)

Part II Religion and people (with chapters on settlement evolution, types of settlement, towns, demography, and the dead population)

Part III Religion and exploitation (with chapters on the agricultural life, pastoral geography, industrialisation, and consumption)

Part IV Geography and circulation (with chapters on religious journeys and migrations, journeys of the deceased, the geography of pilgrimage, commerce generated by religion, and the impact of religion on transport

Part V Geography and lifestyle (with chapters on food, work and the calendar, influence of the dead, and the religious way of life)

---

spatial distributions of denominations within the United States (see Chapter 3).

There are no signs of the emergence of a British school of geographers of religion, and British geographers have made relatively few contributions to the field. Perhaps inevitably one focus of British geographical inquiry has been the religious geography and conflict in Northern Ireland (e.g. Jones 1960; Boal 1969, 1972, 1976). Gay's (1971) book on the *Geography of Religion in England* has not inspired subsequent studies, except perhaps Piggott's (1980) study of the religious geography of Scotland. An emerging research focus is the role of the Church in alleviating poverty, illustrated by recent work on the Church Urban Fund by Pacione (1990, 1991).

GEOGRAPHICAL CONTEXT

There are clearly many ways in which religion and geography interact, in both theory and practice (see pp. 2–7). It turns out that the study of geography and religion is by no means a new field of interest (see pp. 7–16), and recent work represents a rediscovery of one of the oldest focal points of geographical study.

In a recent review of progress in the field Kong found that

> whilst some essays of integration and reviews have been written, each has covered only a selection of the themes that have engaged geographers of religion. To date, no attempt has been made to synthesise the many varied themes, to allow for an eventual evaluation of research trends.
>
> (Kong 1990: 361)

One aim of this book is to provide the sort of synthesis of existing work that will allow more clear identification of fertile areas of future study.

Definitions

A simple definition of the 'geography of religion', offered by a church historian not a geographer, is 'the description and analysis of religious phenomena in terms of the science of geography' (Barrett 1982: 828). It is sufficiently vague as to allow almost any type of study to fall within its remit!

Geographers, following Isaac (1965) and Stump (1986a), find it convenient to distinguish two main approaches:

1 *religious geography*; which 'focuses on religion's role in shaping human perceptions of the world and of humanity's place within it; its primary concerns are the role of theology and cosmology in the interpretation of the universe' (Stump 1986a: 1), and
2 the *geography of religions*; which is 'concerned less with religion *per se* than with its social, cultural and environmental associations and effects. This

18

approach views religion as a human institution, and explores its relationships with various elements of its human and physical settings' (Stump 1986a: 1).

Most recent work in the field has tended to adopt the second approach. Tuan (1976: 271), for example, defines religion as 'the impulse for coherence and meaning' and points out that the strength of this impulse varies, so geographers should note how the variation is 'manifest in the organisation of space and time, and in attitudes to nature'.

This balance of attention towards the *geography of religions* should not overshadow the fundamental importance of *religious geography*, which Sopher (1981: 518) sees as 'the geography and cosmology that are in the mind of the religious man, the way in which, informed by his faith, he sees his world and his place in it'. Tuan (1968a) calls this field *geoteleology*, while Stump (1986a) prefers *geosophy*. Isaac (1965) observed that the early geography that incorporated religious ideas would now be classed as religious geography.

A recurrent debate through the 1960s and 1970s was whether or not geographers of religion had a duty to focus on visible dimensions of religious activity. The debate revolved most clearly around the impact of religion on landscape, which Erich Isaac maintained ought to be the prime focus of geographical studies. In his view, 'the task of the geographer of religion is to separate the specifically religious from the social, economic and ethnic matrix in which it is embedded and to determine its relative weight in relation to other forms in transforming the landscape' (Isaac 1962: 12).

Present status

In some senses, the academic study of geography and religion is now well established because it has been the focus of major texts, substantive reviews and major conference sessions.

Two landmark books have charted the variety of studies published in the field. Sopher's *Geography of Religions* (1967) remains a masterly overview of an emerging field, written by a very active researcher and writer. It was well received, has been extensively cited since, and offers an important baseline against which to measure subsequent developments. Gay's *Geography of Religion in England* (1971) is more of a description of religious groups and traditions, with limited spatial analysis. As the title suggests, however, its focus is entirely on England. It has not provided 'a jumping-off ground for future studies' (Gay 1971: preface) as he had hoped. Both books have yet to be paralleled, even though they are over two decades old, and both are long out of print.

Substantive reviews of progress in the field have also been published by Sopher (1981), Levine (1986) and Kong (1990), which indicates a field growing in significance and seeking to establish a credible intellectual niche for itself within mainstream geography. Theme issues of journals such as *Journal of*

Cultural Geography (1986, volume 7) and *National Geographical Journal of India* (1987, volume 33), devoted to geography and religion, testify further to growing international interest and the emergence of core interest areas and methodological approaches.

A further measure of growth and credibility is the establishment in 1976 of an International Working Group on the Geography of Belief Systems (Singh 1987a). Manfred Buttner (1987: 225-6), the first elected chair of the group, insisted that 'with this foundation, the broad definition of the tasks of the geography of religion asserted itself'. The Association of American Geographers now has a Geography of Religion Speciality Group that publishes a newsletter (*Geography of Religion and Belief Systems*) and organises sessions on the geography of religion at AAG conferences. Themes such as the emerging interest in reciprocity between geographical and religious experience, exploration of new methodological approaches, and consideration of future avenues of religio-geographical debate were discussed at the 1991 Miami conference (Cooper 1991b). The formation in 1988 of a Fellowship of Christian Geographers to bring together a loose coalition of British geographers with a common faith and a shared concern to explore how that might be incorporated into geographical research (Olliver 1989), represents yet a further recognition of a legitimate niche for geography and religion on the academic landscape.

Lack of coherence

Most work in the field falls into one of the four clusters identified by Sopher (1981) – denominational geography, the landscapes and spatial organisation of particular religious groups, the development of sacred centres, and pilgrimage. Stump (1986a: 2) notes that 'even these clusters contain considerable substantive and methodological diversity'. In a prophetic review of geography and religion from a humanistic perspective, Tuan (1976: 271) concluded that 'the field is in disarray for lack of a coherent definition of the phenomenon it seeks to understand. ... A field so lacking in focus and so arbitrary in its selection of themes cannot hope to achieve intellectual maturity.'

Despite the notable advances of the last two decades, few of the geographers who have contributed most to the development of the field are convinced that it has yet reached maturity. Sopher (1981) bemoaned the fact that 'a decade and more of modest increase in the volume of geographic writing on religions and religious institutions has not brought consensus on the nature of the pertinent field or even agreement whether there can be such a field at all'. This lack of coherence is echoed by Stump (1986a: 1), who concluded that despite some interesting publications in the early 1980s 'the study of religion remains a diverse and fragmented endeavour within geography'.

Some critics lay the blame for lack of unity and real progress on an incorrect focus within the field. One champion of this camp is Levine (1986: 437), who points out that 'studies in the geography of religion have often been mere catalogues of artefacts and mentifacts'. He insists that 'it is essential to go beyond this kind of narrow empiricism and to try to appreciate religion as a profoundly social force. It is also necessary to go beyond an idealism which sees religion as motivator in landscape change but fails to appreciate the social nature of religion'.

The development of a coherent geography of religion has also been inhibited by the great diversity of approaches used, and by the lack of consensus about what are the most relevant themes to explore or questions to answer. This partly reflects the lack of an unambiguous definition and widely agreed terms of reference within which to operate.

NICHE WITHIN GEOGRAPHY

The academic niche which students of geography and religion have tried to carve out is rather ill-defined and elusive, reflecting the lack of consensus mentioned above. The niche is more sharply focused in terms of sub-fields within geography; there is no firm evidence of a cohesive niche in terms of paradigms.

Sub-field within geography

Gay (1971: preface) notes how 'during the course of its development geography has greatly extended its boundaries but the field of religion has always been at the outermost margin'. Given the close associations between religion and culture mentioned above, it is no surprise that most practitioners place the study of geography and religion squarely within cultural geography. This is so particularly in the United States, where cultural geography has long enjoyed a position close to the hub of geographical inquiry, and is widely taught at degree level.

Many would agree with Sivignon (1981) that religion and language belong to cultural geography because they constitute the foundation of institutions that allow the structuring of societies. In much the same vein, Levine (1986: 428) claims that 'geography of religion is a sub-field of cultural geography which has lacked theoretical debate'. Religion receives only one mention in Ron Johnston's *On Human Geography* (1986: 60), in a section entitled 'The cultural superstructure' which deals with 'institutions, such as family and organised religions that socialise individuals into accepting their roles within society and legitimate certain divisions of labour'.

Regardless of where the geography of religions is located on the landscape of geographical inquiry, it is of paramount importance that practitioners face up to their responsibilities in addressing the 'big' questions which society faces. Gilbert White (1985: 10), for example, insists that 'geographers have a responsibility to help the human family come to see itself through ... teaching ...

21

research ... [and] policy analysis'. Richard Morrill (1984: 6) argues that one of geography's main responsibilities is to society, in which 'race, ethnicity, language, religion, social class, political allegiance, or even life-style differences are viewed as the legitimate expression of people's values and preferences'.

Paradigm

Students of geography and religion have operated within most of the paradigms within modern human geography (Cloke, Philo and Sadler 1991), and this has been both a strength and a weakness. On the positive side, studies have not been constrained to fit comfortably into a pre-existing philosophical strait-jacket, and this freedom has encouraged the exploration of many different themes in many different ways. On the negative side, however, failure to function within an agreed paradigm has probably perpetuated the field's lack of cohesion and certainly prevented the compilation of an accessible corpus of relevant work.

Modernism

Humanism – with its emphasis on human experience, awareness and behaviour, and its attachment to space and place – has been a fertile patch in which to work. Entrikin (1976: 617) suggests that 'humanistic geography is best understood as a form of criticism' based on empathy, intuition, introspection and other non-empirical methods of study, which emphasises aspects of people that are most distinctively 'human' (meaning, value, goals and purposes). Adopting a definition of religion as 'the impulse for coherence and meaning', Tuan (1976: 272) proposes that 'a humanistic approach to religion would require that we be aware of the differences in the human desire for coherence, and note how these differences are manifest in the organisation of space and time, and in attitudes to nature'.

David Ley (1981: 253), a prominent and persuasive proponent of humanistic geography, stresses that it is 'a theoretical perspective, not a distinctive empirical sub field', which treats human values seriously and focuses on intentionality of action and contingency of context. Thus, he argues, 'geographic facts are not fatalistically predetermined; they are the outcome of both constraint and choice, or processes of negotiation by geographic agents' (Ley 1980: 15). To Ley and other humanistic geographers what really matters is the socio-cultural lifeworld, which is not determined by environment but has autonomy of its own; where values and consciousness are embedded in social contexts, shape meanings and mould human creativity, and find tangible expression in symbolic and structural ways. Religion, expressed at both personal and corporate scales, must surely be a major factor in this rich socio-cultural tapestry.

Anne Buttimer (1976: 277) picks up similar themes in arguing for a phenomenological approach which 'should allow [the] lifeworld to reveal itself in its own terms.' Her central thesis is that 'the phenomenological notion of intentionality suggests that each individual is the focus of his own world, yet he may be oblivious of himself as the creative centre of that world' (ibid.: 279). This notion of 'existential freedom – engagement in, yet transcendence of, one's milieu' (ibid.: 289) suggests many fertile (but as yet largely unexplored) avenues for the geographer of religion to pursue, arising from the influence of beliefs on behaviour, via values and attitudes. Similar frustrations underlie Gale's call for a geography that focused more openly on the links between personal and social values, described 'the character of those social institutions that would truly represent human wants, desires and needs' (Gale 1977: 269), and recognised that social decision-making is based largely on 'short-term truths, context-dependency, and partial information' (ibid.: 272).

Ian Wallace (1985, 1986) takes Ley's arguments a logical step forward and proposes the adoption of a Christian theological perspective on humanist geography. This would place a firm emphasis on three key dimensions of human existence – the materialism of life (material humanity), the centrality of human relationships (relational humanity, expressed most clearly in the idea of stewardship), and the need for people to recognise that they are accountable (humanity).

The structure-agency debate is also a promising context within which to set studies of geography and religion, although few if any have used it to date. Gregory's (1981) critique of humanistic geography highlights the 'bounded-ness' of space within a 'matrix of contingency'. His four models of historical change (Table 1.2) offer a useful framework for describing relationships between action and structure, which must surely have much to offer to students of geography and religion. Simply replacing the words 'society' and 'social systems' with the word 'religion' in Gregory's definitions of *dialectical reproduction* and *structuration* hints at the essentially reciprocal relationships between people and religion and between geography and religion (see p. 26).

Levine (1986) proposes the adoption of the range of phenomenological and historical materialist approaches which have become commonplace in geography. The former, he argues, would increase awareness of the depth of

Table 1.2 Four models of historical change proposed by Gregory (1981)

1 *reification*; where 'society is . . . external to and constraining upon human agency'
2 *voluntarism*; where 'society is constituted by intentional action' by human agency
3 *dialectical* reproduction; where 'society transforms the individuals who create society in a continuous dialectic; society is an externalisation of man, and man a conscious appropriation of society'
4 *structuration*; where 'social systems are both the medium and the outcome of the practices that constitute them; the two are recursively separated and recombined'

the religious quest in the human psyche, and the latter would throw light on religious institutions in class society.

Postmodernism

Perhaps the most promising intellectual framework within which to set future research in geography and religion is postmodernism, which – in geography as in many other areas within the social sciences – 'has become in some ways a vogue term for anything that is new and different' (Rogers 1992: 249). Although its advent has been warmly welcomed as innovative (e.g. Dear 1988), it has not received universal support (see, for example, David Harvey's biting critique in *The Condition of Postmodernity* (1989)).

Novelty *per se*, of course, is not a catalyst for meaningful change. As Rogers (1992: 250) sees it, postmodernism emphasises 'the absence of single explanations and an inability to predict and control reality. . . . Openness, plurality and possibility are the watchwords of postmodernism.' The postmodern movement is pervasive, and has had impacts on many subjects ranging from philosophy to architecture and planning (Cooke 1990b). It is perhaps best reflected in the break-up of hierarchical and centralised forms of organisation (for example in politics and industry), and the echoes of this in philosophy (in the move away from theories that seek to explain everything) and in society (in the growing importance of issues such as rights of citizenship and belonging) (Rogers 1992).

Dear points out that postmodernism

is basically a revolt against the rationality of modernism ... which searches for universal truth and meaning, usually through some kind of metadiscourse or metanarrative. ... In practice, postmodernism has taken the form of a revolt against the too-rigid conventions of existing method and language.

(Dear 1988: 265)

Use of language – which lies at the heart of all knowledge – to portray meaning and to maintain discourse is a central focus in postmodern debates. Deconstruction is used to expose contradictions and paradoxes within the ways we order and articulate concepts, and it 'shows how language imposes limits on our thinking' (Dear 1988: 266). In essence, any text or narrative is open to different interpretations (that is to say, meaning is partly in the mind of the reader or listener), and postmodern writers stress the inherent conditionality of interpretation and understanding that is revealed through hermeneutics (the theory of understanding and interpretation).

In the final analysis, postmodernists claim, all knowledge is conditional and relative – there are no absolutes by which we can order reality, nor any fully objective ways of representing reality. As Cosgrove (1990: 353) puts it, 'today meaning may be constructed, if at all, only in a discourse of images whose contingency is actively embraced'. Everything about us – what we say, how we say it, what we do, how we think, and so on – is heavily conditioned by language and our mental map of reality.

The visible and invisible manifestations of religion within society, where personal faith and interpretations stand alongside received wisdom and codified patterns of behaviour, must surely be prime contenders for postmodern critiques.

PROSPECT

Sopher's (1981: 510) conclusion that 'there is little of coherence, continuity and common purpose that would make for the genesis of a recognisable field' called 'geography and religion' still holds true today. Indeed, it is difficult to disregard two diagnoses made as far back as 1976 – that the geography of religion is 'still in its infancy' (Shortridge 1976: 434), or even worse, it is 'a field in disarray' (Tuan 1976: 271).

A more positive perspective is offered by Buttner (1980: 86), who reminds us that 'the geography of religion is a scientific discipline which, like all others, is continually changing, whose nature, aims and tasks were seen in a different way in every epoch and which was or is correspondingly conducted by researchers of widely varying backgrounds'.

In diversity lies strength and flexibility to respond to new challenges, and, as in most other fields of study, some of the most inviting prospects appear at the points of interface between disciplines. Kong (1990: 367) identifies a 'reluctance to venture into what are considered tangential areas and the tendency to stop short at perceived boundaries [which] may have unnecessarily limited the type of questions that geographers could be asking'. Some of this reluctance, without doubt, reflects an intellectual uncertainty about where the critical boundaries are.

This boundary dispute, in turn, reflects an ongoing debate amongst geographers of religion, between a minority (championed by Isaac 1962: 17) who regard the study of religion itself as an important focus within the field, and the majority (including Gay 1971 and Cooper 1992) who agree with Kong (1990: 365) that 'geographers of religion should apply themselves as geographers only to the purely geographical side of the subject'. Few would disagree with Gay's (1971 preface) view that 'it is not the task of the geographer to evaluate religious experience and neither is he concerned with assessing the validity or otherwise of religious claims and beliefs', but beyond this point there is much less consensus.

Notwithstanding these difficulties, it is possible to detect two emerging themes that either are, or should be, engaging the attention of geographers of religion in the mid–1990s. These are the issue of reciprocity and personal experience, and the acceptance of spirituality within geography.

Reciprocity and personal experience

A hallmark of recent writing in the field (and a reflection of the injection of postmodern thinking into geography) has been the much greater prominence given to reciprocity of relationships. Early work presupposed and therefore sought evidence of one-way relationships, such as the impact of environment on religion, whereas today 'geographers of religion are being exhorted to pay attention to the reciprocity in the network of relations . . . between religion and the environment [because they are] dialectical and . . . [to realise that] to study unidirectional relationships alone (whichever the direction) would be unrealistic' (Kong 1990: 358).

Whilst reciprocity is an emerging theme within geography and religion, it has been part of the sociology of religion for some time (Demerath and Roof 1976). Paradoxically, it is two decades since Buttner (1974: 166) argued that 'it is time to strike a new direction' by abandoning the idea of one-way relationships and studying reciprocal aspects of inter-relationships. His case study of the Herrnhuter Brethren and their settlements offered an encouraging foundation to build upon, but few accepted his challenge.

Recent geographical research highlights the fact that religion can both inform and be informed by society, and emphasises the need for geographers to regard dimensions other than conventional religion as legitimate topics of study. Cooper (1990, 1992) has welcomed this new emphasis on the reciprocity of meaning between place, landscape and religious experience that roots religious experience and geographical factors in the real world and explores the multi-dimensional matrix of actions and interactions by which that 'reality' is created, refashioned, and expressed. This was a central theme of debate during the geography of religion sessions at the 1991 Association of American Geographers' Conference in Miami (Cooper 1991b), and it seeks to locate active human individuality 'within a broader context of personal experience, and social and material relations pertinent to an individual's religious experience, or that of an identifiable group' (Cooper 1992: 123).

Spirituality and geography

Post-Enlightenment geography, with its emphasis on the observable, countable and measurable properties of phenomena, has no place for spirituality. After all, geography in the Middle Ages and Reformation demonstrated the dangers of putting faith before reason, of allowing evangelical zeal to overshadow objective reality, and of allowing geography to be the handmaiden of theology.

The pendulum seems to have swung too far the other way, and so-called 'modern geography' is founded on a set of assumptions about people and what motivates them that gives no credit to the supernatural, the apparently irrational, or the normative influence of belief systems. There are some signs of attempts to redress the balance. If these are even partially successful, we might expect to see spirituality back on the geographical agenda again in the future.

The presuppositional hierarchy

An important watershed was the identification of a presuppositional hierarchy (Table 1.3) by Harrison and Livingstone (1980). They argue (ibid.: 31) that 'the existence of a large number of created human worlds, loosely connected to the one 'real' world, suggests that any understanding of this world will only be an approximation, since it can be seen only through the interpretative activity of the observer'. As a result, they insist, any academic study – indeed, any interpretation of reality – is preconditioned by its presuppositional context, and all results in science are both directed and structured by these presuppositional influences. No observer is value-free, regardless of how objective they think themselves to be. Because cosmological considerations play a far greater role in shaping attitudes and influencing decision than is usually recognised, suggest Harrison and Livingstone (1980), they should be examined more explicitly within all aspects of human geography.

Given that geography often deals with major moral issues, like environment, poverty and equality, Blachford (1979) has called on geographers to give more overt attention to values and moral issues. As he points out (ibid.: 441), 'values are a vital determinant of behaviour, even though the values on which action is based may not be consciously known by the actor'.

A geography that avoids or ignores metaphysics can at best take a partial view of complex moral questions and issues. It is this vacuum that a number of geographers have identified and sought to fill. Most proposals to date advocate the adoption of a Christian perspective.

Emergence of a Christian perspective

Whilst most geographers would question the idea of a Christian perspective on geography, arguing that any academic subject must be built around agreed-upon foundations for sure and certain knowledge, there are certainly signs of a growing interest in exploring what such a perspective might look like. If

Table 1.3 The presuppositional hierarchy proposed by Harrison and Livingstone (1980)

(a) **cosmology**; fundamental beliefs about the *origin* of reality
(b) **ontology**; presuppositions about the *nature* of reality and the sources of knowledge
(c) **epistemology**; constraints on the *understanding* of reality, delimiting the domain of enquiry and specifying legitimate questions
(d) **methodology**; organisation of the *analysis* of reality, identifying the type of analytical techniques and appropriate instruments to be used

nothing else, such an approach would make explicit many of the presuppositions (Table 1.3) which often implicitly underpin geographical research. But, more than that, a perspective which emphasises issues like humanity, justice, truth and relationships must surely hold great potential, at least within human geography.

David Ley was amongst the first well-known geographers to propose a Christian perspective in his much-quoted paper on good and evil in the city, in *Antipode* (1974). He argued that the root cause of evil in the city is privatistic iniquity, not social inequity, and concluded that 'a fuller, more humane view of man is required, one which acknowledges both his dignity and his depravity. Such a perspective is presented by the Christian view of man' (Ley 1974: 71).

The spirit of the Ley line of argument has been developed more fully by Ian Wallace who proposed that

neither the determinism of positivist social science and a dogmatic neo-marxism nor a subjective indeterminism are adequate sources of knowledge for the study of human action and interaction with the natural environment ... the most coherent account of man's nature and his appropriate relationship to the natural environment is, in fact, a Christian one, rooted in the biblical understanding of creation and salvation history.

(Wallace 1978: 91)

A Christian framework within geography, as envisaged by Wallace (1978: 103), would need to accommodate three things – the humanity of the individual in its diversity, depth, freedom and boundedness; the structured nature of society, in which units from the size of the family to that of the international organisation are accorded legitimate power and freedom; and people's complex relationship with the natural environment, on which they are substantially dependent yet over which they have substantial powers of adaptation and exploitation.

Developing this same theme further seven years later, Wallace (1985: 26) suggested that 'the world can no longer be meaningfully regarded as the home of Man because there is ultimately no ontological basis for our aspirations towards coherent personhood and the social, international, and environmental relationships which reflect it'. Such coherence, he proposes, 'becomes possible only with our acknowledgement of creaturehood' (ibid.: 28), which requires a theological understanding of humanity constituted 'in the image of God', and an acknowledgement 'that the material universe and the realm of values originate in one and the same God' (ibid.: 29).

Swan points out that in contemporary geography,

in the interests of objectivity and scientific analysis, preference continues to be for the observed and the observable, the tangible, the measurable and the quantifiable. ... The tendency is to avoid asking the deeper philosophical questions. Often there is a feeling that to do so as geographers would mean talking to the wind. ... [yet] The study of the real world carries the enquirer into the realms of philosophy and to the threshold of transcendence.

(Swan 1990: 278–81)

This, he argues, makes a religious perspective in general – and a Christian one in particular – not only acceptable, but preferable.

A number of niches for Christian explanation in geography are identified by Bradshaw (1990), including stewardship of nature and resolution of environmental issues, and a better understanding of the geography of poverty and social injustice. He confesses that

the Christian cannot pretend to have a ready-made solution to these problems, but can point to the potential prospect of a community where God's worship is central and human relationships are seen in that context – where there is intended to be no Greek, Jew, male, female, slave or free.

(Bradshaw 1990: 382)

Over a decade ago Doughty (1981) suggested that geographers should devote more attention to environmental theology. My own book (1992) *Caring for Creation; a Christian way Forward* explores the implications for modern society of Old Testament models of stewardship and resource use.

The spirit of this emerging Christian perspective within geography is captured in Olliver's report of the inaugural meeting of the Fellowship of Christian Geographers, where she describes how

Christian geographers are attempting neither uncritically to apply theological dogma, nor to resurrect pre-scientific anachronism, nor to create some kind of spurious sub-field of the discipline. Rather ... [they are seeking] to adopt a scholarly approach to the whole discipline, emanating from a genuine, committed world-view ... recognising the need for interdisciplinary, interphilosophical and intermethodological communication ... [in the] search for truth, hope and peace in our discipline and the needy world we study and in which we live.

(Olliver 1989: 108)

A later meeting of the group (now renamed Christians in Geography) explored geographical themes which might integrate Christian principles – such as justice and equity, urban policy and experience, stewardship of nature, attitudes to property and possessions. Martin Clark (1991: 343) reiterated the view that 'Christians in geography are contending for a recognition that Christian orthodoxy and orthopraxis are viable sources of normative knowledge for the conduct and analysis of human life'.

CONCLUSION

This brief review of geography and religion reveals a field of study that is at one and the same time ancient and modern, emerging and retreating, fertile and sterile, and challenging and unresponsive. Even a selective catalogue of the kind of studies that have been published in the field raises serious questions about why it does not enjoy a more prominent and influential position within mainstream geography. The history of the field is littered with topics left untouched, questions left unanswered (often, indeed, unasked), approaches left undeveloped and directions left unexplored. Yet that is its fascination!

David Sopher, without doubt the field's most respected and quoted author, deserves the last word. In his view,

> questions about the validity and viability of a separate [geography of religion] are not ... of much importance. ... For to the extent that geography is prepared and able to take *man* seriously, to accept as data his symbols, rites, beliefs and hopes in all their cultural actuality, religion broadly conceived must become a central object of the discipline's best endeavours.
>
> (Sopher 1981: 519)

2

REFLECTIONS ON RELIGION

Religion ... is the opium of the people.
Karl Marx [Critique of Hegel's *Philosophy of Right*, 1843–4]

INTRODUCTION

Marxism is sometimes seen as a secular alternative to religion. Indeed some would argue that it is the major secular religion to have emerged in the last century. Karl Marx saw religion as both an expression of human distress and a means of disguising its true causes, and so he argued that it was the opium of the people because it offered them happiness that was not real but an illusion (Hinnells 1984a: 205).

The great architect of Communism would have us believe that people seek an escape from reality via religion, which offers a social anaesthetic from the ills and evils of life. Addictive religion might be, but not for the reasons Marx puts forward. It is not without irony that religion (particularly Christianity, both Orthodox and Western) is witnessing a renaissance throughout Eastern Europe and the former Soviet Union in the wake of the collapse of Communism, as a newly liberated people rediscover the life-changing capacity of religion, and existing believers emerge from their chrysalises of covert worship and fellowship.

Edmund Burke said in his *Reflections on the Revolution in France* (1790) that 'man is by his constitution a religious animal'. The same theme is echoed nearly two centuries later, by Yi Fu Tuan (1976: 271–2), who pointed out that 'religion is present to varying degrees in all cultures. It appears to be a universal human trait. In religion human beings are clearly distinguished from other animals'.

Given that this human trait is universal, and that it leaves indelible finger-prints on so many aspects of society, landscape and environment (as we saw in Chapter 1, pp. 2–7), it is important to reflect on what religion actually means (in theory and in practice), how it is expressed, and how it functions in the modern world. This is our task in this chapter.

Yinger (1977: 67) identifies two main traditions in the scientific study of religion. One emphasises differences between religions (particularly in rite, belief and social organisation) and focuses on the historical, cultural and structural sources of these differences. The other emphasises commonality between religions and reflects the view that all religions rest on a common structure and they all fit into human enterprise in similar ways. The latter (commonality) perspective is adopted throughout this book, where our main interest is on religion as a generic influence on people and environment, rather than on individual religions *per se*.

CONTEXT

Writing on the paradox of religion in contemporary Britain, Davie (1990: 395) reckons 'it is hard to imagine a field of enquiry containing a greater number of apparent contradictions'. One way of trying to keep these inherent contradictions in focus throughout this book is to establish a workable context within which to view religion as a 'universal human trait'. We begin here by reflecting on how best to define and classify religion.

Definition

Finding a universally acceptable definition of 'religion' is as challenging as searching for the Holy Grail. The quest is difficult, and an unambiguous definition is elusive, because generalisations have to encompass a vast range of individual belief systems, cultures and contexts. Indeed, to try to portray religion as a homogeneous or unidimensional entity is to presuppose a uniformity and cohesiveness between individual religions that simply does not exist.

Most people would agree that religion goes beyond superstition, folklore (the beliefs, traditions, legends, customs and superstitions of a nation or race) and mythology (traditions and fables concerning gods and goddesses). But how far?

For students of comparative religion, the difficulties of defining 'religion' are a vocational hazard. Hinnells (1984a: 270), for example, suggests that a religion is 'one of a set of recognisable systems of belief and practise having a family resemblance. The set has no sharp boundaries.' He adds that the process of defining the 'set of religions' is both 'arbitrary and artificial', and that 'definers of religion are prone to the error of reification (misplaced concreteness)'.

If it is misguided to try to fix rigid boundaries around the notion of religion, a pragmatic solution is to adopt a working definition (such as 'a system of thought and behaviour expressing belief in God'). We could do much worse than accept Gay's (1971 preface) advice that 'the geographer should adopt the dictum that "if men define situations as real they are real in their consequence", and then immerse himself in examining from a geographical viewpoint the empirical consequences of religion – its organised institutional expression'.

A selection of definitions of religion (listed in Table 2.1) suggests that no single one captures the full meaning of the word, but some common components are evident. Religion is clearly to do with personal experience and change, and also with shared experiences and community. It is to do with the supernatural (which appears to be beyond the powers of nature, and outside of what we normally experience as human beings), springs from an awareness of the active involvement of a higher being or beings in the day-to-day experiences of life on earth, and usually involves worship of that higher being. It involves belief, and is sometimes beyond reason. It is often reflected in behaviour conditioned by some accepted notion of absolute truth.

Religion is about faith. But how wide does the definition extend? Broek and Webb (1973: 134) suggest that religion 'comprises any form of faith from monotheism to ancestor worship and even magic, insofar as it contains an element of reverence for the supernatural'. Throughout this book we will be adopting a narrower definition of religion, which excludes magic and places some emphasis on God or gods.

A functionalist definition of religion, which emphasises what religion *does* or the *effects it has*, rather than what it *is*, is easier to arrive at. Thus Hinnells (1984a: 270), for example, proposes that religion is 'that which promotes social solidarity or gives confidence'. Durkheim defines religion as 'a unified system of beliefs and practices relative to sacred things which unite into one single moral community all those who adhere to them' (quoted by Shortridge 1982).

The word religion is derived from the Latin *religare*, which means 'to bind'. At its root, therefore, religion is what traditionally binds a society together. Tuan poses the rhetorical question 'What is the meaning of religion?', then

Table 2.1 Some definitions of religion

1 'an attempt to explain nature and the mystery of life' (Wynne-Hammond 1979: 30)
2 'the experience of the supernatural; an experience independent of reason' (Isaac 1960: 14)
3 'man's belief in the supernatural, in what arouses in him a feeling of awe or piety, in what he considers sacred' (Broek and Webb 1973: 134)
4 'a system of faith and worship, centrally concerned with the means of ultimate transformation' (Barrett 1982: 841)
5 'an attitude of awe towards God, or gods, or the supernatural, or the mystery of life, accompanied by beliefs and affecting basic patterns of individual and group behaviour' (Bullock and Stallybrass 1981: 537)
6 'belief in, worship of, or obedience to a supernatural power or powers considered to be divine or to have control of human destiny' (*Collins English Dictionary* 1979: 1233)
7 'solutions to questions of ultimate meaning which postulate the existence of a supernatural being, world, or force, and which further postulate that this force is active, that events and conditions here on earth are influenced by the supernatural' (Stark and Bainbridge 1979: 119)

seeks to answer it by reflecting on what people seek in, from or through religion. In his view,

> the religious person is one who seeks coherence and meaning in his world, and a religious culture is one that has a clearly structured world view. The religious impulse is to tie things together. ... All human beings are religious if religion is broadly defined as the impulse for coherence and meaning. The strength of the impulse varies enormously from culture to culture, and from person to person.
>
> (Tuan 1976: 271–2)

Classification

The simplest classification recognises two categories of people – those who are religious and those who aren't (we might call them 'non-religious'). There are, inevitably, enormous semantic and practical difficulties in allocating individuals into even these two groups; the problems are compounded as more categories are included in the typology.

Figures collected from a wide variety of sources by Barrett (1982: 49) show that in the mid–1980s just under 80 per cent of the world population – an estimated 3,463 million people – were 'religionists' (persons who profess a religion), and the remaining 20 per cent – an estimated 911 million people – were unbelievers or non-religious (agnostics and atheists; see Table 2.2 for definitions). Clearly not all of the religionists allow their religious beliefs to control all aspects of their life and behaviour, but if even only a small percentage of them did, then religion – in its diversity of forms – is a much more potent factor in human geography than politics or other socio-economic variables. For this reason alone, then, study of geography and religion merits much closer and more sympathetic consideration amongst geographers.

We noted above that religion is founded in belief, and that there are many different focuses of belief and ways of expressing belief. The range of belief systems defined in Table 2.2 (the list is illustrative rather than exhaustive) indicates just how broad the list of possible candidates is for inclusion under the definition of 'religion', and clearly not all of them would wish or be suitable for inclusion.

The inherent difficulties of defining religion are reflected in problems of how best to classify different religions. Inevitably the choice of classification scheme rests heavily on the scale and purpose of the study; a study of the pattern of religious variation within a city in the United States, for example, would require a different scheme to a study of global patterns of religious beliefs. At the broad scale, Ahern and Davie (1987) offer a useful typology of religion based on two dimensions – organised/non-organised, and supernatural/ empirical referents. In this scheme 'conventional religion' is classified as organised/supernatural, 'common religion' is non-organised/supernatural,

Table 2.2 Definitions of some common belief systems

---

agnosticism – the view that humans can never be certain whether or not God exists

atheism – disbelief in the existence of God or of any gods, reflected in dogmatic rejection of the possibility of God's existence, scepticism about any religious claims, or pure agnosticism

deism – the belief that God created the world in the beginning, but does not intervene in the course of natural and human affairs

dialectical materialism – the belief that social change is governed by underlying material (economic) factors and not by ideas (the basis of Marxism)

empiricism – the belief that experience is the only source of knowledge

existentialism – the belief that humans are free agents in a deterministic and seemingly meaningless universe, and the basis of religion is subjective (personal faith founded on personal experience)

humanism – entirely non-religious beliefs and value, which look towards political reform and science to improve the human condition

logical positivism – the belief that recognises only positive facts and observable phenomena

monotheism – belief that there is one, but only one, God

pantheism – belief that the whole of reality is divine (i.e. God and the universe are the same thing)

polytheism – belief in, or worship of, many gods

rationalism – belief that religion has no basis in reason, which emphasises reason as opposed to experience or emotion

scepticism – denial that there are any grounds for reasonable belief in religious matters

theism – belief in a single God on whom the world depends for its existence, continuance, meaning and purpose

---

Source: Based largely on Hinnells (1984a).

'surrogate' religion is organised/empirical and 'invisible religion' is non-organised/empirical. This book, like most geographical studies, focuses almost exclusively on the 'conventional religion' group, which emphasises the formal or institutionalised expression of religious belief.

The four-fifths of humanity that, according to Barrett (1982), is religionist can be broken down into the major religions represented among them. This is by no means an easy task, because, as Brook stresses,

> the religious composition of the world population can be estimated only approximately. A statistical record of the religious structure is kept in a number of countries, but the officially estimated number of adherents of one or another religion is often much higher than the actual number of believers, since persons polled during population censuses not infrequently give their past adherence to a religion with which they are no longer affiliated. Only active opponents of religion are considered atheists, while people who are indifferent to it are classified as believers.
>
> (Brook 1979: 513)

Other problems include the lack of a standard way of defining membership between many religious groups, and in some countries (like the former

Communist states) it was common for people to state on government censuses and polls that they are atheist. There is no way round these logistical and terminological constraints, and the data summarised by Barrett (1982) are widely accepted as the best available estimates. The estimated strengths of the major world religions in 1985 are shown in Table 2.3.

The evidence suggests that Christianity has by far the most adherents (nearly one in three people around the world), followed by Islam, Hinduism and Buddhism. These major world religions have interesting geographical distributions (as we shall see in Chapter 3), although it must be stressed that the data (Table 2.3) and distributions are still-frames from a moving film of religious change through time and space.

Classification by belief

A common way of classifying religions is on the focus of their beliefs, particularly concerning deities (gods or goddesses). The most basic division is between religions that believe in God (*theistic religions*), and those which do not (*non-theistic religions*).

Table 2.3 Estimated strengths of the major world religions, 1985

Religion	(a) Number (million)	(b) Per cent
1 Christians	1,548.6	32.4
2 Muslims	817.1	17.1
3 Non-religious	805.8	16.9
4 Hindus	647.6	13.5
5 Buddhists	295.6	6.2
6 Atheists	210.6	4.4
7 Chinese folk religion	188.0	3.9
8 New-religionists	106.3	2.2
9 Tribal religion	91.1	1.9
10 Jews	17.8	0.4
11 Sikhs	16.1	0.4
12 Shamanists	12.2	0.3
13 Confucians	5.2	0.1
14 Baha'is	4.4	0.1
15 Shintoists	3.3	0.1
16 Jains	3.2	0.1

Source: Summarised from Barrett (1982: Table 4).
Note: The figures – based on estimates for mid-1985 – show (a) absolute strength; the number of people who profess a particular religion (in millions), and (b) relative strength; the percentage of total world population (4,773 million) represented by that religion. The religions are ranked by number of followers.

Theism, derived from the Greek theos (God), involves belief in at least one God who created the Universe and continues to rule human life. This God (or gods) is regarded as *transcendent* (eternal and infinite) as well as *immanent* (present and active in time and space), reveals itself to believers, and can be approached through prayer and worship (Bullock and Stallybrass 1981: 631). Examples include Christianity, Islam, Judaism and Sikhism. Deism, on the other hand, denies the possibility of divine self-revelation.

Non-theistic religions (such as Theravada Buddhism) have no belief in a person or being called God, although they might not explicitly deny the existence of God. Examples include the *primal* (or *animistic*) *religions* that believe that inanimate objects (such as mountains, stones, rivers and trees) possess souls and should be revered.

The theistic religions can be sub-divided on the basis of how many gods they believe in. Thus there are:

1. *monotheistic religions*, which believe in and worship a single deity (for example Christianity, Islam and Judaism), although there are fundamental differences in how they interpret this one God, and
2. *polytheistic faiths*, which believe in and worship more than one deity (for example, Hinduism).

This type of classification by belief is widely used by students of religion, and it does offer a meaningful basis for sub-dividing religions that reflects fundamental differences in what they believe. It is less successful in distinguishing between variants within religions (between Christian denominations, or between the two major branches of Buddhism, for example), and in detecting changes through time within particular religions (such as the sweeping changes in European Christianity attendant on the Reformation during the sixteenth century).

Classification by geographical scale

An alternative basis for classifying religions is on the basis of geographical scale, and Sopher's (1967) distinction of universal and ethnic religions has many advantages for geographical studies.

*Universal (*or *universalising) religions* – such as Christianity, Islam and the various forms of Buddhism – aim at world-wide acceptance by actively seeking new members. The ultimate goal of the three universal religions is to convert all people on earth. Believers are commanded to share their beliefs with non-believers, and each religion engages in missionary activities (proselytising) and admits new members through individual symbolic acts of commitment (Broek and Webb 1973).

*Ethnic (*or *cultural) religions*, in contrast do not seek converts. Each is identified with a particular tribal or ethnic group. *Tribal (*or *traditional) religions* – typical of Africa, South America, parts of South-east Asia, New Guinea and northern Australia – involve belief in some power or powers beyond humans, to which they can appeal for help. Examples include the souls of the departed, and spirits living on mountains, in stones, trees or animals. More broad-based *ethnic religions* include Judaism, Shintoism (the indigenous religion of Japan), Hinduism and the Chinese moral-religious system (embracing Confucianism and Taoism), which mainly dominate one particular national culture.

Unlike the universal religions, where diffusion is a primary objective, the spread of ethnic religions is slow because they do not proselytise. Although in the historic past Judaism engaged in missionary activity, in principle (and largely in practice today) membership is reserved for the in-group by inheritance. In other ethnic religions, individuals are not accepted until they are fully assimilated into the community. India and China, for example, gradually absorbed foreign tribes into their dominant culture, which expanded accordingly (Broek and Webb 1973).

The historical and geographical relationships between universal and ethnic religions are very interesting, and are discussed in Chapters 3 and 4. We can, however, note three spatial patterns in passing. One is that the universal religions are widely distributed (see Figure 3.1). Christianity has an almost global pattern at the close of the twentieth century, Islam is dominant through much of Africa and Asia, and although Buddhism transcends cultural and political boundaries it has a pronounced concentration in South-east and East Asia). Second, ethnic religions are often confined to particular countries – this is so for Hinduism (which is virtually confined to India), Confucianism (China), Taoism (China) and Shintoism (Japan). Third, traditional religions still persist in many less developed parts of the world.

Expression

One of the fascinations of looking at different religions is the diversity of ways in which beliefs are expressed. This adds something very special to the cultural fabric of an area. Few Westerners who visit ornate oriental temples are not touched by the sights, sounds and smells which make it such a unique experience. Similarly, witnessing elaborate religious ceremonies can be a moving and uplifting experience, even for the non-believer. But these are simply the visible manifestations of religious ways of life which reflect beliefs in different ways.

Ninian Smart (1989), who has written widely on the world's religions, identifies seven key dimensions of old and new religions that provide a framework for analysis and comparative study. These are ethics, ritual, narrative in myth, experience, institutions, doctrine and art. The visible expression of religion is not confined to buildings and ceremonies!

De Blij and Muller (1986) highlight four properties that are shared by all religions, although they are expressed very differently and in turn have different visible manifestations. The first is a set of doctrines of beliefs, which relate to the god or gods recognised by the faith, and which shape the attitudes, values and behaviour of believers. Second, they use religious literature (such as the Bible and the Koran) which embodies the normative codes. The third property is the rituals through which beliefs are given expression. Some mark important rites of passage in people's lives (such as birth and death, or marriage), whereas others are expressed at regular intervals (e.g. thanksgiving prayers at meal times or at night). Finally, many religions (especially Christianity and Islam) have created large, complex organisational structures and may command great wealth.

Each religion has an ethical side (dealing with personal conduct) and a ceremonial side (dealing with worship). Kant refers to these as the 'invisible church' and the 'visible church' (Fickeler 1962: 95). Both leave strong imprints on human society, shape behaviour and attitudes, and are worthy of geographical study.

Religion and postmodernism

As part of the same tide of philosophical change that has started to embrace geography (see Chapter 1, pp. 21–5), students of religion are starting to explore the implications of postmodernism. Milbank (1990) and others, who view religion through social science spectacles, propose that all human culture embodies concepts of reality which shape what we see, how we attach values and how we behave, and which are conditioned largely by experience. Language, the postmodernists argue, gives us an identity and a basis for attaching values. We can never stand fully outside language, which embodies in its vocabulary and the structure of its sentences the keys to how we see the world around us. Language is more than just a device for labelling objects or concepts – it is a filter through which we create meanings and values, which in turn shape what we see and how we behave. Consequently we are all shaped by culture and language, both of which change from place to place, from time to time, and over time; there is no pure meaning that stands outside culture. Postmodernists argue, therefore, that religious behaviour must be understood not in terms of rational functionalism, but as a reflection of human creativity filtered through language and culture. One implication of this is that no interpretation (in religion or in any other field) can be regarded as final, because meaning is never neutral – it is always contingent on social and cultural circumstances. In turn, it follows that 'received wisdom' in religion as elsewhere cannot be 'settled truths', it is always contingent. Such a postmodern perspective is likely to pose profound challenges because it calls into question every form of religious orthodoxy.

DIMENSIONS

Edmund Burke, in his *Reflections on the Revolution in France* (1790), emphasised that 'man is by his constitution a religious animal'. Barrett's (1982:

49) figures – that four out of five people on the planet are 'religionists' – seem to bear him out! What is important, of course, is *what* people believe and how those beliefs influence behaviour.

Belief systems

Religion is founded on faith, which Barrett (1982: 827) defines as 'firm or unquestioning belief in something for which there is no proof'. To that extent, religious behaviour will inevitably appear irrational and eccentric to non-believers.

Different religions embrace different beliefs, and express them in different ways. This is not an appropriate place to review the belief systems of individual religions; it will suffice to point out three common views of the fundamental beliefs embodied within most religions. These are that religion means belief in a supreme being or beings, that it is a unified system of beliefs about life and death, and that it is a system of intellectual beliefs and feelings. Some religions (Islam, for example) have a highly structured system of beliefs, whilst others (such as Hinduism) have no formal system of beliefs. In many religions beliefs are normative; in Christianity religious creeds are accepted by believers as a norm that they abide by.

Table 2.4 International variations in belief in God and in life after death

| | (a) God/universal spirit | | | (b) Life after death | | |
	Yes	*No*	*Don't know*	*Yes*	*No*	*Don't know*
Australia	80	15	5	48	40	12
Canada	89	7	4	54	29	17
France	72	23	5	39	48	13
USA	94	3	3	69	20	11
Britain	76	14	10	43	53	22
West Germany	72	20	8	33	48	19
Benelux	78	20	2	48	40	12
India	98	2	0	72	18	10
Italy	88	7	5	46	36	18
Japan	38	34	28	18	43	39
Scandinavia	65	25	10	35	44	21
Africa (sub-Sahara)	96	2	2	69	17	14
Latin America	95	5	0	54	37	9
North America	94	3	3	67	21	12
Far East	87	7	6	62	21	17
Western Europe	78	16	6	44	39	17

Source: Sigelman (1977).
Note: The figures show a summary of results from surveys in 1947 (11 countries), 1961 (6), 1968 (13) and 1975 (5 continents), where samples of people were asked two questions – (a) Do you believe in God or a universal spirit? (b) Do you believe in life after death?

40

Leaving aside questions of theology and the origins of different religions' belief systems, there is some evidence of interesting geographical patterns of belief. Sigelman (1977) reviewed the evidence collected in multi-nation surveys of religious beliefs between 1947 and 1975, which is summarised in Table 2.4.

The poll results show some interesting patterns. Quite remarkable is the broad cross-national consensus concerning the existence of a divine being, between three-quarters and four-fifths of the respondents in a typical nation said they believe in God or a universal spirit (inevitably there are many different notions of what this God or spirit might be). Sigelman's results indicate that belief in God has declined in some countries (such as Norway and Sweden) in the post-war period, whilst it has remained constant or increased in other countries (such as the Netherlands, France and the United States). Belief in God appears to be much more widespread in the Third World nations of Africa, Latin America and the Far East than in Western Europe. The level of belief in North America is closer to the Third World pattern than to the nations of Western Europe, despite economic and cultural similarities. Levels of belief in life after death are clearly much lower than levels of belief in God, and there is much greater uncertainty about it (shown in the higher percentage of don't knows) in every country and continent.

The polls in the USA and Canada indicate an almost universal belief in God, which is quite striking for countries where Church and state are formally (indeed, constitutionally) separated. Such beliefs are not monolithic, however. Social surveys of how Americans imagine God reveal strong variations between different sub-groups within the population. Nearly 1,600 adults were interviewed by the US National Opinion Research Center in March 1983, and the most common images were of God as creator (82 per cent), healer (69 per cent), friend (62 per cent), redeemer (62 per cent) and father (61 per cent). Females were found to have stronger views of God than males, educated people were less likely to accept any of the images, and older persons adhere to them more than the young.

Studies have also focused on the links between religious beliefs and broader conceptions of people's relationships with the world around them. Empirical studies of what US college students perceive to be the 'basic, permanent question for mankind', described by Yinger (1969), uncovered four areas of 'ultimate concern'. The first was major social issues (such as establishing peace, overcoming poverty and reducing population pressure). Next came interpersonal relations (removing barriers between people, understanding others) and individual creativity and happiness (promoting individual creativity and happiness, balancing needs of individual freedom and social order, using technology creatively). In fourth place was the issue of meaning, purpose, relationship of man to God (addressing questions such as What is the meaning of life? What are basic purposes? Where are we going? What is the relation of man to God? What is the soul of man?). Although the traditional 'religious' area appears only in fourth place, the important thing is that *it is there*.

Yinger (1969) regards religion as an externalised adaptation that serves both the individual and society, by providing ready-made solutions to adaptive problems in people. He identifies four particular cognitive needs which religion seems to cater for. These are conservation (the search for permanence amidst a world of change), representation (how to represent one's own thoughts and the physical and social environment), relations (how we relate phenomena in the world about us in a systematic manner) and comprehension (the ability to grasp relations and the underlying reasons for them). Elsewhere he argues that three beliefs are common among adults in any society, and they provide a direct context for religion (Yinger 1977: 68). They are a widespread interest in problems of meaninglessness, suffering and injustice; a sense that these are persistent and intractable problems; and a conviction that, despite their enduring quality, these problems can finally be dealt with by our beliefs and actions. Religious identity both influences and is influenced by these types of universal beliefs.

Relevance of religion

Religious beliefs, for most believers, are not just theories. They offer codes of moral behaviour that provide a guide for action and a particular lifestyle. This is what Kong (1990: 367–8) describes as the *latent* function of religion (it provides a socially cohesive force), in contrast to its *manifest* function (of explaining that which is outside humankind and mysterious to it). Even many recently secularised societies preserve features of past religious traditions, and it is often rather difficult to determine where religious factors end and secular ones begin.

One way of distinguishing between religious and secular catalysts of behaviour is to focus on the importance people attach to their religious beliefs. It is fair to assume a link between importance of belief and normative behaviour; people who say that their religious beliefs are important to them are more likely to take those beliefs and creeds more seriously and allow them to influence their attitudes and behaviour.

Sigelman's (1977) multi-nation survey of religious beliefs included the question 'How important to you are your religious beliefs?' The results show some pronounced but not entirely unpredictable patterns. Religion emerged as 'very' or 'fairly' important in India (where 95 per cent of respondents put it in those two categories), sub-Saharan Africa (86 per cent), the Far East (87 per cent), and Latin America (80 per cent). Less than half of the people interviewed in Britain (49 per cent), West Germany (47 per cent), Japan (46 per cent) and Scandinavia (45 per cent) described religion as 'very' or 'fairly' important to them.

A different way of estimating the importance which people attach to their religious beliefs is to look at how strong their participation is in religious practices. It is possible, using Barrett's (1982) data, to sub-divide the one in three people around the world who claim to be Christians, into various sub-groups (see Table 2.5). The estimated 70 million crypto-Christians make no public profession of their religion, and keep it concealed usually because of the threat of persecution. Nominal Christians (109 million) are – as the name suggests – Christians in name only, and religious beliefs play no part in their day-to-day existence. This still leaves an estimated 1,323 million affiliated Christians (with whom the churches keep in touch and who are on the records of the institutional churches or organised Christianity), who constitute 30 per cent of the world population. Inevitably, not all of those who are affiliated actually practise (defined as attending church), and this excludes a further 305 million. Most of the thousand million or so practising Christians attend church at least once a month, many doing so weekly.

The global picture is interesting, but the scale is too broad to allow any

Table 2.5 A profile of world Christianity in 1980

		Millions	*Per cent*
World population		4,374	100
(a)	*Religion*		
	Non-religionists	911	20.8
	Religionists	3,463	79.2
(b)	*Christianity*		
	Non-Christians	2,941	67.2
	Christians	1,433	32.8
(c)	*Public profession*		
	Professing Christians	1,362	31.1
	Crypto-Christians	70	1.6
(d)	*Church affiliation*		
	Nominal Christians	109	2.5
	Disaffiliated	−15	−0.4
	Affiliated Christians	1,323	30.3
	Doubly affiliated	−36	−0.8
(e)	*Practising/non-practising*		
	Practising Christians	1,018	23.3
	Non-practising Christians	305	7
(f)	*Church attendance*		
	Non-attending Christians	305	
	Annual attenders	66	
	Occasional attenders	63	
	Festival attenders	60	
	Radio/TV listeners	58	
	Monthly attenders	132	
	Fortnightly attenders	234	
	Weekly attenders	280	
	Daily attenders	125	

Source: Summarised from Barrett (1982).

meaningful evaluation of how and why people practise religion (rather than just professing it). Studies carried out in Sweden and the United States (Moberg 1982) throw some light on such questions. The results show that religious salience (prominence) is highest amongst evangelical groups, members of minority religions faced with prejudice or discrimination, and strongly religious people in secular or anti-religious societies.

Strength of religion (sometimes referred to as 'religiosity' or 'religious vitality') is notoriously difficult to measure. Membership of churches or religious institutions is not suitable because different religions and denominations use different definitions of membership. People are formally admitted to some religions and denominations, while others do not have formal entry requirements or procedures. An alternative is to try to measure religious practice, perhaps by measuring attendance at church, temple or mosque. But attendance is only one dimension of religious vitality and it ignores people of faith who do not attend regularly.

It is clearly difficult to quantify the importance of religion to believers, and this should not be surprising given the heterogeneity of beliefs between (and sometimes even *within*) religions, of ways of expressing beliefs, and of contexts in which religion might be a major determinant of behaviour and attitudes. Perhaps the soundest strategy is to examine patterns and levels of religious involvement, where actions speak louder than words. Yet even here there are problems. Most studies explain variations in religious involvement on the basis of a 'deprivation-compensation hypothesis', founded on the assumption that people get involved with religion (expressed communally) to fill the gaps that other parts of their lives cannot fill.

Religious involvement is much more multi-dimensional than that, argue Bibby and Brinkerhoff (1974), who stress the need to consider other factors too. One is what they term *socialisation*, meaning church involvement by the offspring of new recruits, for whom church-going is often likely to be an end in itself rather than a means to an end (at least initially). Social pressures encourage *accommodation*, as a result of which some people get involved simply to keep others (such as a partner, family or close friends) happy. *Cognition* (rational decisions based on choice not deprivation, and involving intellect, intuition and/or experience) draws others into religious involvement.

Other clues as to why some people become involved with religion and others do not come from a multidimensional scaling study by Russell (1975). This uncovered four axes along which sixteen world religions were described – local-foreign, Godless-fundamental, powerful and powerless, and a measure of social adaptability. The presence of the latter factor reinforces the social context within which religious involvement must be evaluated.

DYNAMICS

Religion in a generic sense, and individual religions, are not static but change through time. In Chapters 4 and 5 we will examine some of the geographical evidence for the geographical dispersion and diffusion of religions at different spatial scales, but here we need to reflect on some of the temporal dynamics of religious change.

Evolution of religions

If we look back through history, look around us today, and glance into the future, we see signs of how religion and religions change through time. Some examples illustrate how wide-ranging the trend is.

Traditional religion, so beloved of social anthropologists, often shows that early peoples had a well-developed capacity to think soundly on 'religious' matters and that the evolution of religion and society were closely bound together (Evens 1982).

Contemporary Islam is not uniform throughout the Muslim world. The mosaic of Islamic forms and practices — whose evolution reflects social, political, economic and class changes through time and space — embraces great diversity, including the feudal aristocracy in northern Lebanon, the working class Sufi brotherhoods of Egypt and the new bourgeoisie of Algeria and Morocco (Gilsenen 1990).

The Americas, throughout history a melting-pot of cultures, provide a fertile seed-bed in which new religious forms have been encouraged to grow. Sernett (1981) concludes that the emergence of Afro-American religion must be interpreted in terms of the interplay of many factors. These include the effect of different environments on religious forms and institutions, the prospect of African religious survivals, the role of clandestine slave religion within the plantation ecology, and the effects of a transition from a rural to an urban landscape on the black religious population in the United States.

Eyre's (1985) study of Biblical symbolism and the role of fantasy geography among the Rastafarians of Jamaica offers a particularly fascinating case study of religious enculturation. As he points out, Rastafarians 'have used biblical geography to invest themselves with sanctity, status and a global significance' (Eyre 1985: 145) because the basic outlines of their geography are drawn from an understanding of the biblical book of Genesis, reinterpreted through black Jamaican experience. Rastafarians see Africa as their original and ultimate homeland, themselves as Black Israel in exile, and the (white dominated) world around them as Babylon.

> Rastafarian fantasy geography has woven strands from biblical Israel and 'real' Africa into a picturesque, dynamic and functional 'map' for mental and behavioural orientation. Specific features of the Israelite wilderness encampment, as given in the Old Testament and many of its levitical

45

laws, are consciously interpreted and incorporated into Rastafarian communities and the spatial organisation within their yards. Dress, colour of buildings, flags and other paraphernalia are adopted and adapted from African models.

(Eyre 1985: 145–6)

Paradise exists in Jamaica, in the form of holy ground, especially the rich ganga land in the parish of St Ann. Ganga (*Cannabis sativa*: marijuana) – 'herbs' to Rastafarians, who believe it to be the herb referred to in the Bible (Genesis 1: 29) – is used universally among them as a sacrament.

Contemporary change

Change is not confined to the past, and today there are various signs of religious change and evolution. Three prominent ones are the emergence of new ideologies and new religious movements, and the rise of fundamentalism.

Emergence of new ideologies

There is no doubt that towards the close of the second millennium many developed countries have witnessed a decline in formal religion. The timing, speed and causes of this process of secularisation (which we look at in detail below) have varied from place to place, but one major – and fairly universal – catalyst has been the tide of socio-economic change that has swept around the post-war world.

The world population now faces a series of critical ethical questions and challenges, and new forms of spirituality are starting to emerge. Many leading contemporary thinkers are lending their voices to a call for a new ethical vision. Whilst there is no firm consensus about what this new vision might look like, it is widely agreed that it would need to encompass the pressing social issues of our times. Logically, therefore, it would include the ethics of nature (made necessary by global environmental problems), the ethics of life (reflected in debates about euthanasia, abortion and genetic engineering), the ethics of development (arising from the growing gap between rich and poor) and the ethics of money (arising from the problems created by the excesses of capitalism and collapse of socialism).

To the secular mind the search for such a new ethical vision must abandon the territory traditionally occupied by formal religion, which – it is argued – has proven so incapable of addressing such urgent questions. Religious leaders of many persuasions, however, would argue that to do so would be to abandon the whole of religion on the basis of (often anecdotal) evidence of religious practice. What's more, those who propose that traditional religions be abandoned in this way are usually judging from without rather than from within.

New religious movements

A second major change in recent decades has been the emergence of new religious movements. Stark and Bainbridge (1979: 124) define religious movements as 'social movements that wish to cause or prevent change in system of beliefs, values, symbols, and practices concerned with providing supernaturally-based general compensators'.

They propose that such movements can be placed along a spectrum representing the degree to which a religious group is in a state of tension with its surrounding socio-cultural environment – 'A church is a religious group that accepts the social environment in which it exists. A sect is a religious group that rejects the social environment in which it exists' (Stark and Bainbridge 1979: 123). This is a more specific meaning for the word 'church' than the two most common meanings – church as a building, or church as 'a body of people meeting on a Sunday in the same premises primarily for public worship at regular intervals' (Brierley 1991: 27).

There is now a growing body of empirical evidence about the dynamics and controls of religious movements, which provides a context within which to appraise contemporary developments. A number of common attributes shared by a range of religious movements – including numerous early Christian, Apostolic Poverty, Croat, Czech and German movements – have been identified by Murvar (1975). Three unsurprising shared attributes in movements that have evolved from earlier religions or denominations are the reinterpretation of original doctrine, the revitalisation of an original religious message, and a challenge to the worthiness of religious office-holders. Anti-imperialism, reflecting opposition to the imperialisation and politicisation of original religious doctrines, has also been a potent force in the evolution of many movements. The other prominent attribute is an orientation towards reform, because the demand for radical social change lies at the heart of many religious movements.

Rise of fundamentalism

One particularly prominent sign of religious activism and revival in recent decades has been the rise of fundamentalism. Fundamentalism is rooted in interpreting religious texts as literal wherever possible. Fundamentalist believers within a particular religion are therefore generally very active in proclaiming those revealed truths, both to non-believers and believers who doubt the literal truthfulness of the texts. Eagerness to encourage or persuade others to shed their doubts or accept the truths, and frustration at the speed with which they are prepared to do so, lies at the heart of many religious disputes and conflicts. We will explore this theme further in Chapters 5 and 6.

The most tangible recent expression of religious fundamentalism was the 1979 Islamic Revolution in Iran, which completely altered the political, cultural and social landscape of the country by placing religious rulers in control, adopting Islamic principles as the basis of civil law, and promoting religion to centre-stage of every facet of the life and work of the country.

In the United States, the rise of religious fundamentalism is reflected in the rise and size of the Moral Majority. This right-wing Christian coalition, which is a vocal and articulate expression of strong religious fervour and concern to preserve what it accepts as unequivocal biblical truths, has wielded enormous political clout and influenced policy decision-making in many areas (including abortion, gay rights and education). It is one sign of the impact which religious fundamentalism has had on social change in the United States, manifest particularly in the Race Relations and Civil Rights movements (Parsons 1966). Empirical studies of what makes individuals support the Moral Majority highlight the significance of a Christian Right orientation, cultural ethno-centrism, and authoritarianism; factors such as education, age and religious television appear to have secondary influence (Johnson and Tamney 1984).

SECULARISATION

Religion continues to play a leading role in shaping values and attitudes in many countries, particularly in traditional societies. In India, for example, religion pervades all aspects of public and private life. Some world religions are in fact doing better than simply maintaining their market niche. A graphic example has been the growing strength of Islam in the Muslim world, encouraged by a resurgence of revolutionary fervour, active government support, and new power based on oil revenues (de Blij and Muller 1986).

Yet there is no doubt that, during the twentieth century, traditional religious influences have been declining in a growing number of countries. Religion is being eclipsed in many modern industrial nations by the emergence of secular society, which places much greater emphasis on personal autonomy and rationality than on received religious wisdom, customs and ethics. Barrett (1982) highlights secularisation and the rise of secular quasi-religions as one of the most important religious trends of the twentieth century.

Secularisation has been defined as 'the discarding of religious faith' (Jordan and Rowntree 1990: 197), 'indifference to or rejection of religious ideas' (de Blij and Muller 1986: 205), and the 'desanctification of thought and behaviour' (Zelinsky 1973: 95). The *Collins English Dictionary* (1979) defines secularisation as 'to change from religious or sacred to secular functions; to dispense from allegiance to a religious order'. Glasner (1977) and others argue that it is the process whereby religious thinking, practice and institutions lose social significance. In short, secularisation means the decline of religion or the disengagement of society from religion (Fenn 1969).

There is no simple answer to the question of why secularisation occurs. The failure of organised religion to adapt and to meet the needs of modern society is doubtless partly to blame. However (as we saw in Chapter 1), religion,

culture and society are tied together by reciprocal threads; there is no simple
cause-and-effect relationship between religion and society. One hallmark of
late-twentieth-century life, particularly in developed countries, has been the
emergence of a culture founded on materialism, consumerism and the pursuit
of personal goals and happiness. God has been eliminated from modern
enquiry, which favours rational scientific explanations. Brook (1979: 513)

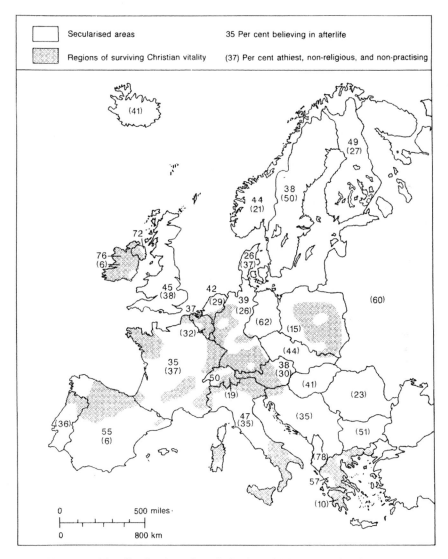

Figure 2.1 The distribution of secularisation within Europe in about 1980
Source: After Jordan and Rowntree (1990).

describes the declining influence of religion as a result of the progress of science and technology in the post-war period, of great advances in the natural sciences, and of the rapid rise in cultural and educational standards. Historians of science trace the roots of modern secularisation back to the scientific revolution of the eighteenth century, when a new-found interest in experiments and natural science started to undermine the religious culture that had given rise to scientific enquiry (Pratt 1970).

Bullock and Stallybrass (1981: 564) note that the word 'secularisation' is used in three different (but compatible) senses. One is to refer to the ending of activities that formally link religion and state. These include state support for religious bodies, of religious teaching in national schools, of religious tests for public office, of legislative protection for religious doctrines (such as the prohibition of contraception) and of censorship and control of literature, science and other intellectual activities in order to safeguard religion. Individuals are then free to deviate openly from religious dogmas and ethics. The second sense in which the word is used is to describe the decline of widespread interest in religious traditions, so that religious bodies cease to enjoy popular respect or attract many practising supporters. Examples include the rise of humanism amongst intellectuals, and systematic official attempts in Communist countries to suppress religion as antisocial. The third, and most extreme, context of secularisation is the end of all interest in religious questions and attitudes, including mysticism (Pratt 1970).

The rise of secularism has not been uniform across either space or time. The evidence shows that recent religious decline has been fastest in urban and industrial areas (Jordan and Rowntree 1990) and in Christian countries (de Blij and Muller 1986), particularly the United States and the traditionally capitalist countries of Europe. Figure 2.1 shows one interpretation of the geography of secularisation in contemporary Europe, which clearly suggests that Christianity has ceased to be of much importance to most of the population. The map shows some residual regions where the vitality and relevance of Christianity survive, and there is little doubt that these religious refuges are shrinking in size and declining in number. There is no universal explanation of the pattern or pace of secularisation.

Knudsen (1986) provides an interesting case study of the progress of secularisation within Norway, using the distribution of votes for the Christian People's Party as evidence of the strength of religious culture. His empirical study uncovered religious strongholds in western Norway, in rural areas and among people of low socio-economic status, and showed how 'urban places represent bridgeheads for the process of secularisation – a process that threatens the religious periphery' (Knudsen 1986: 1) where traditional values survive. The evidence also showed that secularisation takes place on different geographical levels, reflecting the results of different processes.

There is no single process of secularisation, and neither is it a simple linear change from secular to sacred society. It is an incremental progression, with different symptoms at different stages. Brook suggests that

> at first it is expressed in the fact that adherents of a certain religion do
> not observe its precepts and have a poor knowledge of the subtleties of
> the official religious teachings; then in the growing number of people
> who, although they consider themselves adherents of a church, do not
> believe in God; and finally in the steadily increasing number of
> convinced atheists.
>
> (Brook 1979: 515)

Some stages in the secularisation process are reflected in tangible ways, such as
falling church membership and declining church attendance. Membership and
attendance of Christian churches in Britain have declined at a fairly constant
rate of about 1.3 per cent a year over the last two decades, and in 1985 church
membership stood at 15 per cent and weekly church attendance at 10 per cent
of the population (Bradley 1987).

But statistics on secularisation don't tell the full story. Indeed, they *can't* tell
the full story because religion cannot be defined simply in quantitative terms,
using statistics such as the numbers of believers or churchgoers. De Blij and
Muller (1986) point out that the world religious map and national religious
tables must be interpreted carefully, because it is easy to draw the conclusion
that populations in areas shown as one particular religion adhere strictly to that
religion. This is simply not the case, because many people practise religions
other than the dominant one, and many others practise no religion at all.
Church or religious membership figures give a false impression, too, because
they are usually much higher than the number of active and practising
members. If anything, the statistics tend to underplay the rise of secularism in
the modern world.

To speak of secularisation as a single process is highly misleading because it
often reflects radical shifts in the orientation and expression of religion rather
than irreversible wholesale religious decline (Perrin 1989). Decline in established religions is often associated with its perceived failure to meet modern
needs, but new reform-orientated religious movements sometimes emerge to
take their place. In Britain, for example, the continued growth of the House
Churches (a nonconformist Christian movement that is presently small, but
growing rapidly) is eclipsed in global statistics by the continued decline of the
established Church of England. Knudsen (1986: 6) reports that in Norway,
'while the traditional rural religion is losing ground, the urban free churches
as a group experience a slow but steady growth'.

Martin is critical of

> the conventional history of secularisation [which] selects a particular
> historical baseline, employs over-simple sets of comparisons, often cares
> little for the coherence of its use of concepts, or their precise specification,

and is capable of organising material in terms of ideas of historical evolution rooted in the presuppositions of rationalistic or Marxist philosophy.
(Martin 1973: 82–3)

He finds no conclusive evidence – in art, science, and religion (church attendance, baptism and belief) – of religious decline in England in recent centuries, and concludes that secularisation is not a straightforward concept in a pluralistic society like modern Britain.

Perhaps it is more realistic to retain the word 'secularisation' to describe the process of religious decline, but abandon as unworkable any notion that it is possible to identify 'secularised' countries. There is no discrete point along the spectrum spanning sacred and secular societies, at which we can meaningfully or unambiguously state that a particular country is 'secularised'. Figure 2.1 shows areas classed as secularised, apparently on the basis of a sizeable (but unspecified) percentage of the population being classed as 'atheist, non-religious and nonpractising'. Just how meaningful this sort of distinction is in practice is hard to tell.

The processes, patterns and consequences of secularisation represent a largely unexplored patch, for geographers in particular, and even a modest investment of research in this area promises rich returns. It is a theme that is growing in importance, spreading in impact and emerging in its implications. It deserves more study.

RELIGION AND STATE

Religion clearly influences people's values and behaviour in a wide variety of ways. Most of what we have focused on in this chapter up to now relates to organised religion and how individuals respond to it by choice. In reality, other dimensions of the religious imprint can also have a strong influence on both individuals and society. One of the most prominent and pervasive of these is the relationship between religion and state.

Established religion

Religion and state have often been closely associated if not formally tied together. Throughout pre-Reformation Europe, for example, the Roman Catholic Church had immense and far-reaching political power, and political decision-making was heavily influenced by religious views and values. In post-Reformation times, the established (Protestant) Church of England has enjoyed a privileged position as the country's official church, headed by the monarch. In the post-war world, Islamic states have rooted political and legal systems and priorities firmly in the teachings of the Prophet Mohammed, and no distinctions exist in practice between religion and state. The establishment of the Jewish state of Israel in 1948 (to which we return in Chapter 6) was the outcome of the Zionist political movement (based on religion but in reality a secular nationalist movement), and the country's national identity has been delineated in fundamentally religious terms (Weissbrod 1983).

Religious pluralism

In many countries, however, ties between religion and state are much looser. This generally reflects religious pluralism and/or historical developments. North America is a classic example of a religious and cultural melting pot, inherently pluralistic, in which religious diversity is inbuilt and largely transcends ethnic or regional variations of social type or culture. Parsons (1966) suggests that this pluralism results from the interplay of many factors, including immigration, religious tradition (particularly Calvinism), secular-isation and individualisation, the liberating influence of free public education, anti-foreign sentiment, upward mobility, and the integration of blacks and whites.

The pluralism is positive, not passive, argues Parsons (1966) because it actively embraces rather than simply tolerates Roman Catholics and Jews, as well as the core Protestant community.

The roots of this pluralism can be traced back to the wisdom of the country's founding fathers, who enshrined a key principle in the United States Constitution. The First Amendment of the Constitution states that Congress (and by later extension the states) 'shall make no law respecting an establishment of religion, or prohibiting the free exercise thereof' (Gamwell 1982: 273). This is interpreted to mean that government cannot do anything that promotes either a particular faith or religion in general. Contemporary religious pluralism in the United States is more varied than the founding fathers could ever have imagined. Gamwell (1982: 273) notes that it was originally based on diversity of Christian communities, but now embraces Mormons, Jewish communities, and groups claiming some religious heritage from Islam and non-Western religions.

As Gibbs (1991) points out, the First Amendment clause was not designed to promote clashes between believers and nonbelievers, but in practice it tightly constrains what a person can say, where, and to whom. Clashes stemming from this formal separation of Church and state have become more common and more acrimonious in the United States particularly since the fundamentalist Moral Majority was formed in 1979, and they have raised complex questions about freedom of speech and religious liberty. The debate has been most visible and hostile in America's public schools, centring on issues like the place of prayers in classrooms. The ongoing debate in American public schools between fundamentalist creationists and those who believe in mechanical biological evolution has also produced more heat than light, and led Strahler (1983: 87) to conclude that 'the clearest solution in accordance with the First

Amendment is to exclude the transnatural realm entirely from the science curriculum of all public schools'.

Civil religion

Paradoxically, although state and religion are separated by constitutional requirement, God is by no means absent from the American public scene. The motto 'In God we trust' appears on every coin, on every dollar bill and in most state constitutions. Schoolchildren pledge allegiance to one nation, under God. Presidential speeches (certainly under George Bush) close with the benediction 'God bless America' (Gibbs 1991). Presidential inaugurations and opening of Congress sessions are accompanied by prayers and speeches by clergy (Parsons 1966). John F. Kennedy, a devout Roman Catholic, closed his inaugural address in 1961 with the claim 'here on earth, God's work must truly be our own' (Gamwell 1982: 285), and the well-known evangelist Billy Graham was amongst those who prayed at Bill Clinton's inauguration in January 1993. These forms of public expression of religion in the United States in some ways transcend conventional definitions of what religion means, partly because the term 'God' is used in a generic rather than a religion-specific sense.

Bellah (1967) has coined the term 'civil religion' to describe an elaborate system of practices and beliefs born of America's unique historic experience, which exists alongside formal religion but remains free from any single religious sect. It is, he argues, the one religion in the United States that is broad enough for all citizens to accept. His logic is this:

> although matters of personal religious belief, worship, and association are considered to be strictly private affairs, there are, at the same time, certain common elements of religious orientation that the great majority of Americans share. These have played a crucial role in the development of American institutions and still provide a religious dimension for the whole fabric of American life, including the political sphere. This public religious dimension is expressed in a set of beliefs, symbols, and rituals that I am calling the American civil religion.
>
> (Ballah 1967: 3–4)

The God of American civil religion is a curious construct of the American mind, who is 'not only rather 'unitarian', he is also on the austere side, much more related to order, law, and right than to salvation and love. ... He is actively interested and involved in history, with a special concern for America' (Bellah 1967: 7). In this way, Bellah proposes (ibid.: 8), civil religion has 'served as a genuine vehicle for national religious self-understanding' and (ibid.: 18–19) 'is concerned that America be a society as perfectly in accord with the will of God as men can make it, and a light to all the nations. ... it is a heritage of moral and religious experience from which we still have much to learn as we formulate the decisions that lie ahead'.

Bellah's paper on American civil religion has been welcomed as 'an inspiration to scholars of religion and society . . . [which is] of exceedingly great importance and value' (Stauffer 1975: 394) but criticised because 'it assumes that a substantial degree of moral and civil religious consensus is attainable in contemporary America . . . [and] it assumes that civil religion is a functional requisite of all societies' (ibid.: 393). None the less it opened up two decades of fertile debate and creative exploration of the idea and essence of civil religion, particularly amongst sociologists, although interest in the theme appears to have subsided since the early 1980s (Mathisen 1989).

CONCLUSION

In this chapter we have explored some of the dimensions of religious beliefs and practices which underpin and influence human behaviour. Whilst it is difficult to arrive at an unambiguous definition of the term 'religion', the evidence suggests that religion serves to unite believers and it gives them a meaningful way of interpreting what is going on around them. Classifying formal religions into theistic and non-theistic types helps to account for some observed differences in beliefs and practices, and distinguishing between universal and ethnic religions provides a framework for explaining how and why religions grow and spread (through space and time).

Religion is not a static phenomenon, and a geography of religion which over-looks the significance of religious dynamics (embracing historic evolution and contemporary change) ignores much of the richness and vitality of the subject. One of the purposes of this chapter has been to highlight important themes which have been relatively neglected by geographers. The emergence of new religious movements and the recent rise of religious fundamentalism pose particularly interesting challenges and are worthy of more detailed geographical study. Similarly, the increasing secularisation of many industrial and post-industrial societies – with secular values replacing sacred ones – reflects continued evolution (in this sense, of post-religious society) and deserves more close geographical examination. The broader theme of religion and the state (encompassing established religion, religious pluralism and civil religion) rarely surfaces in geographical explanations of socio-economic and cultural patterns and differences within and between countries, yet its implications could be widespread and highly relevant.

3

DISTRIBUTIONS
Spatial patterns of religion

Nothing is so fatal to religion as indifference, which is, at least, half infidelity.

Edmund Burke, Letter to William Smith, 29 January 1795

INTRODUCTION

Spatial patterns have traditionally captured the geographical imagination, and the study of the distribution of religion at different scales is doubtless what most other disciplines expect geographers to be engaged in. It is the most logical link between geography and religion, lends itself most readily to geographical analysis and interpretation, and is an area largely neglected by other disciplines. Moreover, it is a long-established focus within geography. Recall (from Chapter 1, pp. 8–14) the great interest during the 'golden era' of the sixteenth and seventeenth centuries in the distributions of different religions. It has continued to inspire geographical research and writing during the twentieth century. Many of the early publications on geography and religion, especially between about 1900 and 1960 and particularly from the European schools, were largely descriptive and most of them laid great emphasis on describing distributions of religions. Classics include Fleure's (1951) paper on 'The geographical distribution of the major religions', Le Bras's (1945) paper on religious geography, and Deffontaine's (1948) *Geographie et religions*.

The field has evolved greatly since this descriptive phase, and it now encompasses detailed empirical studies of how religions spread and take root in new areas, how religions survive in different places, and how religion can exert powerful influences on the character of culture regions. We shall examine these topics further in the next two chapters. It is important, however, first to review what sort of work has been done on spatial patterns of religion, because – as well as being interesting in their own right – present-day distributions can give valuable clues about historical evolution and contemporary change. Analysis of patterns might also suggest factors that help to explain those patterns. It is also useful to have a geographical framework established, because patterns of religion doubtless influence other aspects of human geography, including landscape (see Chapter 7) and pilgrimage (see Chapter 8).

DATA

Whilst the data shown in the maps and tables have an air of precision about them, we must be careful not to overlook the immense (and largely intractable) problems of data quality, reliability and availability. Recall (from Chapter 2, p. 35) Brook's (1979: 513) comment that 'the religious composition of the world population can be estimated only approximately'. Some countries have much more and better quality information on religion than others; indeed, for some countries, best guesses are all that exist. Where hard data do exist, there are considerable variations between countries in reliability and spatial resolution. Not all data refer to the same time-period, too. Definitions and classifications are not always consistent from one country to another, so this adds further distortion.

The most accessible and widely used statistics on contemporary religious distributions are contained in Barrett's (1982) monumental *World Christian Encyclopedia: A comparative study of churches and religions in the modern world, AD 1900–2000*, from which most of the tables in this chapter have been compiled. Inevitably when data are drawn from different sources (as in this chapter), some incompatibilities and inconsistencies must be allowed for. Given that our main focus is on broad patterns and characteristics of the distribution of major religions, these inconsistencies are of an order of magnitude that can be tolerated.

Two scales of analysis are used in this chapter – the global and the national – largely for convenience. Inevitably the national scale can be treated in more detail, although there are few countries for which reliable statistics exist. The data sources for the United States are the best national collections in the world, and this allows us to review a range of studies that describe different aspects of the spatial patterns of religion in North America.

GLOBAL PATTERNS

The world is a tapestry on which are woven some striking patterns of religious variation. Ideally, we might wish to be able to examine different dimensions of this religious tapestry, such as the degree of religious plurality in different places, or perhaps broad patterns of strength or character of variables such as religious commitment, adherence or activism. Data to allow such analyses of religiosity are simply not available, however, and we must be content with an examination of the distribution and relative strengths of major religions.

Continental scale

The data show some striking variations in the religious composition of different continents (Table 3.1). Although roughly one in three people on earth is classed as Christian, the spatial distribution is uneven. Thus a high percentage of the population in Europe (84 per cent), the Americas (91 per cent) and Oceania (84 per cent) is Christian, whereas the figure drops to 8 per cent in Asia and 45 per cent in Africa. Conversely, the great majority of Muslims (72 per cent) are in Asia, and most of the rest (26 per cent) are in Africa. Perhaps not surprisingly both Hinduism and Buddhism (both over 99 per cent) are overwhelmingly confined to Asia. Judaism, by far the smallest (numerically) of the five main world religions, has a much more dispersed pattern than the others.

Religious diversity

The continental data (Table 3.1) also yield some clues about large-scale variations in religious diversity. Whilst they do contain members of other major religions, Europe, Oceania and the Americas are so heavily dominated by Christianity that to all intents and purposes they can be classed as Christian. Africa, on the other hand, is not so dominated by one religion; both Christianity and Islam are dominant in roughly equal measure. Asia presents a radically different religious profile, and – at this coarse continental scale at least – it is very pluralistic. Hinduism, Islam, Buddhism and Christianity are all very strong there, though smaller-scale patterns doubtless exhibit greater homogeneity in particular areas.

Disaggregating the continental figures, and adding the numbers of non-religious people (which in many cases are very large), gives perhaps a better picture of religious diversity at this broad scale. What emerges is in some ways quite surprising (Table 3.2).

Table 3.1 Summary of the global distribution of world religions in 1980

| Religion | Number of people (millions) | | | | | | |
| | | | | America | | | |
	Europe	Asia	Africa	North	Latin	Oceania	Total
Christianity	420.9	250.2	236.3	227.2	392.2	21.5	1,548.3
Islam	9.2	588.7	215.8	2.6	0.6	0.1	817.0
Hinduism	0.6	644.0	1.2	0.7	0.6	0.3	647.4
Buddhism	0.2	294.7	0.01	0.2	0.5	0.02	295.63
Judaism	1.5	7.1	0.3	7.9	1.0	0.09	17.89
Total population	499.9	3,055.5	520.4	260.8	419.2	25.79	4,781.5

Source: Based on data in Barrett (1982).

Table 3.2 Distribution of the main world religions, by continent, in 1980

	Number (millions)	Per cent
(Former) USSR		
Christian	102.2	36.3
Non-religious	83.1	29.6
Atheists	60.6	21.6
Muslims	31.5	11.2
Jews	3.2	1.1
Population	281.2	100
South Asia		
Hindus	644.0	40.1
Muslims	534.9	33.3
Buddhists	150.9	9.4
Christian	125.9	7.8
New-religionists	66.8	4.2
Population	1,605.2	100
Oceania		
Christian	21.5	83.7
Non-religious	2.9	11.3
Atheists	0.5	1.9
Hindus	0.3	1.2
Muslims	0.1	0.4
Population	25.7	100
North America		
Christian	227.2	87.1
Non-religious	19.0	7.3
Jews	7.9	3.0
Muslims	2.6	1.0
Atheists	1.0	0.4
Population	260.8	100
Latin America		
Christian	392.2	93.6
Non-religious	12.9	3.1
Spiritists	6.7	1.6
Atheists	2.4	0.6
Tribal religionists	1.2	0.3
Population	419.2	100
Europe		
Christian	420.9	84.2
Non-religious	49.4	9.9
Atheists	17.4	3.5
Muslims	9.2	1.8
Jews	1.5	0.3
Population	499.9	100

continued

Table 3.2 (Continued)

	Number (millions)	Per cent
East Asia		
Non-religious	618.9	53.0
Chinese folk-religionists	179.3	15.3
Buddhists	143.4	12.3
Atheists	123.4	10.6
New-Religionists	38.1	3.3
Population	1,168.8	100
Africa		
Christians	236.3	45.4
Muslims	215.8	41.5
Tribal religionists	63.9	12.3
Non-religious	1.3	0.2
Hindus	1.2	0.2
Population	520.4	100

Source: Summarised from Barrett (1982).
Note: The figures for each continent show the number of adherents (in millions) and the percentage of the total population of that continent, in 1980. Religions are ranked within each continent on the basis of absolute strength; only the top five religions in each continent are shown.

The non-religious and atheist groups figure prominently in most continents. Although absolute numbers are usually quite small (the glaring exceptions being East Asia and the former USSR), in relative terms they are usually strong – non-religious are ranked second in the former Soviet Union, Oceania, North America, Latin America and Europe. Half of the people in the former Soviet Union, six out of ten people in East Asia, and roughly one person in eight in Europe is non-religious or an atheist. This lends weight to the argument about secularisation (see Chapter 2, pp. 48–52), but in the absence of historical data it is not possible to establish whether this is an emerging or persistent feature of the world religious landscape. The religious profile of the former Soviet Union is also far more diverse than one might initially assume on the basis of seven decades of anti-religious Marxist dominance. In the post-*perestroika* era of the Commonwealth of Independent States and Eastern Europe there are clear signs of religious revival, so the 1980 profile is likely to change a great deal in the coming decades. The persistence of tribal religionists in Africa and Latin America, and the emergence of new religionists in South and East Asia, also raise interesting questions about the dynamics, controls and influences of religious change (to which we return in Chapter 5).

Christianity

Christianity can be singled out for special treatment for two reasons – it has more followers than any other religion (Table 3.1), and it is better documented, particularly in terms of statistical information. We have seen already that nearly one in three of the world's 4.8 thousand million population is classed as Christian, and that Christians are found in large numbers in most places. The largest concentrations of Christians are in Europe and Latin America, where over half of the world's 1.5 thousand million Christians live, accounting for around 17 per cent of the global population (Table 3.3). About one person in seven in North America and Africa is classed as Christian, accounting for nearly another half a billion individuals (just under a tenth of the world population).

Like all other major religions, Christianity is not monolithic. Through history Christianity has been split into branches by religious schisms and differences of interpretation of religious history, traditions and texts (in this case the Bible). Some landmarks in this evolutionary process will be examined in Chapter 5, but it is important to note here that the numerical strength (both absolute and relative) of different Christian sub-groups (sects) varies from place to place (Table 3.4).

The Eastern Orthodox Church is particularly strong in the former Soviet Union, and in parts of Europe and Africa (particularly North Africa). Roman

Table 3.3 Relative and absolute strength of Christianity, by continent, based on estimates for mid-1985

	(a) Adherents *(millions)*	*(b)* Per cent of Christians	*(c)* Per cent of world population
Africa	236.3	15.3	4.9
East Asia	22.4	1.4	0.5
Europe	420.9	27.2	8.8
Latin America	392.2	25.3	8.2
North America	227.2	14.7	4.8
Oceania	21.5	1.4	0.5
South Asia	125.9	8.1	2.6
USSR	102.2	6.6	2.1
Total Christian	1,548.6	100	32.4
World Population	4,781.1		100

Source: Summarised from Barrett (1982.
Note: The figures show (a) the total number of Christians in mid–1985, (b) the percentage of all Christians who were living on that continent in mid–1985, and (c) the Christian population on that continent as a percentage of total world population in mid–1985.

Table 3.4 Distribution of the main Christian sects, by continent, in 1980

	Christians	*Catholic*	*Protestant*	*Orthodox*	*Anglican*
USSR	102.2	3.9	4.4	63.0	0.001
South Asia	125.9	72.8	23.0	3.2	0.3
Oceania	21.5	7.5	7.6	0.5	5.6
North America	227.2	86.5	94.5	9.3	7.6
Latin America	392.2	391.0	13.5	0.4	1.2
Europe	420.9	251.1	79.0	35.8	33.5
East Asia	22.3	3.6	7.7	0.07	–
Africa	236.3	89.7	62.9	32.7	–

Source: Summarised from Barrett (1982).
Note: Figures show the total number of Christians (in millions) by continent in 1980, with a break-down by Christian sect. The four sects shown account for most Christians; the balance includes non-conformists, including Baptists, Methodists and Pentecostals.

Catholicism – altogether much larger and more widely dispersed than the Orthodox Church – has its strongest presence, at least numerically, in South America and Europe. In South America almost all Christians belong to the Roman Catholic Church; in Europe well over half do.

Protestantism, a product of the Reformation in sixteenth-century Europe, remains numerically quite strong in Europe (where it accounts for nearly one in five of all Christians) but has its strongest base in North America (where it accounts for over 40 per cent of Christians). About a quarter of the large (and growing) number of Christians in Africa is associated with the Protestant churches. The so-called Anglican Communion – representing the Church of England, the Church of Ireland, the Episcopal Church in Scotland, the Church in Wales, the Episcopal Church in the United States, and other churches that are in full communion with each other – has most (70 per cent) of its members in Europe.

Global distribution

A number of maps of the distribution of world religions have been published, mostly in general geography textbooks. They often differ from one another in detail, particularly in terms of the classification of religions that is used, and the boundaries that are shown. There is, however, a broad consensus over the overall global pattern. Figure 3.1 shows a typical map of the global distribution of religions.

Distribution

The map (Figure 3.1) amplifies some aspects of the pattern outlined on the basis of continental statistics, and it contradicts other aspects. North America

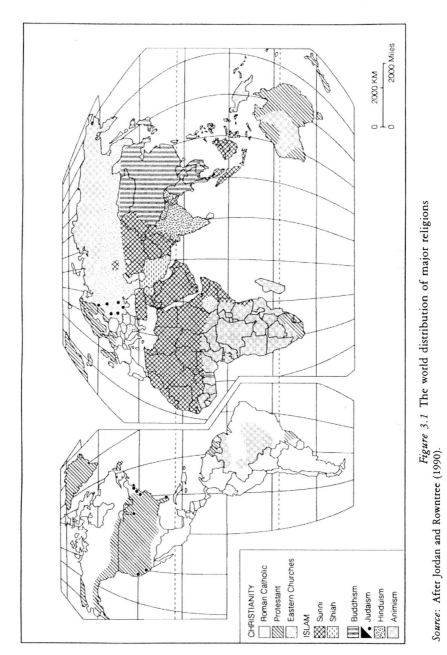

CHRISTIANITY
☐ Roman Catholic
▨ Protestant
▦ Eastern Churches

ISLAM
▩ Sunni
▨ Shiah

■ Buddhism
● Judaism
▨ Hinduism
☐ Animism

0 ———— 2000 KM
0 ———— 2000 Miles

Figure 3.1 The world distribution of major religions

Source: After Jordan and Rowntree (1990).

is Christian (predominantly Protestant), with some large Jewish populations in major cities. South America is mainly Roman Catholic, with residuals of animism in remoter parts of Brazil. Europe is almost entirely Christian – Protestant in the north and Roman Catholic in the south. Northern Africa is dominated by Islam, and much of central and southern Africa is classified as animism. There are a small number of Christian strongholds in different parts of Africa. Asia has a much more diverse mosaic of religions than any other continent, embracing animism in the far north, a large swathe of Eastern Orthodox Christianity through the former Soviet Union, the Islamic-dominated Middle East, Hindu India and the Buddhist and oriental religious strongholds of China and South-east Asia. Australia is mainly Protestant, with a large tract of animism surviving in the interior. Figure 3.1 contradicts the continental statistics mainly in Africa, where the large Christian presence is not reflected on the map.

As well as reading the map in terms of what religion is dominant in a particular place, we can use it to indicate in broad detail where the main religions are strongest. Christianity, with the most followers (Table 3.1), also has the widest distribution around the world. Not only is it dominant in many countries (Figure 3.1), it also has a strong presence in countries dominated by other religions (Table 3.2). Islam dominates northern Africa and south-west Asia (Figure 3.1), although it too has many followers in other countries (including North America, the former USSR and Europe according to Table 3.2). Buddhism remains strong in South and East Asia (Figure 3.1), but has far fewer followers elsewhere (Table 3.2). Hinduism comes third after Islam in terms of number of followers world-wide (Table 3.1), but most of the two-thirds of a thousand million Hindus is concentrated in South Asia (Table 3.2), almost without exception within the Indian sub-continent (Figure 3.1). Judaism is widely scattered around the world beyond the state of Israel (Table 3.2 and Figure 3.1).

Limitations

Inevitably the map (Figure 3.1) gives the impression that religion within any one of the shaded units is relatively uniform, which of course is clearly not the case. The map shows dominant or prevailing religion only, and it gives no impression of how competitive the situation is between leading and other religions. In this sense it also masks considerable variations in the strength of the absence of religion, not just in terms of the distribution of atheists and non-religionists but increasingly also in terms of the emergence of secular society. The map reveals nothing about another important religious variable, too, and that is religious vitality or adherence. It would be misleading to assume that each religion shown in the distribution was followed equally faithfully by all of its believers in all places, or that each religion was followed as faithfully as the rest. Similarly, the distribution masks some quite significant variations in how religion is expressed, both within and between religions. Sunni Muslims are not the same as Shiah Muslims, and their religious beliefs are not expressed in exactly the same ways. Roman Catholics, Protestants and Eastern Orthodox believers are all classed as Christian, but whilst they share some common beliefs and values, they express them in very different ways.

The map is also misleading in the sense that whilst large areas might be shown to have a particular religion dominant, what really matters is the population distribution – which is naturally not uniform within or between countries. Thus, for example, the large area in Australia classified as animism in reality accounts for a relatively small number of people. Conversely, the few large North American cities classified under Judaism account for up to 7 million individuals.

The mosaic of world religions raises interesting questions about how this pattern came into being, and what factors influenced it. Clearly, some components of the distribution are largely endemic. Animism, for example, is common amongst traditional societies and the archaeological evidence suggests that it was present in most cultures before more modern forms of religion took hold. Other components reflect religious persistence in or close to areas where those religions first appeared. Hinduism has dominated India since its birth, and Buddhism retains its foothold in the area where it first spread and became important. A third set of components reflects the spread of major religions from original source areas over time. Christianity is a good example – from origins in the Middle East, it now spans the globe. We will look further at this question of origins, diffusion and dispersion of religions in Chapter 4.

NATIONAL PATTERNS

The world map shows only the general pattern of religious distribution and the mosaic is more detailed, varied and relevant at smaller scales. For example, India appears on most maps of world religion as Hindu, and there is no denying the overarching importance of Hinduism to the country. But other religions (particularly Islam) survive and prosper there too. Inevitably, different scales of analysis reveal different patterns and distributions, and we must be careful not to draw too many definitive conclusions from the world map.

Few countries have sufficient high-quality national data to allow religious distributions to be examined in great detail, and there are vast areas of the globe for which no detailed studies have been carried out at all. Indeed, the availability of national-scale data on religious distributions remains the exception rather than the rule. Plugging this gap is a useful and potentially very fruitful challenge for contemporary geographers that would doubtless repay the effort.

Most of the published studies focus on one of three countries – India, Britain and the United States. We will explore the first two here, and devote the following section to the United States, which has by far the most comprehensive coverage and has been the focus of some pioneering distribution studies.

India

Religion and India are closely related in many ways, because South Asia was the birthplace of some of the world's most influential and important religions and the area has functioned as a religious cross-roads throughout history. Yet, perhaps surprisingly, the overall composition and distribution of religion within India have remained remarkably stable. Information on religion in India is collected regularly in national censuses, which provide a rich source of data for geographical studies (although the accuracy of the data is sometimes open to question).

One of the earliest detailed geographical studies of religious distributions in India, based on census data, was reported by Brush (1949). It is particularly illuminating, because it describes the situation in 1941, before Independence. India was granted independence from Britain in 1947, and Pakistan was created at the same time; the partitioning was heavily influenced by religious criteria. Brush produced a series of maps of the distribution of people within (the original) India, classified by religion. Figure 3.2 shows a much-simplified summary of his results, indicating those areas where Hindus, Muslims or aboriginal tribes account for over half of the 1941 population. There were relatively few areas dominated by traditional religion (Figure 3.2b), and these were small, scattered and located away from the main centres of population. The bulk of India was predominantly Hindu (Figure 3.2a), and it is no accident that modern India was carved out of this Hindu homeland, encompassing all of southern India and most of central (modern) India and the Indo-Gangetic Plain south of the Himalayas. The two strong Muslim enclaves, to the north-west and to the east of the Indo-Gangetic Plain, provided the nucleus of post-Independence Pakistan. Former West Pakistan is now the Islamic Republic of Pakistan, and in the late 1980s almost all (97 per cent) of its population was Muslim. Former East Pakistan is now the People's Republic of Bangladesh, with a slightly more mixed religious composition (83 per cent Muslim and around 16 per cent Hindu in the late 1980s).

Population redistribution along religious lines quickly followed partition in 1947. Changes in the religious composition can be detected in the 1951 census, only four years after the event (Mamoria 1965). The 1951 census results show that 85 per cent of the population of (post-Independence) India was Hindu, 9.9 per cent was Muslim, 2.3 per cent was Christian and 1.7 per cent was Sikh (with much smaller numbers of Jains, Buddhists, Parsis and tribals). Even between the 1941 and 1951 censuses there is evidence of a considerable increase in the number of Hindus in India, which Mamoria (1965) attributes to a balance of trade between Hindu immigration from Pakistan and Muslim emigration to Pakistan. The 1951 data also indicate a strong Hindu dominance

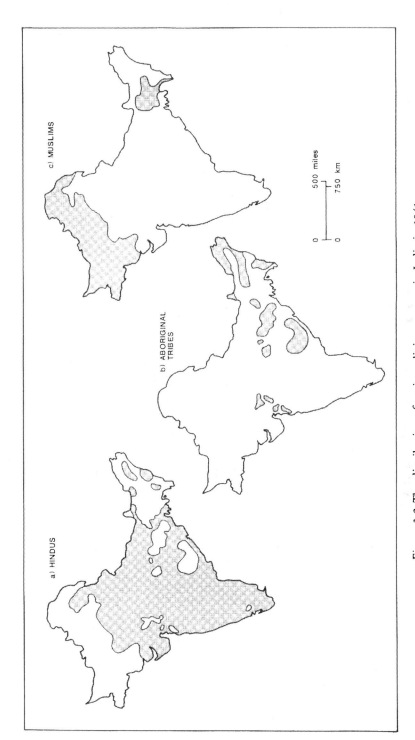

Figure 3.2 The distribution of major religious groups in India in 1941

Source: Based on information in Brush (1949).

in the central and southern parts of India and in Madras, with Muslims concentrated in the north and east. Christians were largely confined to the south (where missionary work first started). Sikhs were still heavily concentrated (90 per cent of them) in the Punjab, their original homeland. Buddhists were mainly restricted to Sikkim and the adjoining hills, and tribals were scattered in sparsely populated hills and jungles.

More recent census data shed some light on the dynamics of population changes between the major religious groups in India. One useful source is Roy's (1987) analysis of changes in the religious composition of India between 1901 and 1981 (Table 3.5). This shows that Hindus remain far and away the largest group, accounting for more than four-fifths of the total population and outnumbering the second group (Muslims) by more than seven to one. Most of the main religious groups have continued to grow rapidly through the twentieth century, particularly Muslims who accounted for over 11 per cent of the total population in 1981 (compared with less than 10 per cent in 1951 according to Mamoria (1965)).

The diversity of India's religious population begs much closer scrutiny and more detailed analysis than the rather summary treatment given by Brush (1949). Dutt and Davgun (1979) illustrate the benefits of using multivariate statistical approaches to the delineation of religious regions, using factor analysis on census data. Their empirical analysis grouped the population of India into five major religious classes (mainly Hindu, mainly Muslim, mainly Sikh, mainly Buddhist and mixed), the distribution of which could then be mapped. The resultant maps are, not surprisingly, not too different from the earlier and simpler maps (such as Figure 3.2). They show Hindus spread out all over the country, Sikhs concentrated in the Punjab, Muslim majorities in the north-western Kashmir and Lakhadweep, a strong Buddhist presence in the Western Kashmir area, and mixed religious areas in the north-eastern and south-western tip of India. They conclude that 'the "heart" of India is essentially Hindu, while only the islands, the north-western, eastern and southern peripheral limbs have sizeable non-Hindu concentrations' (Dutt and Davgun 1979: 213).

Table 3.5 Religious composition and change in India, 1901–81

Religion	1981 population (millions)	Per cent of total 1981 population	Per cent decadal growth
Hinduism	549.7	82.63	24.14
Islam	75.57	11.36	30.69
Christianity	16.17	2.43	16.83
Sikhism	13.07	1.96	26.15
Buddhism	4.71	0.71	22.52
Jainism	3.19	0.48	23.17

Source: Summarised from Roy (1987).

No discussion of religion in India is complete without some reflection on the Hindu caste system. As Schwartzberg emphasises, although the caste system is gradually weakening in the cities,

> it continues to exert a major influence in shaping the lives of the nation's 375 million villagers, including a large number of Sikhs, Muslims, and Christians, to whom caste is theoretically anathema. . . . Caste rules cover virtually all major aspects of life; whom and when one may marry; what, how, and with whom one may or may not eat; what forms of respect one ought to show to, or receive from, members of higher or lower castes; what rituals one shall perform and how; and, for most castes, what forms of work one may or may not undertake . . . there can be no understanding of Indian society, economy, or political life without an appreciation of the pervasive role of caste.
>
> (Schwartzberg 1965: 477–8)

With that in mind, Schwartzberg (1965) tried to map the distribution of castes in part of the Indo-Gangetic Plain, using 1931 census data (the most recent and comprehensive data source available). He focused on ten of the 28 castes with more than a million members (Table 3.6), each of which had different distributions reflecting needs and opportunities. 'Each caste has its characteristic distribution, whether viewed regionally or intraregionally on a village basis. The distributions are not random but are related to specific facts of history and

Table 3.6 Characteristics of some Hindu dominant castes in India, 1931

Caste	Traditional occupation(s)	Status	Per cent literate
Brahmin	Priestly and other learned professions	Very high	22.0
Rajput	Warriors and landowners	High	9.2
Jat	Cultivators	Medium high	3.6
Kurmi	Cultivators	Medium	3.8
Lodh	Cultivators	Medium	1.3
Ahir	Cowherds and cultivators	Medium	1.5
Sheikh	Occupationally not specialised	Varies with region	12.0
Chamar	Leather workers and agricultural labourers	Very low	0.4
Pasi	Toddy tappers and cultivators	Low	0.4
Nai	Barbers and factotums	Low to medium	2.5

Source: Schwartzberg (1965).

to the distinctive role played by the caste in the total socio-economic system' (ibid.: 495). Thus Brahmins (the priestly caste), for example, tended to congregate in cities and towns, whilst maintaining their ties with ancestral villages. Jats (yeoman cultivators) were found to be generally the most numerous and most powerful caste throughout the Punjab plain; they appeared to be willing to provide for themselves or go without some services traditionally provided by certain specialised castes. The Chamar (untouchables; they are really a non-caste, below the caste system) were a widespread group who perform services such as removing dead animals from the village, flaying them and tanning their hides, making and repairing shoes and other leather objects.

Britain

A number of aspects of the geography of religion have been studied, although there are many more gaps than pieces in the jigsaw of understanding. Paradoxically, we know probably more about patterns in the nineteenth century than we do about twentieth-century patterns, thanks to a comprehensive nation-wide Census of Religious Worship carried out in 1851. No survey has been carried out on that scale since.

The 1851 Census of Religious Worship provides a fascinating snapshot of the mid-nineteenth-century geography of religion in Britain (Table 3.7). Sixty per cent of the population in England and Wales attended church or chapel on census day, and nearly half of them were Anglicans. Most of the rest attended Nonconformist chapels. Relatively few (about 2 per cent of the population) went to mass in Roman Catholic churches, and even less (just over 1 per cent) attended other churches.

Gay (1971) makes extensive use of the 1851 Census evidence in his *Geography of Religion in England*. The book includes some informative maps of the distribution of various indicators of religious practice, based on Gay's

Table 3.7 Summary of the 1851 Census of Religious Worship for England and Wales

Denomination	Attendance as per cent of total population	Attendance as per cent of all attenders
Anglican	29.5	48.6
Old Dissent	12.8	21.1
Methodism	15.2	25.0
Roman Catholic	2.1	3.5
Others	1.2	1.8
Total	60.8	100

Source: Summarised from Coleman (1983).

own Ph.D. thesis. The spatial pattern of religious practice in England, based on levels of church attendance, shows a marked north–south gradient (Figure 3.3). Levels of attendance were on the whole much lower in the northern counties, and low attendance was also prominent in the metropolitan region and in growing industrial areas.

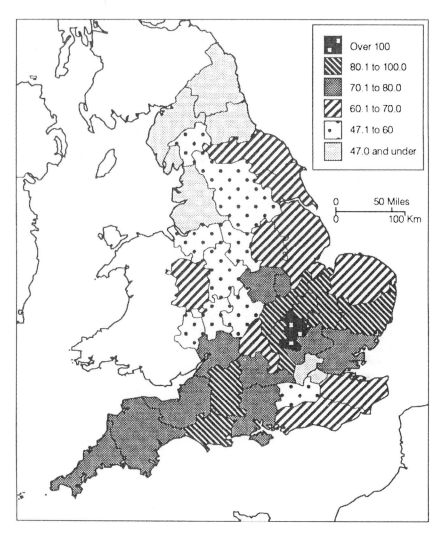

Figure 3.3 Geographic variation in religious practice in England in 1851
Note: The map shows variations in Index of Attendance, which is the total church attendance on Census Sunday (30 March 1851) expressed as a percentage of total population for that area. Some people attended two or three church services on the day, and an Index value of 100 would mean that just over half of the total population attended.
Source: After Gay (1971).

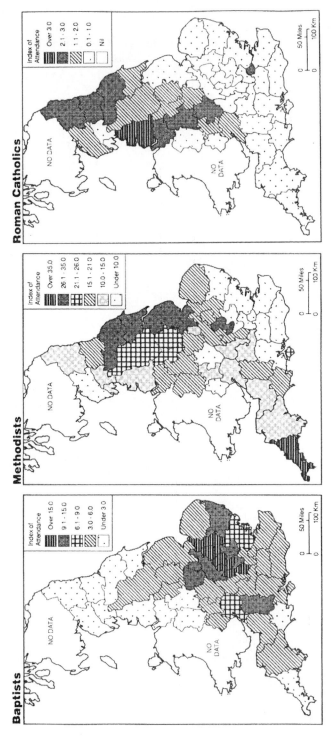

Figure 3.4 Distribution of Roman Catholics, Baptists and Methodists in England in 1851

Source: After Gay (1971).

Interesting patterns emerge when denominations are viewed separately (Figure 3.4). Roman Catholic attendances were confined almost exclusively to the North and Midlands, with a small but noticeable outlier in London. Baptists had quite the opposite distribution, with highest levels of attendance in the south and east. The Methodists, on the other hand, show a much more variable pattern with highest attendance in Cornwall in the south-west, and high levels also in the north-east. Hempton's (1984) study of Methodism and politics in British society highlights some of the broader political implications of this national pattern. The map evidence (Figure 3.4) suggests that each denomination had its stronghold areas, and these largely reflect regional tradition and socio-economic make-up.

Maps of overall levels of attendance, or attendance by denomination, reveal interesting patterns around the country. They reveal little of the religious character of different areas, however. In 1860 a Dr Hume produced a Religious Map of England, based heavily on the 1851 Census returns, which tried to portray variations in the religious landscape of the country (Figure 3.5). The pattern is quite striking, with Wales monopolised by chapel-going Dissenters and much of southern England dominated by church-going Anglicans (with pockets of Dissenters and irreligious in the south-west). Most of northern England was dominated by either the irreligious or nominal Anglicans, and a boundary zone between north and south had a strong but not dominant Anglican presence. Individual large towns and cities sometimes reflected the dominant religious complexion of their surrounding area, but sometimes showed strikingly different allegiances. A good example of the latter is the towns with sizeable Roman Catholic populations, within mainly irreligious areas.

Coleman's (1983) analysis of the patterns of attendance in southern England revealed great diversity both within and between counties. Absolute levels of Anglican attendance were high by national standards, but patchy, and he found great unevenness in the incidence of Nonconformity. Anglicanism seemed to be weaker in the upland areas (such as the Weald, the Mendips, Dartmoor and much of Cornwall) and in industrial areas (such as old textile areas of Wiltshire and east Somerset, and mining districts of Devon and Cornwall). Urbanisation also appeared to influence Anglican attendance, which was low in and around the major towns (such as Portsmouth, Southampton, Bristol and Plymouth), whereas Roman Catholic attendances were relatively large (but small in absolute terms) in larger urban districts and fast-growing towns (where strong labour demand drew Irish immigrants into the region).

A contemporary map of religious adherence in England and Wales would doubtless show quite marked differences from the 1851 picture, with local and regional changes superimposed on an overall decline in levels of church or chapel attendance. Such a map would also highlight the increasing diversity of religion in Britain, reflecting the injection of new (as in non-indigenous)

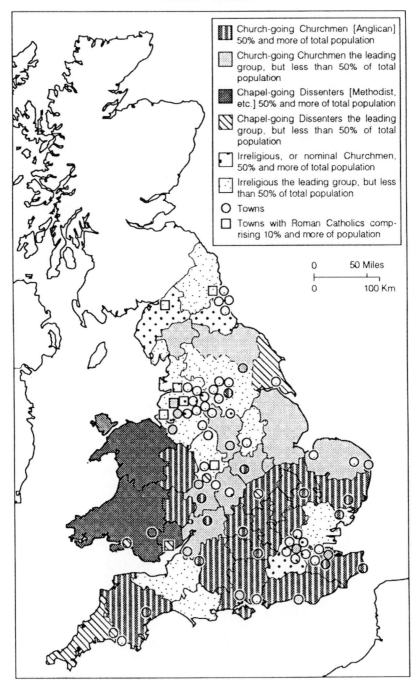

Figure 3.5 Distribution of religious adherence in England and Wales in 1850
Source: After Gay (1971).

religions and cultures through immigration and diffusion. The 1851 census makes no mention of Jews, Muslims, Hindus and other religious groups that would now claim a niche on the key of maps like Figure 3.5.

Studies of the distribution of Jews in the United Kingdom (who represented just over 1 per cent of the population) show a pronounced concentration in major cities. Waterman and Kosmin (1986) show that about twice as many Jews live in London as in the provinces, and Greater Manchester is the major provincial centre. Some cities (like Leeds) have declining Jewish populations, whilst others (like Glasgow) are relatively stable. Secondary provincial centres appear to be emerging in retirement communities along the affluent South Coast (including Brighton and Hove).

Whilst the contemporary geography of religion across Britain as a whole remains largely unexplored, two areas that have been studied are Scotland and Northern Ireland. We will look at Northern Ireland in Chapter 6, because the province's religious geography and its political history are intimately intertwined.

Piggott's (1980) study of the geography of religion in Scotland emphasises regionality and diversity within this small country, and it highlights the patterns and causes of recent declining church membership. The nineteenth century saw dramatic changes in the Scottish religious landscape, with the founding of new denominations and massive growth and migration of population. These trends can be detected in the results of the 1851 Census of Religious Worship, which allowed Piggott to compile maps of attendance at Roman Catholic, Church of Scotland, Free Church and Mormon churches. Surveys show that church attendance in Scotland was highest in the mid-1950s, when up to 60 per cent of the adult population were church members. Total membership and attendance have declined significantly since then (in 1971, for example, about half the adult population were church members), which Piggott accounts for in terms of net population decline, migration and suburbanisation.

THE UNITED STATES

More studies have been undertaken into the geography of religion in the United States than in any other country, partly because of a long-standing institutional interest in religion there and also because of the interest of geographers (particularly within cultural geography). Another major factor is the availability of high-quality data sources, which has enabled detailed studies of how religion varies in both time and space in the United States. Historical research is illustrated by Gaustad's (1976) *Historical Atlas of Religion in America*, and recent patterns are documented in the *Atlas of Religious Change in America, 1952–1971*, edited by Halvorson and Newman (1978). Sources of contemporary information include *The Encyclopedia of American Religions* (Melton 1989).

Although the American data are both more abundant and more detailed than that available elsewhere, it is still far from perfect because some religious groups are unable or unwilling to provide statistics and definitions of membership are not uniform from group to group. Everyone who has been baptised (including infants) is counted as a member of the Roman Catholic Church, whereas only those who have been confirmed (generally over 13 years old) are counted by the Protestant churches. As we noted in Chapter 2, 'membership of' does not necessarily mean 'active involvement in' a particular church. 'Membership' spans the spectrum from passive adherence through to militant participation, so membership figures by themselves can be misleading.

Numerically, Christianity is by far the strongest religion in the United States (see Table 3.1) and it is not surprising that most of the detailed studies of religious patterns have focused on aspects of Christianity. Inevitably, most of the case studies reviewed in this section echo this overwhelming emphasis.

This abiding interest in American religiosity is interesting in the light of the traditional separation of religion and state and in the light of widespread support for civil religion (see Chapter 2, pp. 54–5). Many studies have demonstrated strong levels of religious commitment or intensity in America, which Stump (1984b: 144) regards as 'somewhat paradoxical ... for it clashes with very poor levels of demonstrable knowledge of Christianity'.

Diversity

One prominent hallmark of religion within the United States is its diversity. This melting pot of a country boasts an almost unrivalled variety of religions, reflecting both historical factors (particularly migration) and contemporary socio-economic processes. Zelinsky has noted that

> in terms of religion, the United States has been a land of phenomenal diversity from the very beginning, and this heterogeneity fails to show the slightest sign of slackening. ... The best account available lists some 251 religious bodies able and willing to offer statistics on church membership, but some scores of additional groups undoubtedly exist. Much of this variety was imported from Europe. ... The multiplicity of religious bodies is visible not only at the national or regional scale but even within a single small city, village, or neighbourhood.
>
> (Zelinsky 1973: 94)

He argues that such diversity is

> a ringing affirmation of the dogged individualism of the American. The freedom to choose one's church, to float freely from one group to another, or even to invent an original sect of one's own is deeply cherished, and is closely akin to other forms of personal mobility and self-assertion.
>
> (Zelinsky 1973: 94)

The large number of religious bodies that Zelinsky was referring to includes many small groups and sects with local or regional rather than national distributions. The 'big league' is dominated by a small number of large Christian denominations, which have followers distributed throughout the country. The profile of Church membership in 1970 (Table 3.8), for example, shows the overwhelming dominance of the Protestant and Roman Catholic churches, who between them accounted for just under 92 per cent of the national total.

A particularly valuable source of data on religion in the United States is the Church Membership Study, which has collected county level statistics for the entire country in 1952, 1972 and 1980 (Newman and Halvorson 1984). These have been by far the most comprehensive studies of their type, although they relate only to Christian churches. The unique national archive has enabled detailed studies of religious change through time, spatial patterns of religious membership, and correlations between religion and a wide range of socio-economic variables. The census data show great stability in the number of Christian denominations over the period 1951 to 1971 (Table 3.9), although it is not immediately obvious whether this represents religious constancy or a persistent focus within successive studies on the same groups. A dramatic growth in the number of members of the Christian churches is evident in the data – total membership rose by more than half over the three decades.

Table 3.8 United States church membership, 1970

Religious group	Members
Protestant	69,740,000
Roman Catholic	47,872,000
Jewish	5,870,000
Eastern Orthodox	3,745,000
Old Catholic, Polish, National Catholic and Armenian	818,000
Buddhist	100,000
Total	128,145,000

Source: Broek and Webb (1973), Table 6.3, p. 150.

Table 3.9 Summary of United States church membership, 1952 to 1980

	1952	1972	1980
Number of denominations	35	35	35
Number of adherents (millions)	71.2	103.0	110.0

Source: Newman and Halvorson (1984).

Distribution

The surveys of church membership in the United States, along with compatible information from the Canadian census, 'constitute a rich source of information for mapping' (Shortridge 1982: 177) which has been quarried extensively since the early 1960s.

Gaustad's (1976) *Historical Atlas of Religion in America* provides a baseline against which to measure recent changes in the strength and distribution of major religious groups. Maps and data in the atlas show that large denominations are spread across the whole country, although there are some interesting contrasts between the cities and the countryside. American Jews are almost entirely concentrated in cities, and Roman Catholics, Episcopalians and Unitarians are also predominantly urban. The Baptists, on the other hand, tend to be more heavily concentrated in rural areas, along with other smaller sects (such as the Mennonites, including Amish) and fundamentalist groups derived from Puritan settlers (Broek and Webb 1973: 150).

Most studies focus on church membership rather than any measures of religious activism. There is considerable scope for detailed analysis of patterns and controls of variables like church attendance. Broek and Webb (1973: 149) report that on average about two-fifths of the adult population of the United States attends worship services in a typical week (a higher national average than in most European countries). The proportions vary from three-fifths for Roman Catholics to one-fifth for Jews. They outline but do not illustrate some interesting patterns in church-going. For example, the proportion that attends worship services in a typical week is highest in the Midwest and lowest on the West Coast. Attendance levels are highest in rural areas and lowest in big cities, too.

The distribution of religion within a country such as the United States mirrors other socio-economic distributions, at a variety of scales. Shortridge (1982: 178) anticipates correlations between religious diversity and factors such as social stratification, wealth and individuality, and he points out that 'where one or two faiths dominate, one could expect strong influences on such things as demography, place names, land patterns, food preferences, and general social attitudes'.

The nation-wide church membership data have been used in numerous studies, many of them designed to identify geographical patterns in religious denominations. Several different approaches have been used, and the results have often been interpreted in terms of cultural regions.

Denominational pattern

Zelinsky's (1961) paper 'An approach to the religious geography of the United States; patterns of church membership in 1952', published in the *Annals of the Association of American Geographers*, was the first systematic attempt to

78

map religious variations across the United States using empirical data. It remains a classic of its type, and rightly enjoys a reputation as one of the most valuable and widely quoted papers within geography and religion. It has spawned a number of subsequent studies, which have confirmed and complemented its results rather than contradicting them.

County-level data from the 1951 Church Membership Study were used to produce a series of maps of the nation-wide distribution of individual denominations. These were then brought together in a composite map that delimited seven religious regions (Table 3.10). Variations in the size and distribution of religious groups were explained largely in terms of spatial movements and success in attracting new members and holding old ones, rather than as a result of differential fertility and mortality. Zelinsky concludes that

> the contemporary map of American religion indicates four major sets of migrational events: 1. the colonisation of the Atlantic Seaboard by regionally distinctive groups; 2. their westward progression into the interior; 3. the concentration of post-colonial immigrants from Europe, Canada, and Latin America in particular localities; and 4. the rather less visible interregional migration of members of different groups.

> (Zelinsky 1973: 97–8)

The Zelinski (1961) map has been widely used in textbooks, and some writers have modified the detailed boundaries a little and proposed further subdivisions of the major units. De Blij and Muller (1986), for example, suggest that a French Catholic area centred on New Orleans should be recognised, as well as the mixed denominations of the Peninsular Florida (with a large Spanish Catholic group in metropolitan Miami).

Gastil (1975) has slightly updated the original Zelinsky map to take account of more recent census data, and his regionalisation is shown in Figure 3.6. This

Table 3.10 Religious regions within the United States

1   New England (defined by the Puritan heritage, a dominantly Protestant region)
2   Midland (a mixed region, with Jews important in New York City, Catholic dominance in most cities, and Methodism dominant among the strong Protestant presence)
3   Upper Middle Western (dominated by nineteenth-century immigrants and their faiths; mixed Catholic, Lutheran and Evangelical)
4   Southerns (heavily Protestant, primarily Southern Baptist; parts of the South are the most homogeneous religious areas in the country)
5   Mormon (mainly Utah and southern Idaho; in some counties over 99 per cent of adherents to religious groups are Mormons; population is nearly all church-affiliated)
6   Western (area of low religious affiliation)
7   Spanish Catholic

Source: Based largely on Zelinsky (1961).

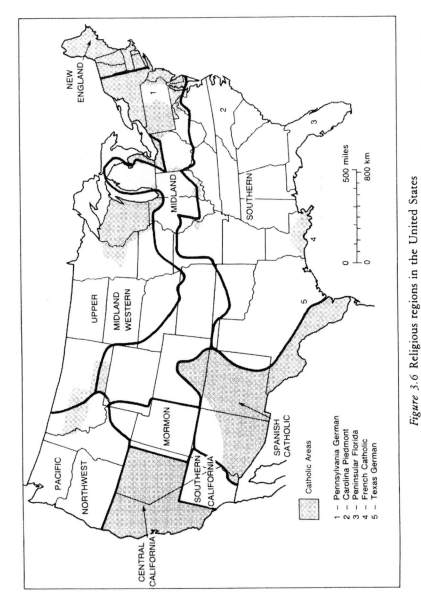

Figure 3.6 Religious regions in the United States

Note: The map shows a modified version of Zelinsky's (1961) classification of religious regions, based on church membership data.

Source: After Gastil (1975).

recognises a strongly Catholic area in New England, and a broad region extending from the Middle Atlantic in the east to the Mormon region in the west with a mixture of denominations dominated by no single church (although Methodism is the largest single group). The Upper Middle West is dominated by Lutheran churches, and the Mormon region (centred on Utah) provides a distinctly separate religion unit. Baptists are the leading denomination in the South, where – together with other conservative fundamentalist denominations – they have given rise to the so-called 'Bible Belt'. Spanish Catholics dominate the South-west. No single denomination dominates the West, which Gastil proposes should be sub-divided into two regions – the Pacific South-west Region (which is strongly Catholic, with a large Jewish population in the Los Angeles area), and the Pacific North-west (with even lower religious affiliation and Protestant dominance).

The pattern (Figure 3.6) is interesting, and many writers have pointed to the close correlations between some of the more prominent religious regions and culture regions. But the map must be treated with caution, because – like the world map shown in Figure 3.1 – it shows dominant religion and can mask significant variations within and between delimited religious regions. The Methodist Church, for example, is important at the national scale but dominant in only a handful of areas. Because of data limitations the maps completely overlook the 13 million or so Americans associated with black churches, along with the 8 million associated with Eastern Orthodox churches, many native American churches and Jews (Gastil 1975).

Numerous studies have sought to update Zelinsky's (1961) pioneering paper, most usually by plotting census data by county without then generalising the patterns into religious regions. Results from one such study are shown in Figure 3.7, based on United States county data. The map is shaded according to the dominant church or denomination. Interpretations of the national pattern usually place heavy emphasis on migration history (Broek and Webb 1973, Jordan and Rowntree 1990). Thus, for example, the distribution of Roman Catholics partly reflects waves of immigrants from Europe and other parts of the Americas. A concentration of Catholics along the Mexican Border in Texas, New Mexico and Arizona might reflect the legacy of the Spanish-Mexican influence, along with recent immigration from across the border. Similarly, the Roman Catholic enclave in the coastal region of Louisiana betrays the area's French heritage. Large numbers of Catholic immigrants from Ireland and central and southern Europe have swamped the original Protestant stronghold of New England.

Utah stands out as *the* Mormon state, which is not surprising given that nine-tenths of all church membership in most counties belongs to the Church of Latter-Day Saints (the Mormons). This distinct religious region also defines the Mormon culture region (see Chapter 5, pp. 159–63).

The Protestant churches also exhibit distributions that seem to owe as much to history as to contemporary socio-economic factors. The South is strongly

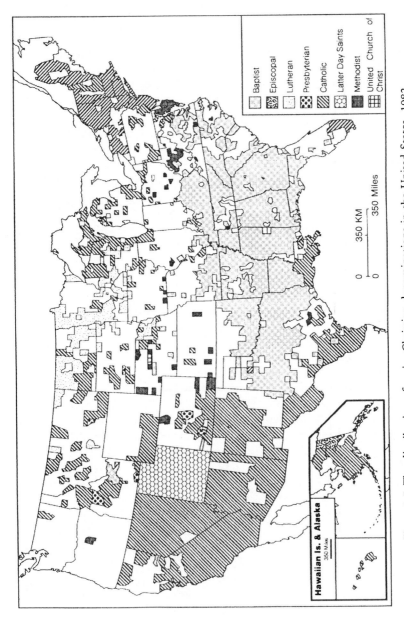

Figure 3.7 The distribution of major Christian denominations in the United States, 1982

Note: The map shows areas where 50 per cent or more of the total church membership is claimed by the particular church or denomination.

Source: After Jordan and Rowntree (1990).

dominated by Baptists (Figure 3.7), and Lutherans dominate parts of the Midwest farm belt. Congregational churches are still strong in New England, and are scattered throughout the Midwest. The most widely dispersed of the Protestant denominations are the Methodists, Presbyterians and Episcopalians. The main centre of Methodism runs through the Middle Atlantic states and the southern part of the Midwest to the Rocky Mountains, whilst the main centre of Episcopalians stretches from their original core area in southern New England to Virginia (Broek and Webb 1973: 150).

Multi-dimensional classification

The studies of denominational patterns highlight the diversity of religion within the United States and they provide one illustration of just how complex the spatial mosaic is. But, without undermining just how useful such maps are, some critics have argued that they do not portray the picture very clearly because they are effectively one-dimensional. Most such maps (Figure 3.7 being an example) show dominant denomination in an area, which gives no indication of the degree of pluralism in that area or of the level of religious activism there.

An alternative approach is to go beyond the simple mapping of church membership statistics, and try to characterise patterns of religiosity in more meaningful ways. Shortridge (1976, 1977) has championed this school by attempting a new regionalisation of American religion, based not on denominational patterns but on the basis of religious philosophy, ideology, intensity and diversity.

Shortridge (1976, 1977) analysed data from the 1971 national survey of church membership. The initial database included membership information on 53 denominations representing 81 per cent of the estimated 125 million Christian church members in the country. He tried to portray, in a multivariate way, how many churches were present in each county, and how strong religious belief or commitment was. As a first step, individual churches were divided into two groups – liberal (such as some Baptist, Methodist and Lutheran sects), and conservative (such as the Lutheran Church–Missouri Synod, the Southern Baptist and the Free Methodist). Second, Shortridge developed measures of four important dimensions of religion:

1 Catholics as a percentage of total population;
2 liberal Protestant adherents as a percentage of all Protestants;
3 a measure of religious commitment or intensity (church members as a percentage of the total population);
4 an index of religious diversity (the degree of local numerical dominance by a given religious body).

Next, the geographical distribution of each measure was analysed, using the nation-wide county-level census data. For example, Liberal church adherence as a percentage of total Protestant adherence was found to be highest in New England and Pennsylvania and in a belt running through the Middle West. Not surprisingly, this variable scored lowest in the southern Bible Belt. Religious diversity was greatest along the Pacific Coast states, in peninsular Florida, and in much of the Middle West and Central Plains. Diversity was particularly low in the Mormon West, southern Louisiana, much of the South, the Spanish borderlands, and the northern Plains. The map of religious commitment was complex and not easy to interpret, but it highlighted three areas of high commitment – Mormon oases in Utah and Idaho, lowland South (including the Spanish borderlands and French Louisiana) and the upper Middle West. Shortridge (1976: 431) points out that all three are inherently conservative places, where church is seen as a symbol of ethnic heritage and group solidarity.

Step four was to use the four measures to create a regionalisation of American religion. Five 'religious types' of area were identified, using an optimisation type of clustering analysis:

1 transition;
2 intense Conservative Protestant;
3 diverse Liberal Protestant;
4 Catholic;
5 super Catholic.

A map was then constructed, showing the distribution of each 'religious type'. It showed three areas of 'intense conservative Protestantism' (in the South, Utah and adjacent states, and parts of the northern Plains) and two main areas of 'super Catholicism' (in French Louisiana and parts of the Spanish Southwest). The Middle West and the West Coast emerged as 'diverse, liberal Protestant' areas. Religious types 1 and 4 were generally found not to occupy large relatively homogeneous areas. 'Transition type' occurred in the western two-thirds of the upland South, and the 'Catholic type' was found mainly in parts of the north-east and upper Middle West. The northern Plains area emerged as a religious melting pot, with no particular 'religious type' dominant.

The summary map shows a lower southern 'Bible Belt' of strong conservative Protestantism, a northern belt of diverse, liberal low-intensity Protestantism, and several 'super Catholic' areas in which the Roman Catholic Church is strongly dominant. As Shortridge (1977: 150) concedes, this statistical regionalisation echoes patterns shown on Zelinsky's (1961) map and updated versions of it. This, he insists, reinforces the concept of religious regions and illustrates their stability even when multivariate definitions are attempted.

A second and much more narrowly focused multi-dimensional study is Pillsbury's (1971) statistical characterisation of the religious geography of Pennsylvania. The study used Principal Components Analysis (PCA) to build multi-denominational religious regions within the state, although the approach could be adopted nation-wide. Five primary religious areas were identified within Pennsylvania – the late immigrant-dominated western coal fields, the Pennsylvania German Susquehanna basin, the New England-influenced northern border counties, the late immigrant-dominated central cities, and a strip of conservative Protestant-dominated counties between the German and western coal field areas. Pillsbury concluded that the PCA method produced similar results to Zelinsky's (1961) study, and his methodology was at least as effective as Zelinsky's and required far less intuitive interpretation on the part of the investigator. It is perhaps surprising that more studies have not adopted this type of approach, particularly given the availability of such detailed church membership data at county level for the United States. Herein, doubtless, lies an inviting and potentially very rewarding avenue for further geographical research.

Dynamics

One of the problems of compiling maps of religious distributions is the impression given that patterns are unchanging through time. Naturally this is not so, and a number of studies have focused on different aspects of the dynamics of religion within the United States.

The existence of comparable statistics on church membership for 1952, 1971 and 1980 opens up the prospect of examining how membership has changed over the three decades. Newman and Halvorson (1984) used the membership archives for this very purpose, and they found – contrary to expectation – that for both the 1952–1971 and 1971–1980 periods, the broadly representative denominational data showed remarkable stability rather than change. This was so, they point out, despite the fact that one in five Americans changed place of residence each year. Such pronounced stability suggests that Americans do not carry their denominational affiliations with them when they move, but that they adopt the religious organisations of their new environment. The results contradict a widely held assumption that a highly mobile population leads to religious mixing and, in turn, decreases the sharpness with which religious regions can be defined. This so-called 'religious convergence' argument is not borne out by the facts, which in fact illustrate the opposite.

Regional culture in the United States appears to be not only strong, but also persistent. Support for the notion of regional divergence, rather than convergence, is offered in other studies of changing patterns of religious affiliation. Stump (1984a, 1984b) found a twentieth-century trend towards regional divergence between the main Protestant groups in the United States. The evidence showed that Baptists in the South, Lutherans in the upper Midwest and Mormons in the West all dominated their regions more thoroughly in the early 1980s than they did at the turn of the century. Such persistence is particularly interesting given the common assumption that the US national culture is rising

at the expense of traditional regional cultures, and the widespread talk about
the growing relevance of civil religion within American life.

A relatively understudied aspect of American religion is the Protestant–
Catholic distinction, which has traditionally been associated with social and
economic distinctions too. Shortridge notes how

> the religious labels were, in part, mere symbols for a growing national
> tension completely nonreligious in nature. The traditional American self-
> image of Jeffersonian democracy, the ideal of an agrarian nation peopled
> by independent and individualistic men and women, was labelled as
> Protestant. Catholicism, in contract, came to symbolise a threat to this
> vision. Catholics were urban and involved in labour unions; they were
> seen as the pawns of big business.
>
> (Shortridge 1978: 57)

The distribution of Catholics in 1971 shows their virtual absence in the South,
with only two large clusters outside the Louisiana French and recent migrants
to Florida. Shortridge concludes that 'on most socio-economic yardsticks the
Protestant-Catholic distinction has essentially disappeared' (ibid.: 59) and
although 'economic divisions have declined and disappeared . . . cultural ones
remain strong. Intermarriage rates, although increasing, are still low' (ibid.:
60).

American Jews also figure prominently in the religious scene, although maps
like Figures 3.6 and 3.7 overlook them entirely. Newman and Halvorson (1979)
examined the changing distribution of Jews between 1952 and 1971, and
found that they had increased in number substantially in areas of the country
not traditionally associated with Jewish residence. Despite this shift, the Jewish
population remained highly concentrated in metropolitan area counties.
Regardless of their size, Jewish communities were overwhelmingly situated in
areas characterised by high degrees of religious pluralism.

One of the primary mechanisms by which religious distributions change
through both time and space is denominational switching (individuals and
families transferring their membership from one church to another). Stump
(1987) has examined regional variations in such switching among white
American Protestants, using logistic regression on data for 1974 and 1984.
Significant regional differences in the determinants of switching were
uncovered, extending beyond patterns of religious affiliation and participation
into the factors that influence religious behaviour, which are themselves shaped
by the regional context of the switcher. The main conclusion of the study was
that

> patterns of switching within the nine regional cohorts differ most strongly
> in the degree to which they are influenced by religious conversion,
> childhood affiliation and interdenominational marriage. Such differences
> appear to derive from regional contrasts in the social influence of religion,

the role of church affiliation in social identity, attitudes towards religious experimentation, the prominence of particular denominations, and levels of denominational diversity.

(Stump 1987: 448)

Different motives appear to operate in different regional settings. Sometimes the decision to switch reflects a religious motivation, such as an experience of conversion. Sometimes it reflects social concerns, such as a desire to belong to the church attended by one's spouse or associated with one's ethnic or status group.

Clearly, many interesting geographical questions concerning spatial and temporal variations in religious stability and change have yet to be asked, let alone answered. We return in Chapter 5 to explore some of the dynamics of religious change, but it might be instructive here to reflect on a few brief case studies of particular religions or religious areas within the United States.

Case studies

These case studies deal with the Bible Belt, the unchurched, and civil religion within the United States. They review some interesting studies that complement the more mainstream geographical interest in the distribution of church members and the definition of religious regions.

The Bible Belt

One of the most widely discussed dimensions of American religion is the existence and significance of the so-called Bible Belt in the South. The expression was first coined in the mid–1920s and has been part of the American vocabulary ever since 'to designate those parts of the country in which the literal accuracy of the Bible is credited and clergymen who preach it have public influence' (Tweedie 1978: 865). Whilst there is broad agreement that the Bible Belt is located in the rural South, the actual area where fervent religious fundamentalism flourishes is rarely defined geographically. Indeed, it is shrouded in 'locational cloudiness' (Heatwole 1978: 50) since 'neither church attendance records nor church membership statistics are valid indicators of fundamentalist religious orientation' (Tweedie 1978: 866) because of the wide range in theological, social and cultural attitudes found within individual denominations.

Heatwole (1978) tried to delimit the Bible Belt on the basis of 24 Christian denominations that profess a belief in the literal interpretation of the Bible, using data from the 1971 census of church membership. His map (Figure 3.8) shows the percentage of Bible Belt population by county, nearly two-thirds being members of the Southern Baptist Convention. Twelve fundamentalist groups account for 98 per cent of the Bible Belt population used in compiling

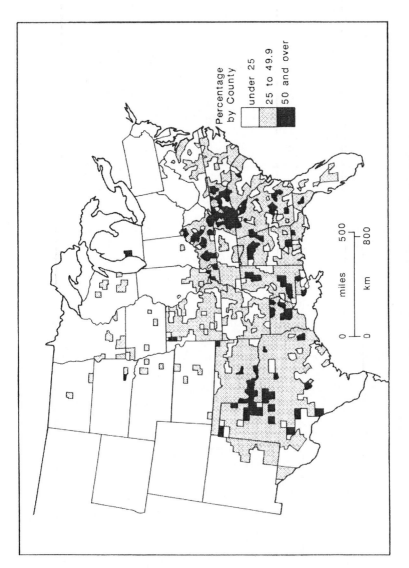

Figure 3.8 The distribution of Bible Belt population in 1961

Note: The map shows the percentage of population on the southern United States, by county, which belongs to a fundamentalist Bible Belt church. The Southern Baptist Convention accounts for most of the pattern.

Source: After Heatwole (1978).

the map. The pattern is far from uniform, and there are many gaps in the distribution. Such variations reflect the presence of non-fundamentalist groups (especially the Methodists), and the lack of significant Bible Belt populations in most urban counties in the South (attributed largely to in-migration of non-Southerners and diffusion of secular ideas amongst the better educated city folk).

An alternative approach to the delimitation of the Bible Belt was used by Tweedie (1978). Recognising the role played by religious television and televangelism in galvanising the fundamentalists who give the Bible Belt its name, Tweedie mapped Sunday audience estimates for the five leading evangelical, fundamentalist religious programmes (Figure 3.9). On this basis, he concluded that 'viewed as a whole, the Bible Belt appears as a broad zone stretching from Virginia to northern Florida in the East and from the Dakotas to central Texas in the west. Within this Belt are two cores, an Eastern and a Western' (Tweedie 1978: 873). The Eastern core is centred on the five major television markets in Virginia and North Carolina, and the Western core hinges on Little Rock and Tulsa. Connecting the two cores is a zone of moderate response. Tweedie (1978: 875) stressed that the core areas of religious television cut across religious regions based on denominational patterns, such as those identified by Zelinsky. He concluded that 'The Baptist South certainly is a major part of this Bible Belt, but areas of strength also include parts of the Methodist-dominated Midwest as well as portions of the predominantly Lutheran Dakotas.'

The unchurched

Geographers of religion in the United States have, quite understandably, devoted most of their attention to the distribution of particular religions and church membership. As a result, Heatwole (1985: 1) argues, hardly any research has been done on the unchurched (people who are neither members nor involved in any religious institutions), who thus remain 'a largely overlooked segment of the region's social fabric'.

Heatwole (1985) set about redressing the balance by looking at the geography of the unchurched in the South-east United States, using 1980 data. In some ways this theme represents the negative of the Bible Belt snapshot, because many of the gaps in the Bible Belt church membership map (Figure 3.8) are occupied by the unchurched! Heatwole's results show that

> a substantial portion of the South-east's population is unchurched and that the unchurched percentage generally is on the rise. The situation is not the same everywhere, however. Unchurchedness appears to be greatest in Florida, northern Virginia and Appalachia. Unchurchedness has no proven cause(s), but several factors appear to be spatially associated with it. These include: recent in-migration; rugged terrain;

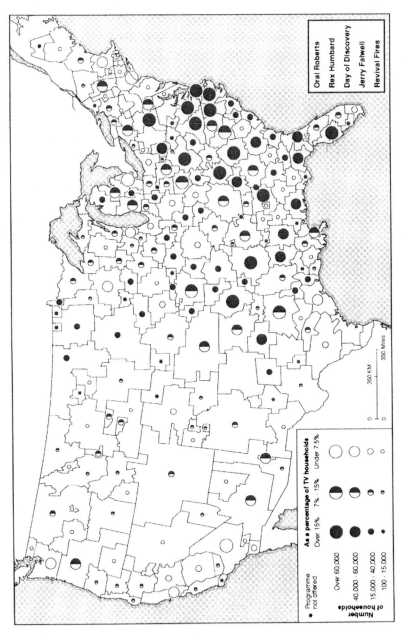

Figure 3.9 Sunday audience estimates for the five leading independent religious television programmes in the United States, 1973

Source: After Tweedie (1978).

low population densities; high population densities; liberalism; poverty; and lack of ethnic or other minority group identity.

(Heatwole 1985: 12)

Looking at patterns of religious absence, as opposed to religious activism, would appear to be a promising theme that deserves much more attention from geographers. It could throw some much-needed light on the geography of secularisation, and would doubtless cross-fertilise with studies on civil religion in the United States.

Civil religion

We have already seen in Chapter 2, pp. 54–5) that civil religion is a prominent feature of religion in the United States, reflecting a deep-rooted and persistent interest in patriotism in a country where Church and state are formally separate. Sociologists and historians have devoted studies to civil religion since Robert Bellah first coined the term in 1967. However, as Stump (1985: 87) observes, 'little attention has been devoted . . . to regional variations in the practice of civil religion, despite abundant evidence of persisting regional variation in many other aspects of American religious behaviour'.

The only study yet published on the geography of civil religion in the United States, by Stump (1985), explored patterns in the popular observance of the American Bicentennial. He analysed where most Bicentennial events were held, using a Register compiled by the American Revolution Bicentennial Administration during 1975 and 1976. The results show that the strongest commitment to public celebration of the Bicentennial occurred in a large region stretching from the western prairie states across the Great Plains into the northern Rocky Mountains. Suggested causes of the prominence of civil religion in that region include the importance of religion generally in providing a sense of community in a region characterised by cultural isolation, the effect of rural and small-town conformity, and the influence of a regional culture that is closely tied to the traditional mainstream of American culture. Stump concludes that

the interior North may represent a previously unrecognised religious region, in which there exists a particularly high commitment to public celebration of American civil religion. This commitment is expressed both in the large number of Bicentennial events, relative to the region's population, and in the distinctively celebrative quality of those events.

(Stump 1985: 93)

CONCLUSION

Spatial patterns of religion are one of the most obvious focal points of geographical studies, but, as the examples described in this chapter show, they can also be very fruitful and highly informative. Inevitably some countries have better data sources than others, so the treatment is not uniform from country to country. Equally inevitably, the most detailed studies are those dealing with small areas. Notwithstanding these practical constraints, the published studies reveal interesting patterns which hint at underlying processes of diffusion and change. By far the most detailed and penetrating geographical studies have focused on the United States, and in recent decades some of the finer spatial detail of the nation's rich religious tapestry have been described and mapped. There is abundant scope for further geographical research into spatial patterns of religion in all countries.

4

DIFFUSION
Religious beliefs and organisations

Persecution is a bad and indirect way to plant religion
Sir Thomas Browne, *Religio Medici* (1643)

INTRODUCTION

The religious patterns described in Chapter 3 raise interesting questions about how religions evolve, and what processes are important in distributing them through space and time. At different spatial scales it is possible to both describe and delimit a complex mosaic of religious variations. The global pattern begs fundamental questions about how and when Christianity became so widely dispersed, for example, and about why Hinduism has not been more widely embraced outside India. Maps of national patterns, in India, Britain and the United States, raise more questions than they answer – such as what role is played by immigrant groups in the dispersal of religions?, what factors allow some religions to survive in an area while other religions decline?, and how does religious mixing and enculturation in diverse societies change the original religions and sometimes give rise to new religious forms and expressions?

Such questions of religious dispersal, survival and change are addressed in this chapter and the next. Here we focus mainly on patterns, mechanisms and processes of religious diffusion in an attempt to better understand how religions disperse across geographical space. In Chapter 5 we turn to the dynamics of religious change and review the relevance of factors like religious adherence, the role of ethnicity and migration, the topic of religious persistence, and the eminently geographical issue of religious culture regions.

Any consideration of how religions spread and change must concede the difficulties of distinguishing between religious change and cultural change. As Tyler rightly points out,

> many of the major religions of the world have become so inextricably linked with particular racial groups, cultures, political systems and lifestyles, that it is difficult to imagine one without the other. It is hard to imagine Thailand without Buddhism, or India without Hinduism, for

example. Christianity has become intricately bound up with the lifestyle of Western culture.

(Tyler 1990: 12)

At the outset, therefore, we must recognise that to talk of the diffusion of particular religions as if they were separate from (indeed, as if they were *separable* from) the culture in which they are embedded is folly. Thus, for example, any attempt to examine the spread of Islam or the recent rise of Islamic fundamentalism that does not reflect the fundamental and wholesale integration of religious belief and culture, politics and lifestyle is doomed to be superficial and simplistic. Throughout this chapter it is taken as axiomatic that religion and culture are usually so closely intertwined that one reflects the other.

GLOBAL PROCESSES AND PATTERNS

The point was emphasised in Chapter 3 that present-day distributions of religions are merely snap-shots in a continuously unfolding moving film. Even within the twentieth century there are plenty of signs of change and evolution of the geographical patterns of religion. Inevitably the trends are not uniform in either space or time; some religions are declining in some places at the same time as they are expanding in other places. Contrast the significant growth of Christianity in Africa, where the number of believers trebled between 1940 and 1990, with its continuing decline in Europe over the same period (Tyler 1990).

At the global scale, two factors are particularly important in accounting for the distribution of the major religions at any point in time – the places where religions originated, and the processes by which they were dispersed and diffused. A number of studies have focused on these themes, and a useful source for students of geography and religion is the *Historical Atlas of the Religions of the World* edited by al-Faruqi and Sopher (1974).

Even a simple summary of the distribution of major religions towards the close of the twentieth century indicates that many of them have spread considerably beyond their country of origin (Table 4.1). Thus, for example,

Table 4.1 The spread of major religions around the world

Religion	Country of origin	Main concentration today
Buddhism	India (Nepal)	South-East Asia, China, Japan, Tibet
Christianity	Israel	Europe, the Americas, Australasia
Hinduism	India	Mainly India
Islam	Arabia	Gulf States, Africa, Bangladesh, South-East Asia
Judaism	Israel	Israel, Russia, USA, Western Europe
Sikhism	India (Punjab)	Punjab, Africa, Europe

Christianity has spread from an initial source area within present-day Israel to span the globe. Hinduism, in contrast, remains largely confined to its country of origin (India). Given the enormous impact of religion on many dimensions of human geography (see Chapter 1, pp. 2–7), the question of how and why religions spread at different speeds and in different ways is surely of more than marginal interest to geographers.

Origins

One particularly striking aspect of the geography of religions is that all of the main world religions originated within a relatively small area in what is today south-western and southern Asia. The fact that they originated there at different times only serves to heighten the curious coalescence of religious birthplaces. Late nineteenth- and early twentieth-century attempts to explain such puzzling geographical phenomena relied heavily on environmental determinism.

Environmental determinism

Environmental determinism is founded on the assumption that human activities are controlled or determined by the environment. The adoption of such a perspective in explaining the origin of the major world religions is not new. As far back as 1795, according to Buttner (1980: 93), the German geographer Kasche pointed to the influence of climate on religion. Kasche (1795: 35) wrote that 'a raw or mild climate often has an influence which is often ignored but which is none the less remarkable. The gentle climes of Italy bring forth none of the phantoms, ghosts and apparitions that frighten the superstitious inhabitants of the colder North.'

Geographers ever since have speculated about why the major world religions evolved initially in the Middle East and in India, and whether the particular environments there might help to explain the origin and initial development of the religions. Semple (1911) argued that early nomadic desert dwellers of the Middle East could see the movement of stars and planets through clear skies, which must have impressed on them order and progression and suggested that a single guiding hand created that order (hence the origin of monotheism in the Middle East). She also stressed that the imagery and symbolism of a religion are significantly affected by its place of birth, so that

> the Eskimo's hell is a place of darkness, storm and intense cold; the Jew's is a place of eternal fire. Buddha, born in the steaming Himalayan piedmont, fighting the lassitudes induced by heat and humidity, pictured his heaven as Nirvana, the cessation of all activity and individual life.
>
> (Semple 1911: 41)

Huntington (1951: 18) suggested that 'every religion is at least modified by its surroundings, especially those of its birthplace'. Like Semple, he also argued that objects of worship are frequently determined by geographical factors. Thus the Rain God is particularly important in India (where rains are uncertain), the ancient Egyptians worshipped the River Nile (for similar reasons), and Christianity originated in a dry region where sheep-herding was a major occupation and this led to the widely used biblical metaphor of the 'Good Shepherd'.

Deterministic interpretations of the origin and evolution of religion – as caused primarily by the interplay of environment and culture – are also offered by geographers such as de la Blache (1926) and Fleure (1951). The theme is echoed in more recent writing, too. Thus, for example, Bhattacharya argues that

> conception of God, and Hell and Heaven, taboo or sanction of particular food and drink, worship of one or many gods, punctuation of the year with various fasts, feasts and festivals, origin and development of holy places, adoption of sacred symbols and above all the distribution of principal religions of the world have been very largely influenced by the various earth factors directly or indirectly. . . . Each principal religion had its cradle land, from where it was disseminated . . . and each religion . . . reflected the geographical spirit of the cradle land. . . . Islam is cardinally a desert religion, Christianity a religion of Mediterranean region and the Indian religion a product of the prolific Monsoon environment.
>
> (Bhattacharya 1961: 13)

Wynne-Hammond insists that

> there is little doubt that representations of the gods, ideas of heaven, objects of worship and even styles of temples vary according to physical surroundings. In countries of seasonal or uncertain precipitation, the rain god is a common deity; in societies dependent upon fishing there is often a god of the sea. . . . Hindu temples and images seem to reflect the exotic vegetation of Asia whilst Islamic mosques are dark and cool to keep out the heat of the sun.
>
> (Wynne-Hammond 1979: 30)

Cradle lands

The cradle lands of the main religions are well established through detailed historical and archaeological research (Figure 4.1). Northern India provides the core area of Hinduism in the Punjab, and Buddhism (an offshoot of Hinduism) in the Ganges Plain. From here both religions spread through the Indian subcontinent, but Hinduism (an ethnic religion) extended little further whilst Buddhism (a universal religion) dispersed across much of central and eastern Asia. Judaism and Christianity originated in Palestine, and Islam (partly based on both Judaism and Christianity) began in western Arabia. Both Christianity

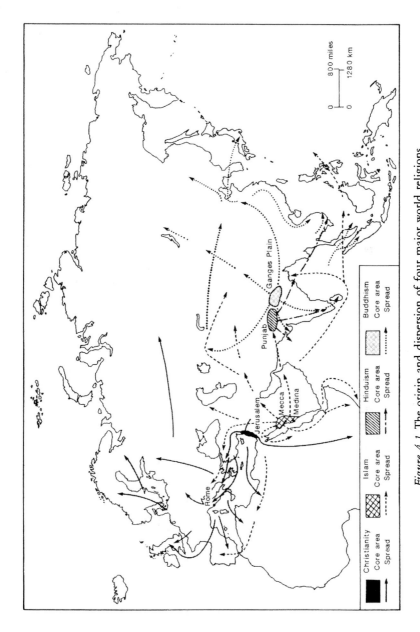

Figure 4.1 The origin and dispersion of four major world religions

Source: After Jordan and Rowntree (1990).

and Islam – the great universal monotheistic religions – dispersed widely through the old world. Christianity gained a particular stronghold in Europe and Islam spread through north and east Africa, as well as further east into central and southern Asia.

Broek and Webb (1973) describe the two areas where the main religions originated as 'religious hearths'. It is convenient to adopt their terminology in the following two sections, which describe the so-called Indus–Ganges and Semitic Hearths. They emphasise two important locational properties of these religious heartlands. First, the religious hearths closely match the core locations of the major ancient civilisations in Mesopotamia and the Nile and Indus Valleys. This makes cultural evolution of religion a distinct possibility (though, as they point out, spatial correspondence does not in itself establish cause–effect). Second, and equally important, the religions emerged on the margins not the centres of the great civilisations. This hints at a more complex interplay between religion and culture, involving factors such as innovation and cultural diffusion, religious adaptation, and exchanges of ideas, beliefs and values along migration and trade routes.

Dispersion

Whatever the reasons for the emergence of religions within such a small area, the fact remains that many religions have spread far beyond their original homeland. Paradoxically, many religions are stronger today in countries other than their source areas.

Many religions have changed a great deal as they have spread and grown, so that the form they display today is often far removed from their original form. Through dispersion the main religions have come into contact with and been influenced by different cultures and customs, some have divided into sub-groups (sects), and many have changed forms of worship and organisation. Modern Christianity, for example, is different to what it was like in the first century after Christ. Similarly, Hinduism has evolved a great deal over nearly thirty centuries.

Part of the explanation of religious distributions lies in the relationships between ethnic and universal religions (see Chapter 2, pp. 34–9) (Broek and Webb 1973). Each of the major universal religions (Christianity, Islam and Buddhism) grew from within earlier ethnic religions, so they started with similar roots but evolved into much more prolific branches.

One of the particular strengths of universal religions, as far as survival and growth are concerned, is their adaptability to local cultures. A religion that is adaptable can be modified to better suit new conditions it encounters, both as it spreads through space and as it survives through time. The flourishing universal religion is thus able to assimilate dimensions of ethnic religion, which increases its attractiveness to new converts and promotes its prospects of long-term survival. This is illustrated in the global pattern of the three main

Christian groups – Protestants, Roman Catholics and Eastern Orthodox (Figure 3.1) – which clearly reflects the interaction of universal and ethnic factors.

The universal religions have an inbuilt dynamic towards expansion and diffusion, because they deliberately seek new converts. Thus, missionary zeal and endeavour must also be considered in the search for an explanation of contemporary religious patterns. Tyler comments that

> Christianity is the most aggressively ambitious religion, in the sense that it endeavours to spread the Gospel of Christ throughout the world. . . . Islam, and to a lesser extent Buddhism, are also missionary religions, but neither has had the same success in converting the 'unfaithful' to its beliefs as has Christianity.
>
> (Tyler 1990: 12)

The history of most of the major religions is well documented (al-Faruqi and Sopher 1974, Broek and Webb 1973, Jordan and Rowntree 1990), and these sources allow us to paint thumbnail sketches of how each arose and spread. Before we look at the Indo-Gangetic and Semitic hearths in later sections of this chapter we need to examine some of the principles that underlie the diffusion of religions at various scales.

PRINCIPLES OF RELIGIOUS DIFFUSION

Religion is in many ways like any other set of ideas or values that can be spread among and between groups of people, often separated by considerable distances. Geographers have long been interested in the diffusion of innovations, and this is an area within geography and religion where principles and concepts can be borrowed from other areas of geographical inquiry. Most of the general principles that apply to the diffusion of innovations – like new agricultural and industrial technologies, new architectural styles or consumer preferences – should apply equally well to the diffusion of religious beliefs and practices.

Abler, Adams and Gould (1972: 389) highlight two key principles of spatial diffusion that transcend both the context and the content of the diffusion. The first is that anything that moves must be carried in some way. This means that the processes, speeds and dynamics of this movement must be understood if we are to understand how and why diffusion occurs, and it is not enough simply to be aware of the outcome (usually the spatial patterns) of the diffusion. The second key principle is that the rate at which some things move over geographic space will be influenced by other things that get in the way. As a result, we must recognise the existence and operation of both carriers (which promote diffusion) and barriers (which inhibit diffusion). Many spatial patterns cannot be understood simply in terms of the dynamics of the transfer processes. Barriers, which are sometimes obvious and visible but often subtle and hidden, can radically influence the speed and direction of diffusion – often as much as the diffusion processes themselves can.

A standard typology of diffusion processes recognised two main types:

1 *Expansion diffusion*; in which the number of people who adopt the innovation grows by direct contact, usually *in situ*. For example, an idea is communicated by a person who knows about it to one who does not, so the total number of knowers increases through time.
2 *Relocation diffusion*; this involves the initial group of carriers themselves moving, so they are diffused through time and space to a new set of locations. Migration is a classic relocation diffusion mechanism, because those who migrate take their beliefs, values, attitudes and behaviour with them to new places. Missionaries who deliberately introduce religion into new areas fall into this category.

Expansion diffusion can be further sub-divided into:

1 *Contagious diffusion*; this is diffusion through a population by direct contact. Diseases spread this way. Such diffusion always expands, and it is strongly influenced by the frictional effect of distance. This operates like a series of concentric waves moving over the surface of a pond after a stone has been thrown in – places close to the points of diffusion normally adopt the innovation first, and more distant places adopt after a time-lag during which intervening places have adopted. In human terms, ideas are passed to people close to those who already have them. Much religious diffusion is of this contagious type, and takes place by contact conversion as a product of everyday contact between believers and nonbelievers.
2 *Hierarchical diffusion*; here the idea or innovation is implanted at the top of a society and appears to leap over intervening people and places. Innovations are adopted or received from the top of the hierarchy down. New fashion trends, for example, emerge amongst important people (the leading designers) and appear first in major centres (like Paris, Rome and New York). They are subsequently transmitted to others lower down in the hierarchy, eventually to reach the clothes' stores in small towns. Hierarchical diffusion of religion has occurred through history when missionaries deliberately sought to convert kings or tribal leaders, in the hope that their people would follow.

The most common type of diffusion process for most innovations, including religious ideas and practices, is contagious expansion diffusion. Traditionally this has taken place mainly as the physical relocation of people as carriers of the innovation (in this case a new religion), but modern telecommunications has opened up the prospect of using radio and television to spread religious messages across much bigger areas more quickly (Stump 1991). Such processes underlie the evolution of televangelism in the United States (see Figure 3.9).

The speed or ease with which individual people adopt an innovation is a significant determinant of rate and patterns of diffusion. It reflects two major factors – the existence and impact of barriers such as culture, language and competition from existing (often indigenous) religions, and the character of the people exposed to the innovation. Four major categories of innovation acceptors can often be recognised:

1 innovators (who adopt first);
2 the early majority (who quickly adopt the innovation after the innovators);
3 the late majority (declining numbers of new adopters through time); and
4 laggards (who adopt eventually, when most others have already done so).

An ideal diffusion cycle passes through several distinct stages (Clark 1984). At the start of the diffusion, only a very small number of adventurous early innovators adopt it. Then, often suddenly, the innovation takes hold and the number of adopters rises quickly as the early majority comes in. Through time the more conservative late majority progressively adopt the innovation and the diffusion slows down. Eventually the resistant laggards adopt, and progressively the diffusion process ceases as the number of non-adopters dries up.

Few innovations are so important or universally embraced that every single person in an area adopts them, and most innovations are voluntarily adopted by a large majority at best. Religion falls into this category, and universal religions engage in diffusion much more readily (and deliberately) than ethnic religions. This largely explains the significantly larger areas dominated by the universal religions (Figure 3.1), and the much larger number of followers they have (Table 2.3). Militaristic conversion, in which adoption of the new religion is enforced rather than voluntary, represents a mixture of contagious expansion and relocation diffusion. ` \ Hierarchical diffusion `

RELIGIONS OF THE INDO-GANGETIC HEARTH

This important religious source area is based on the lowland plains of the northern edge of the Indian subcontinent that are drained by the Indus and Ganges rivers. Hinduism, Sikhism and Buddhism were born there (Jordan and Rowntree 1990: 201). Hinduism had no single founder, and the reasons why it emerged here around 2000 BCE remain unclear. Buddhism and Sikhism evolved from Hinduism as reform movements, the former around 500 BCE and the latter in the fifteenth century.

Once a religion is born, the quickest way in which it can spread is by diffusion. Throughout history India has been an important cultural cross-roads and a centre from which cultures, beliefs and values were scattered far and wide. Kirk (1975: 33) identified two primary corridors of diffusion within India – the Gangetic and Deccan corridors – and noted that 'India has functioned as a great selective filter in cultural diffusion', particularly after the expansion of the Gangetic civilisation after about 500 BCE. Ideas and cultural systems (including architectural styles, town planning, metallurgy and agricultural techniques) imported from the West were absorbed, reorganised and transformed in India to meet the needs of the physical and cultural environments they encountered there. The cultural package was subsequently transmitted eastwards to Southeast Asia, generally in a much-modified form.

Hinduism

Hinduism was the earliest major religion to emerge in this area, at least 4,000 years ago. It is known to have originated in the Punjab, in north-west India (ancient Hindu literature refers to seven rivers – the Kabul, Indus and five rivers of the Punjab). It later stretched from Afghanistan and Kashmir to Sarayu in the east, followed by a major wave of expansion across the Ganges to occupy the region between the Sutlej and the Jumna. From here it spread eastward down the Ganges and southward into the peninsula, absorbing and adopting other indigenous beliefs and practices as it spread. It was eventually to dominate the whole of the Indian sub-continent (Figures 3.1 and 4.1). Hindu missionaries later carried the faith overseas, during its major universalising phase, although most of the convert regions were subsequently lost.

Hinduism has remained a cultural religion of South Asia, and has not spread by expansion diffusion in recent times. During the colonial period many hundreds of thousands of Indians were transported to other countries, including East and South Africa, the Caribbean, northern South America, and the Pacific islands (particularly Fiji). This relocation diffusion did effectively spread Hinduism far beyond its source area, but it has not had a major lasting effect. Most of the scattered clusters survive, but relatively few non-Hindus were converted to the faith.

Buddhism

Buddhism began in the foothills bordering the Ganges Plain about 500 BCE, as a branch-out from Hinduism. Its founder was Prince Gautama (born 644 BCE), who found Enlightenment while sitting under a pipal tree (later known as the Bodhi or Enlightenment tree). After a period of peaceful meditation by the tree he decided to make known to others the way of salvation he had found. He preached the so-called Middle Way (that those who lead the religious life should avoid the two extremes of self-indulgence and self-mortification), initially in the Deer Park at Isapatana (now called Sarnath, near Benares). Starting with five converts who became disciples (monks), the Buddha soon gathered around him sixty monks who were sent out to preach and teach.

During the Buddha's lifetime his preaching activities were confined to northern India and a few small communities in the west of India. During the next two centuries Buddhism spread into other parts of India, although it was to remain confined to the Indian subcontinent for centuries after that (Figure 4.1). Missionaries and traders later carried Buddhism to China (100 BCE to 200 CE), Korea and Japan (300 to 500 CE), South-east Asia (400 to 600 CE), Tibet

(700 CE) and Mongolia (1500 CE). As it spread Buddhism developed many regional forms. Ironically, it was subsequently to die out in the very area it had originated, and was re-absorbed into Hinduism in India in the seventh century (although it has survived among the mountain people of the Himalayas and on the island of Sri Lanka).

Sikhism

Sikhism originated in Punjab at the end of the fifteenth century in a reform movement initiated by a spiritual leader called Nanak. The evolution and diffusion of the movement have been clearly documented by Dutt and Davgun (1977). Nanak travelled widely as a young man and visited many Hindu and Muslim holy places including Varanasi, Mecca and the sacred places of the Himalayas, South India and Sri Lanka. Challenged by various visions and visitations of God, he started to preach an eclectic mixture of Vedantic Hinduism and Islamic Sufism, founded on the belief that the best way to approach God was by repeating his name and singing hymns of praise. Before long he was being regarded as a holy man (guru), his ideas found widespread support, and he was preaching to large numbers (many of whom had travelled especially to hear him).

The new religion was widely adopted in the Punjab because it offered a fresh spiritual idea which people found attractive, particularly its criticism of the caste system that was so central a part of Hinduism. It grew fastest when peaceful conditions prevailed, which was not always the case (especially because of disturbance by Muslim invaders), and its consolidation and expansion were greatly aided by initial political patronage. During the first two centuries Sikhism remained confined to its source area in the Punjab, mainly because successive gurus were chosen in accordance with family lines.

Between about 1850 and 1971 there was considerable diffusion of Sikhism. Sometimes this occurred by voluntary migration, because the Sikh community was notoriously adventurous. Often the diffusion followed forced migration caused by political unrest. This was so especially with the creation of Pakistan after the partition of India in 1947, which divided the Punjab into an Islamic western half and a dominantly Hindu eastern half. Large numbers of Sikhs embarked on a mass exodus to India from the former West Punjab and other states in Pakistan. Since partition there has been an almost complete shift of the Sikh population from West Pakistan to India. Many of the immigrants settled in Punjab, where nationalism based on both religion and language led to the eventual formation of Punjabi Suba (state) in 1966.

RELIGIONS OF THE SEMITIC HEARTH

Judaism, Christianity and Islam – the three great monotheistic religions – all developed first among the Semitic-speaking people in or on the margins of the deserts of south-western Asia in what is today the Middle East (Figure 4.1). Like the religions of the Indo-Gangetic hearth, these three have family ties. Judaism originated about 4,000 years ago, and Christianity emerged from within Judaism 2,000 years ago. Islam was born in western Arabia about 1,300 years ago.

Many commentators have questioned why it should be that the three great monotheistic religions all developed in the same basic core area but at different times. Environmental factors cannot be ruled out, as the determinists enthusiastically argued before about the 1950s (see pp. 95–6), but it is doubtless too simplistic to seek one single or even one dominant cause or explanation.

Monotheism has spread throughout the world (see Figure 3.1), and between them Christianity and Islam have nearly 2.4 thousand million believers, accounting for half of the world population (see Table 2.3). Christianity and Islam, two dominant universalising religions, have played key roles in the dispersion of monotheism from their initial Middle East heartland. Judaism, the oldest Semitic religion that does not seek new converts and thus remains an ethnic religion, has played a more minor role, at least numerically.

Judaism

Judaism developed out of the cultures and beliefs of Bronze Age people who wandered through the deserts of the Middle East nearly 4,000 years ago. As the first monotheistic religion (committed to one God who, they believed, ruled sovereign over them and had chosen them as a special people), it borrowed ingredients from the great civilisations of the ancient Near East world and established rituals and ethical and civil laws.

> From the Babylonian world, where Abraham was born, the Hebrews derived a body of traditions centring around the concept of law. . . . From Egypt, where the Hebrews first prospered and were later enslaved, they came forth with a fierce dedication to freedom. . . . From their revulsion at the 'abomination' of Egypt, they derived an aversion to the priestly caste system.
>
> (al-Faruqi and Sopher 1974: 139)

Like all major religions, Judaism spread and was quickly dispersed over a wide area. By 586 BCE, when King Solomon's Holy Temple was destroyed, the Ten Tribes that constituted the northern kingdom of Israel had already been resettled in northern Assyria for four generations. This diffusion and scattering were to become a prominent feature of Judaism through the rest of its history. The Jewish Diaspora (dispersion) began some time before 550 BCE, and it was led by Jewish refugees and immigrants who refused to give up their faith when persecuted by pagan neighbours. The Diaspora comprised a network of Jewish communities bound together by deep-rooted loyalty to the basic tenets of Judaism, including the promised land and Jerusalem (with its Holy Temple).

Like other religions, too, Judaism was to experience the emergence of a number of variants through its history. The Samaritans, who built a Holy Temple on Mount Gerizim, continued the religion of the northern kingdom of Israel. Recent studies suggest that during the period from the sixth century BCE through to the second century CE, Judaism was characterised by great diversity. Leading sects included the Sadducees (who controlled the Holy Temple), the Pharisees (an order of pietists) and the Essenes (an ascetic group who shared all possessions and mostly lived in monastic communities).

Judaism spread into Europe by the forced and voluntary migration of Jews, starting with the forced dispersal from Palestine in Roman times that scattered Jews throughout the Mediterranean Basin. Through time most European Jews became concentrated around the present Russian–Polish border in an area that became known as the 'Jewish Pale'. In 1939 well over half the world's Jews were living in Europe and the Soviet Union (almost 10 million). Poland housed over 3 million, and there were other concentrations in the Soviet Union, Romania and Germany (Jordan 1973).

Modern Zionism (the political movement for the establishment of a national homeland for Jews in Palestine) has roots in medieval Jewish migrations to the Holy Land, but the most important catalyst was a series of shocks that shattered the life of Jews in Europe.

> The Russian pogroms of 1880–1905, the Ukrainian massacres of 1917–1922, the Polish persecutions and pressures of 1922–1939, and, above all, the rise of Nazism in 1933 and its attempt to annihilate totally the Jews in its conquered territories from 1939 to 1945 produced social pressures that resulted in the presence of about 650,000 Jews within the borders of Palestine in May 1948, when the British mandate over its territory came to an end.
>
> (al-Faruqi and Sopher 1974: 154–5)

Christianity

Christianity began in Jerusalem when disciples of Jesus of Nazareth proclaimed that he was the expected Messiah, that he had risen from the dead, and that he would soon return, heralding the end of time. The movement spread (albeit slowly) even while Jesus was alive, to the great embarrassment of the Jewish religious leaders and Roman civil leaders of the day. After Jesus's death it spread more rapidly, and the diffusion was greatly assisted by Christian preachers and missionaries. It spread first to Samaria (in northern ancient Palestine), then to Phoenicia to the north-west, and south to Gaza and Egypt (Figure 4.1). Afterwards it was adopted in the Syrian cities of Antioch and

Damascus, then subsequently in Cyprus, modern Turkey, modern Greece, Malta and Rome (al-Faruqi and Sopher 1974: 201).

During the early years Christianity was confined to a relatively small number of believers scattered over a relatively restricted area. It spread fast, and numbers quickly grew. Within the first century there were an estimated million Christians, comprising less than one per cent of the total world population. But within 400 years over 40 million people, nearly a quarter of the total population, had adopted Christianity (Table 4.2). Imperial sponsorship of Christianity in the fourth century accounted for its rapid increase in influence and membership. The proportion of the world population classed as Christian fell slightly between about 500 and 800, mainly because of more rapid population growth within the other world religions, but it has remained at roughly a third throughout the twentieth century.

Each of the diffusion processes described earlier (see pp. 99–101) played a significant role in the dispersion of Christianity. The early spread of Christianity through the Roman Empire was achieved mainly by relocation diffusion aided by the well-developed system of imperial roads. Christian missionaries like Paul travelled from town to town spreading the good news (of the gospel). In later centuries the pattern of Christianity reflected hierarchical expansion diffusion; early congregations were largely confined to towns and cities while the countryside remained largely pagan. Once planted in an area, Christianity spread further via contagious diffusion (contact conversion) (Jordan and Rowntree 1990).

Christianity diffused through Europe along a number of different routes (Figure 4.2), mainly via missionaries initially. Diffusion and adoption were slow during the first 300 years, and most early converts were town dwellers. Progress speeded up after 313 when the Christian Roman Emperor Constantine

Table 4.2 Growth in the Christian population, 30–2000 CE

Year	World pop. (millions)	Christians (millions)	Per cent Christian
30	169.7	0.0	0.0
100	181.5	1.0	0.6
500	193.4	43.4	22.4
1000	269.2	50.4	18.7
1500	425.3	81.0	19.0
1800	902.6	208.2	23.1
1900	1,619.9	558.1	34.5
1980	4,373.9	1,432.7	32.8
1985	4,781.1	1,548.6	32.4
2000	6,259.6	2,019.9	32.3

Source: Summarised from Barrett (1982).

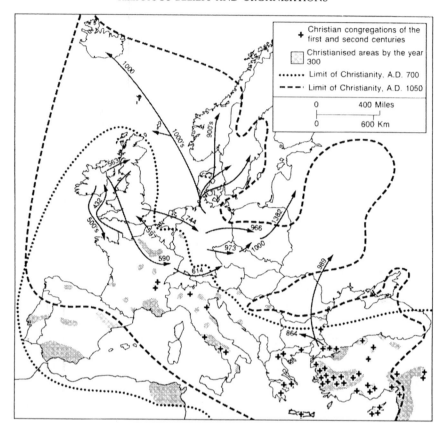

Figure 4.2 The diffusion of Christianity in Europe from the first century CE onwards

Source: After Jordan and Rowntree (1990).

issued an edict of toleration for Christianity that led eventually to its status as state religion (Jordan 1973). The Roman Catholic Church emerged in the fifth century, presided over by the Bishop of Rome (the Pope).

During the fourth and fifth centuries the Roman Church spread rapidly in the western Mediterranean. Roman Catholic missionaries introduced Christianity to northern Europe. Prominent missionaries include Patrick who arrived in Ireland in 432 and quickly started to win converts among the Celts there, and Ninian who used his monastery at Iona as a base for evangelising Scotland towards the end of the fourth century. Between the fifth and seventh centuries Roman Catholicism gained a stronghold throughout Britain.

Monks were an important and effective vehicle in the spread of Christianity around Europe, and monasteries were hubs in a network of diffusion points (al-Faruqi and Sopher 1974: 219). The earliest were the Benedictine monasteries between Cluny and Burgundy, which were influential during the sixth and seventh centuries. The monastery mechanism was even more effective under the Cistercians from the early eleventh century onwards, and the Order of Premontre after 1120. Mendicant friars – itinerant preachers who lived by begging, amongst the best known of whom are the Franciscans and the Dominicans – also played a significant role in spreading Christianity.

While Christianity was winning its battle against paganism in northern Europe, however, Islam was making inroads into the already Christianised Mediterranean region. In the eighth century North Africa was won by Islam, and has remained Muslim ever since. A sizeable area within the Iberian Peninsula (Spain and Portugal) was under Muslim rule for many centuries.

The world-wide dispersion of Christianity coincides with the era of colonial acquisition by European countries. Roman Catholicism was introduced into Middle and South America by the Spanish, after they had invaded the continent (in the mid-sixteenth century). Many Protestant refugees from the seventeenth century onwards emigrated to North America to escape conflict and oppression in Europe, taking their Calvinist brand of Christianity with them and planting it firmly there. Much of Africa and small parts of India were converted by Christian missionaries, who were particularly active there during the nineteenth centuries.

The Reformation in the sixteenth century, which gave rise to Protestantism as an offshoot from Roman Catholicism (see pp. 111–17), served to intensify rather than diminish the enthusiasm of the Christian Church for evangelism. One response of the Catholic Church to the challenges posed by the reformers was to recognise and approve the educational and missionary work of the Jesuits. This religious order was founded by Inigo de Loyola in 1534 as the Compania de Jesus and approved by the Pope in 1540. After the 1560s it was to become the primary agent of expansion of Roman Catholicism. Jesuits introduced Christianity into many areas including Ethiopia, Morocco, Egypt, India, China, Japan, the Philippines, Persia, Tibet, Ceylon, Malaya, Siam, Indo-China and the East Indies (al-Faruqi and Sopher 1974: 227).

Many dissenting Protestant groups sought refuge in the New World from religious tension and persecution they were facing throughout Europe. Relocation diffusion accounts for the presence of numerous Christian groups in the United States. 'The best known are the Congregationalists of Calvinist persuasion (Plymouth Colony, Massachusetts Colony, Connecticut), the Baptists (Rhode Island), the Friends (Pennsylvania) and the Moravians, Mennonites and other German Pietist groups (largely Pennsylvania but also North Carolina)' (al-Faruqi and Sopher 1974: 231–2).

Few detailed studies exist which describe geographical patterns of Christianity, particularly at the national scale. One such study on the Netherlands (Knippenberg, Stoppelenburg and Van der Wusten 1989) identified remarkable regional segregation based on religion. Three religious–cultural groups were delimited (Roman Catholics, orthodox Protestants, and liberals), and the resultant patterns were shown to be hardly unchanged between 1920 and 1985 despite over six decades of economic development and cultural change.

Christianity has remained a universalising religion, with an abiding commitment to active proselytism (the conversion of non-believers). Whilst other major religions including Islam and Hinduism have grown in numbers, this is largely accounted for by population growth rather than wholesale conversion. As Tyler points out,

> Christianity stands alone in having continued to push its territories. Only one-third of the world's population is Christian (or nominally Christian), but, in terms of land areas, predominantly Christian lands cover a disproportionately large area. North and Latin America, Europe, sub-Saharan Africa, Australia, New Zealand, Papua New Guinea and the Philippines are today all predominantly Christian.
>
> (Tyler 1990: 14)

Islam

Islam means 'submission to God', and this strict monotheistic religion was founded by Mohammed in Medina in 622. That year is taken as the start of the Islamic calendar. Mohammed claimed to have been shown the absolute truth in revelations by Allah (God). He wrote it down in the Koran, which Muslims accept as divinely inspired and have adopted as the basis of Muslim law. Islam contains ingredients from both Judaism and Christianity, blended with Arab beliefs and customs (such as the practice of polygamy). By the time Mohammed died in 632, he ruled the whole of Arabia (in both religious and political terms).

Islam spread and expanded mostly by force initially, because conversion of the mainly Christian populations it encountered usually required political control. Within less than a hundred years, Arab Muslims had conquered lands over a vast area – stretching from the Atlantic Ocean in western Europe to the borders of India, and including Spain, North Africa, Egypt, Syria, Mesopotamia and Persia. Today's distribution of Islam reflects a significant retreat from this early core emirate or territory (see Figure 3.1), although the spread of Islam into India, Central Asia, the Sudan and the margins of East Africa has left an enduring legacy. Islam also has a strong presence in South-East Asia. In Sarawak, for example, the religion, customs and beliefs of the Malays display the prominence of Islam, which has significantly shaped their cultural identity (Kling 1989). The Islamic faith has been propagated there through the Malay language, to the mutual benefit of both religion and language (Saharan 1989).

Broek and Webb (1973) suggest that an important factor in the rapid spread of Islam was its emergence at the hub of a series of important trade routes, including caravan trails leading from the Middle East through Central Asia to North China, and across the Sahara to the Sudan. Many Muslim traders were also effective missionaries, acting as multiple diffusion nuclei who travelled widely.

Expansion diffusion accounts for the spread of Islam from its Arabian source area, and relocation diffusion accounts for its subsequent dispersal to Malaysia, Indonesia, South Africa and the New World (de Blij and Muller 1986). Unlike Hinduism, Islam attracted converts wherever it took hold. New core areas soon turned into effective source areas for further dispersion, by a combination of contagious and hierarchical diffusion.

A major milestone in the spread and consolidation of Islam was the Crusades. These Christian wars against Islam – three major and three minor Crusades between the eleventh and thirteenth centuries – were designed primarily to keep open the Christian pilgrimage route to Jerusalem and the Holy Land, which passed through Muslim lands. Al-Faruqi and Sopher conclude that

> the balance sheet of results and consequences of the Crusades is heavily weighted on the side of the tragic. From the Muslim side they were wild surges that had no valid justification and that brought about nothing but bloodshed and grief. . . . The Crusaders did not add to the civilisation of the Muslims, for they were far less civilised and less cultivated. Muslim learning, craftsmanship, arts and culture flowed from East to West. The Crusaders returned to Europe loaded not only with the material products of the Muslim world, with which every church and every nobleman's home was adorned, but with the ideas that later helped to bring about the Renaissance.
>
> (al-Faruqi and Sopher 1974: 255)

In common with other major religions, Islam has evolved internal diversity. The two main sects are the Shiite (or Shi'ah) Muslims and the Sunni Muslims, and the split occurred over the succession of the *imam* (head of the Muslim community). Sunnites claim that the *imam* should be elected, whilst the Shiites insist that he be a descendant of the prophet Mohammed. Roughly nine out of ten Muslims today are Sunni, who are strongest in the Arab-speaking countries and in Indonesia. Other Sunni strongholds include Pakistan and Bangladesh. Shiite Muslims form the majority in Iran, Iraq and Bahrain, and in parts of south and central Asia (Jordan and Rowntree 1990).

Fundamentalist revival amongst the Shiites has caused significant religious and political tension, most evident in the violent feuding between the Sunnites and Shiites (Evans 1989b). The Islamic revival movement is characterised by 'terrorism, intolerance and revolution for export' and 'the Islamising movement at heart is a reaction, sometime virulent, against the ways of modernity exemplified by Western culture' (Walsh 1992: 28).

In recent years Islam has once again started to spread into Europe, caused not by military invasion but by 'a quiet and frequently clandestine movement of Muslims from North Africa, the Middle East and southern Asia, as the dispossessed of those regions come to Europe looking for jobs and new hope' (Evans 1989a: 8). Europe now houses an estimated 7.5 million practising or cultural Muslims, many of them in France, Germany and Britain. Evans speaks of a

> new Islamic activism . . . sweeping through much of Western Europe [in which] mosques, Muslim associations and centres have sprung up in nearly all the major cities. Equal rights and job opportunities, as well as cultural and religious autonomy, have become vital issues to Muslims across the continent.
>
> (Evans 1989b: 10)

Muslims constitute the second largest population group within the former Soviet Union, and their numbers are rising at a rate four times as fast as the Soviet population as a whole (Halsell 1990). Separatist movements quickly emerged in the dying days of Communist rule, and by 1990 the peoples of the Soviet Union's Muslim republics (Azerbaijan, Kazakstan, Kirgizia, Tajakstan, Turkmenia and Uzbekistan) were seeking to regain control of their own destinies. During the early 1990s many of the Soviet Union's 26,000 mosques which had been shut down under Communist rule were restored and reopened, in a visible display of the restoration of organised religion.

Public expression of support for Islam was also suppressed in other Eastern European states under Communist rule. Islam was introduced into Albania at the time of the Ottoman conquest in the late fifteenth century, and its influence grew during the sixteenth century. Christians were heavily taxed, many Orthodox Christians were subdued by organised fighting, and converts were also won by the practice of *devshirme* in which selected young Albanian boys were educated at the Sultan's court, with those who converted to Islam enjoying a more favourable economic and social situation than the rest (Daaniel 1990). When the Communists took control of Albania in 1944 religion was suppressed by a series of direct measures. For example, church property was nationalised, religious schools came under state control, clergy were imprisoned and killed, religious festivals and Muslim practices were banned and Muslim holidays were greatly reduced. Reasons for failure

DIFFUSION OF THE REFORMATION IN EUROPE

One of the most important single episodes in the history of Christianity, which had a significant impact on the pattern and character of this religion and has in turn been translated into geographical patterns on the ground, was the Reformation. Hanneman's (1975) detailed study of how Reformation beliefs and practices were transmitted across Europe provides an excellent case study of how a major religious innovation is diffused in both time and space. It reveals as much about the underlying dynamics of religious diffusion as it does about the spread of a particular schism within Western Christianity.

The Reformation

To understand the meaning and significance of the Reformation we need to reconstruct the history of the time, and examine the behaviour and motives of some leading individuals. The drama of this momentous event is set against the backdrop of early-sixteenth-century Europe, which had embraced Roman Catholic beliefs and practices. One widespread practice that was causing growing concern to some theologians and believers was the sale by the Church of papal indulgences. These were pardons issued for sins forgiven, on receipt of cash contributions to help build the dome of St Peter's Basilica in Rome (the spiritual and administrative centre of the Roman Catholic Church).

A reform movement emerged, initially to counter what was seen as the abuse of papal authority and corruption of the Church. Quickly, however, the reform was to embrace wholesale theological revision that sought to re-establish the Church along what the reformers believed to be the original lines (which they argued had been corrupted over the centuries). This process of re-establishment in practice meant breaking away from the mainstream Roman Catholic Church, and thus emerged the Protestant Church.

Three leading reformers took public stands against the teaching of the Roman Catholic Church, and provide the original centres of innovation of the reform movement (Figure 4.3). The first, in 1517, was Martin Luther (the Augustinian canon of the university at Wittenberg in Germany). Luther's views were regarded as heretical by the Church, and he was excommunicated (expelled from the Church) in 1520. Dissatisfied with the corruption of the Roman Catholic Church, and eager to restore the Church to its original form, Luther looked at the writings of Saint Paul in the New Testament of the Bible for a model of how the early Church operated. He found there two key principles that the Church had long since either lost or abandoned – the doctrine of justification by faith (people's sins are forgiven and their souls are saved if they believe that Jesus is the Son of God and died for their sins), and the doctrine of the common priesthood of believers (all believers are equal before God). Huldreich Zwingli, a Swiss priest, began a similar reform movement in Zurich (Switzerland) in 1518. Zwingli's views mirrored Luther's, but he claimed to have held them before Luther articulated his. The third leading player was John Calvin, a French Catholic layman based in Geneva, who sought to remodel the church along New Testament lines.

Diffusion

Hannemann's (1975) pioneering study of the diffusion of the Reformation focused on the critical early stages (1518–1534) and the geographical spread of

Wittenberg (Luther, 1517)
Zurich (Zwingli, 1518)
Geneva (Calvin, 1530s)
Canterbury (Henry VIII, 1534)

||| Anglican   Lutheran

Calvinist (reformed)   Anabaptist

0   400 Miles

0   600 Km

Figure 4.3 Diffusion of the Protestant Movement in Europe, cira 1570
Source: After Jordan (1973).

this religious innovation across south-west Germany. He examined how a number of indicators of the reform movement (the first evangelical sermon and liturgical changes in communion and mass) spread spatially.

Pace of adoption

Maps compiled from exhaustive archive research show that Luther's ideas were initially accepted spontaneously by many people over a wide area. The innovation was certainly popular, and it travelled far and fast. After the diffusion was launched it was actively promoted in a number of localities, and this served to speed up the pace of change.

Four successive stages in the development of the Reformation movement were identified:

1 opening: the movement began when news spread that Luther was actively opposing the Roman Catholic establishment;

2 launch: the real Reformation began with the preaching of Luther's ideas;
3 diversification: preaching of Luther's ideas was followed by changes in the liturgy, such as the communion under both forms;
4 consolidation: further liturgical change came with the abolition of mass, which symbolised a total break with the Roman Catholic Church and the beginnings of the political Reformation.

Stage 2, the place and date of the first evangelical sermon, was the most decisive of the four developmental stages. This new form of sermon, based on the Bible and rejecting any notion of supreme authority vested in the Pope or bishops, was an important hallmark of the reform movement.

The pattern of adoption of Luther's reformist ideas mirrors the adoption pattern for most innovations (Figure 4.4). Using the number of evangelical preachers as an index of Lutheranism, the relatively rapid rise in adoption of this religious innovation between 1518 and about 1523–4 is clear. From 1525–6 there was a sharp drop, then a progressive slowing down from 1526 onwards, which Hannemann accounts for in terms of the impact of numerous rebellions in several cities in 1523 coupled with a sharp decline in number of students attending universities during that period.

Catalysts and constraints

Hannemann (1975) identifies some key factors that encouraged the rapid diffusion of Reformation ideas and practices. One was the political character of Switzerland and the Upper Rhine and central Germany, which was decentralised. This facilitated the arrival and dissemination of new ideas, including religious ones. A second important catalyst was the good road network in Upper Germany, which made it possible for reformist preachers to move out freely from the cities that provided the initial power bases of the new movement. Conversely, natural barriers like the Black Forest and Lake Constance acted as constraints to the diffusion process, and forced the innovation to spread in some directions faster than others. Thus, for example, the ideas of the Strasbourg reformers diffused principally along a north–south axis rather than evenly throughout the area.

The distribution of wealth and poverty was another significant factor, because Hannemann's study of old tax records showed (contrary to the popular view) that the new religious ideology and practices were adopted more readily in prosperous places. Universities were by far the most important agents for the diffusion of evangelical ideas, and university towns in central Germany, along with Heidelberg and Frieburg, were significant diffusion centres particularly in the early years. Once the new evangelical ideology was planted in a place, the primary route for dispersion was through networks of personal relationships. Nine out of ten Germans at the time were illiterate, so conversations between families, friends and acquaintances allowed the exchange and transfer of ideas and practices.

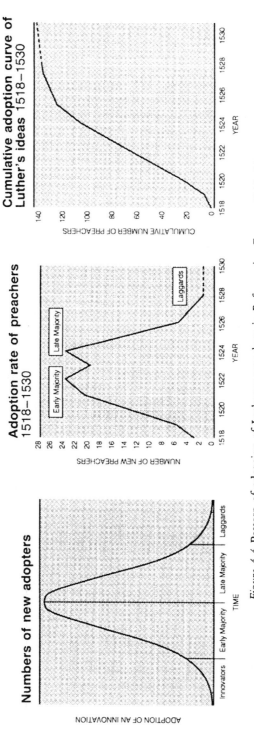

Figure 4.4 Pattern of adoption of Lutheran preachers in Reformation Europe, 1518–50

Source: After Hannemann (1975).

Diffusion processes

Hannemann points out that all three types of diffusion (expansion or contagious, along with relocation and hierarchical) operated simultaneously and in varying intensities during the first fifteen years of the Reformation. Expansion diffusion – the gradual spread of an idea outward from a centre – occurred mainly in the initial phase of the movement, as rumours spread rapidly among the population about Luther's opposition to the Roman Catholic Church. Lutheran preachers also spread initially by expansion diffusion as graduates and friends of Luther at Wittenberg, and Zurich-based carriers of the new faith moved outwards from their original bases.

Relocation diffusion – involving a change of location from which preachers operated – occurred mainly because of constraints imposed by territorial, ecclesiastical and city authorities on the new religion. Some communities lost their pre-reform preachers (often to independent imperial city-states), and once government censorship on evangelical preaching was lifted many such communities welcomed Lutheran preachers from other areas. Relocation of preachers was often done through an agent.

Hierarchical diffusion operated widely, and was the dominant and most important type of diffusion. During the first 15 years acceptance of the reform movement was limited mostly to cities and towns, from which it progressively filtered down into the surrounding countryside. New urban diffusion centres were established when opinion leaders from Heidelberg and Frieburg were forced to relocate to nearby cities (such as Strasbourg) because of political harassment. Luther's ideas were also diffused through a sociological hierarchy, from individuals of high social status to lower social groups. As the top-down diffusion progressed, the number of adopted rose accordingly.

The wider context

Although the Protestant movement (the main outcome of the Reformation) grew at different times in different places, the major turning point came in 1517 when Martin Luther issued a challenge to the Roman Church at Wittenberg, Germany. The movement spread far and wide, along a number of different routes and involving a number of different groups. Jordan (1973) notes how John Calvin was instrumental in dispersing Puritanism from his headquarters in Geneva (Switzerland) to England, along with Presbyterianism to Scotland, the Reformed Church to the Netherlands and Germany, the Huguenot faith to France and lesser Calvinist groups to eastern Europe (Figure 4.3).

The course of the Reformation was also influenced by a group who were much more radical than Luther, Zwingli or Calvin in insisting on the absolutely literal truth of the Bible. The most important of this radical group were the Anabaptists, who denounced infant baptism and argued that only adults could consciously make such a commitment to obey God's commands. The Anabaptist movement began in Zurich as soon as Zwingli had initiated reforms there. Amongst the leaders of this group were Jacob Hutter (died 1536), who founded the Bruderhof (Hutterite brotherhood) in Moravia, and Menno Simons (died 1561), who founded the Mennonites in the Netherlands (from which sprang the American Amish). Their followers – the Hutterites and Mennonites – were in time to migrate to the New World and play a leading role in the persistence of these fundamental reform movements.

Reformation ideology was also instrumental in the founding of the Anglican Church (Church of England) by King Henry VIII in 1534, as another break-away from Catholicism.

CASE STUDIES OF RELIGIOUS DIFFUSION

Whilst Hannemann's (1975) detailed study of the diffusion of the initial stage of the Reformation in Europe throws invaluable light on some of the processes and patterns of religious innovation, it remains the only study published on the topic. Historical and documentary sources were used to great effect in reconstructing how the innovation spread, but the task was massive, the data constraints considerable, and the spatial scale quite restricted.

Smaller-scale and more recent case studies of how religious ideas and practices have diffused, in different countries, raise different questions to the ones Hannemann asked or addressed and they broaden our understanding of the dynamics and controls of religious diffusion. In this section we review a sample of the published studies, most of which have focused on the diffusion of religion within the United States.

The Black Christian Methodist Episcopal (CME) Church

The spread of the Black Christian Methodist Episcopal (CME) Church in the United States between 1870 and 1970, described by Tatum and Sommers (1975), is a classic illustration of relocation diffusion (see pp. 99–101). Although the major white branches of the Methodist Church in the United States reunited after a rift during the Civil War, the black branches did not. Three major black churches coexist in the United States today – the African Methodist Episcopal Church, the African Methodist Episcopal Zion Church, and the Christian Methodist Episcopal Church – each of which has adapted white Methodist ideas and modes of religious expression in slightly different ways.

The Christian Methodist Episcopal (CME) Church was organised in December 1870. Its policies and practices were identical to those of the Methodist Episcopal Church (South), except for its use of two significant black American adaptations – the use of 'old time' preaching and black American music. The church played a vital role in empowering freed slaves and in enabling them to integrate more easily into mainstream (white) American society. Tatum and Sommers (1975: 343) stress that it was an important means through which they 'found meaning in life, developed character, began economic enterprises, embraced basic educational tenets, fostered institutions of learning, developed indigenous leadership, and practised basic forms of self-government'.

The Black or 'Southern' CME, as an ethnic church for black slaves, was originally confined to the southern states. Through time, after emancipation, its members migrated from the South, and they took their church with them and planted it where they settled. As a result, membership of the CME diffused widely and within a hundred years there were churches and active members in most states (Figure 4.5). From its birthplace in Tennessee, the church spread with its migrant members to the Mississippi Delta, Georgia, Alabama and Texas. Further expansion followed, through the urban complexes of the Middle West, West Coast and the East.

Three mechanisms assisted the consolidation of the CME church as it was diffused by member migration. One was the mobility of CME ministers. Itinerant preachers toured a church circuit on foot or horseback, often preaching in homes and uniting black communities. Resident ministers exchanged circuits periodically, some as frequently as each year. This mobility ensured the maintenance of religious vitality and fervour, and prevented the expanding church from becoming inward looking and complacent. A second mechanism was the dissemination of CME teaching through publications, especially *The Christian Index* and regional variants, which attracted new members and helped existing ones to feel part of a cohesive family of believers. Provision of educational opportunities was the third mechanism, and a network of CME institutions of higher learning (including colleges such as Texas College) has broadened the horizons of many church members, promoted the church's existence and black focus, and established the legitimacy of the CME on the American cultural landscape.

Tatum and Sommers (1975: 357) conclude that 'after 100 years on the American landscape, the CME Church remains primarily rural and southern, though it is increasingly found in sections of northern and western cities. It has attracted some affluent Negroes but has been more successful in serving the poor'.

The African Methodist Episcopal Zion Church

The African Methodist Episcopal Zion Church (AMEZC) is one of the largest black denominations in the United States. Although it shares some geographical attributes (such as a strong concentration in the south-east) with the Black Christian Methodist Episcopal (CME) Church, its history and diffusion dynamics are rather different.

Heatwole's (1986) analysis of county level data showed an estimated total nation-wide AMEZC membership in 1980 of 1,092,723 people. The distribution

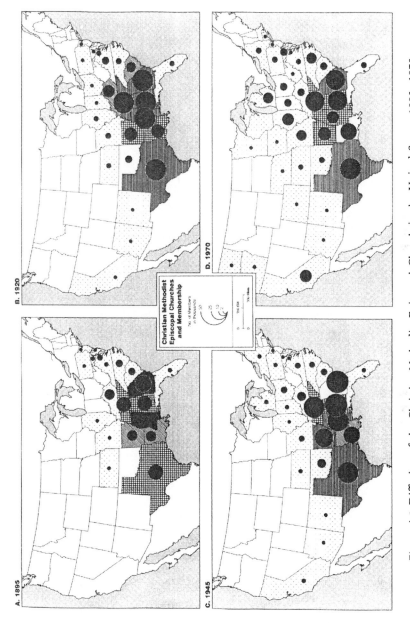

Figure 4.5 Diffusion of the Christian Methodist Episcopal Church in the United States, 1895–1970

Source: After Tatum and Sommers (1975).

is partly clustered and partly dispersed. Major clusters centre on the large AMEZC populations in North Carolina (311,000 individuals; 28.4 per cent of the total), New York (174,00 people; 15.9 per cent) and Alabama (140,000 people; 12.8 per cent). Minor clusters were discovered in the Lower Mississippi Valley, Appalachia and the central valley of California. Elsewhere congregations were widely dispersed, with over 1,800 congregations scattered around 37 states.

'The geography of the AMEZC is like no other church previously mapped', concluded Heatwole (1986: 4), who explained the pattern as the product of three key stages in the evolution of the church. Stage one, 'northern genesis', began at the close of the eighteenth century when some blacks in northern states began to separate from mainstream Protestant denominations. The separation was promoted by the non-inclusive nature of the mainly white Protestant churches, some of which refused to ordain blacks and in many of which blacks were relegated to 'Negroes only' pews at the back. Separatists founded their own black independent churches, one of which was the John Street Methodist Episcopal Church of New York City in 1796.

Unlike the CME church, which grew largely by relocation diffusion, the AMEZC church grew initially by hierarchical expansion diffusion. Missionaries were dispatched from the parent church in New York City to organise other black Methodist Episcopal churches throughout the north-east under their own denomination (the African Methodist Episcopal Church in America). The new church grew as it was adopted by black residents in northern cities, and by 1860 there were an estimated 6,000 members in 49 congregations. AMEZC had little presence in the southern states, where most blacks remained slaves (and many were probably members of CME).

The second phase, of 'southern development' began in 1865 after the defeat of the Confederacy, when the South was incorporated back into the United States. AMEZC missionaries quickly exploited the newly opened opportunity to convert southerners, particularly when most black members of the Methodist Episcopal Church (South) defected to either their church or the Methodist Episcopal Church (North). This fuelled a massive expansion of AMEZC membership from less than 10,000 in 1864 to around 400,000 by 1884. By 1910 AMEZC had grown into a relatively large church concentrated in the rural south. The church became particularly prominent in the Carolinas and Alabama as a result of the placement of missionaries, spatial expansion and numerical increase.

Since 1911, when AMEZC had around half a million members, the membership has more than doubled and it has dispersed significantly. Heatwole (1986) accounts for this phase of 'national maturity' largely in terms of demographic shifts, rather than church policy. A major contributory factor has been the migration of large numbers of blacks from the rural south to urban centres of the North-east, Middle West and Pacific Coast.

120

The spread of Christian Science from Boston

Christian Science is more a system of spiritual healing than a proper religion, and Lamme (1971) has traced the diffusion of this unorthodox Christian sect within the United States. Within a century, Christian Science grew and spread from an initial base in New England in 1870 to enjoy a nation-wide distribution by 1970 (Figure 4.6).

The evolution of this pattern has a definitive beginning in 1866 in Boston, where Mary Baker Eddy was living when she made her 'religious discovery' of a system of spiritual healing based on the principle that mind is the only reality and matter is an illusion. The new theology she proposed, explained in her *Science and Health with a Key to the Scriptures* (1875), was highly unorthodox and it appealed particularly to the relatively affluent. Between 1866 and 1886 the headquarters of her movement was permanently located in Boston, and Christian Science began to appear in other locations throughout the country. By 1910 the distribution was nation-wide, and a national pattern has continued since.

In the absence of membership data, Lamme had to use data on services to map the distribution of Christian Science in 1970. Whilst services are found in all parts of the United States, the pattern varies greatly in density (reflecting historical patterns, population distribution and organisational structures). There were 2,409 congregations in the 50 states and District of Columbia, with concentrations highest in California and Washington. Services were relatively rare in the South and along the Eastern Seaboard, although migrants from northern states guaranteed a strong Christian Science presence in parts of Florida.

Lamme's study illustrates the 1970 distribution, but reveals little about how the expansion occurred or what diffusion processes were involved. Membership appears to be lower in poorer than in wealthier areas, and there are many more services in urban than in rural areas. Christian Science services also seem to be more frequent amongst diverse populations, and where there is evidence of other forms of innovative behaviour.

The spread of the Jehovah's Witnesses in Spain

One of the most dynamic religious groups in terms of disseminating its beliefs is the Jehovah's Witnesses. This unorthodox Christian group was founded as the People's Pulpit Association in 1884 by Charles Taze Russell. Its legal name is the Watch Tower Bible and Tract Society, although since 1931 it has been unofficially known as Jehovah's Witnesses. The pacifist group claims to have up to 200 million followers world-wide. It propagates its own (allegedly infallible) version of the Bible, rejects all other religions as false or evil, and believes that a final battle (Armageddon) is imminent after which the Witnesses will rule the earth with Christ.

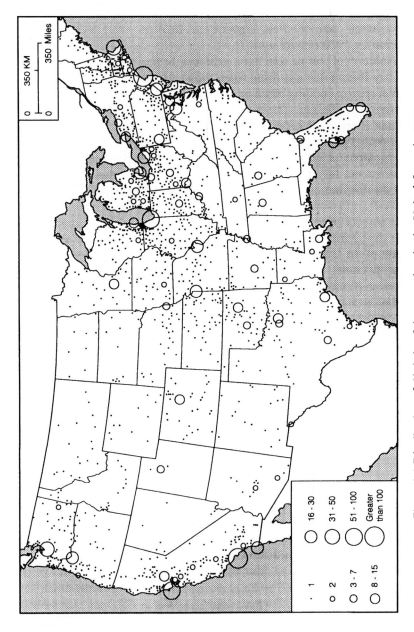

Figure 4.6 Distribution of Christian Science services in the United States in 1970

Source: After Lamme (1971).

The Jehovah's Witnesses was mainly an American movement until 1911–12 when Russell made a world-wide evangelising tour. Evangelism and conversion of non-believers are high on the Witnesses' corporate and personal agendas, and great time, energy and resources are invested in house-calling and distribution of religious literature (including magazines, books, tracts and bibles).

Landing's (1982) study of the diffusion of the Jehovah's Witnesses in Spain between 1921 and 1946 offers little detail of how the process worked, but the overview does highlight some important factors. The Jehovah's Witness movement entered Spain by relocation diffusion. A native Spaniard Witness was sent from Brooklyn in 1921 to begin preaching among the Asturias miners near the city of Oviedo. Presumably this group of potential converts had been carefully identified and deliberately targeted, but we are not told why. The pioneer missionary was recalled to the United States in 1924 and replaced by Spanish-speaking English missionaries. The focus of evangelism was also switched to the large cities of Madrid, Barcelona and Valencia.

The missionaries' messages were well received, and within two years (by 1926) Jehovah's Witness congregations had been founded in the major cities. The Witnesses denounced both Communism and Fascism in their preaching, and met little opposition amongst the largely illiterate population. Most converts were won in the provinces with left-wing populations, and in the industrial and commercial centres of Valencia, Barcelona, Bilbao and Madrid.

Given the prominence which Jehovah's Witnesses attach to disseminating their beliefs, and the enthusiasm and commitment with which they do so in many countries, it is perhaps surprising that more geographical studies have not focused on the diffusion of this particular religious innovation.

The Old Order Amish in Europe and North America

The case studies outlined above deal with the diffusion of particular religious groups within individual countries. Patterns and processes of diffusion are different at the international scale, and it is at this broader scale that much of the detail of the global distribution of religions (Figure 3.1) must be explained. The diffusion of Reformation thinking within Europe (see pp. 111–17) is a good example of an innovation that starts in particular places at particular times but then spreads through both space and time and ultimately has an impact that transcends space and time. On a lesser scale, the diffusion of entire religious bodies – particularly between countries or even continents – illustrates the broader context within which the religious mosaic evolves. This final case study focuses on the diffusion and growth of one small religious sect, the Old Order Amish, as documented by Crowley (1978).

Origins

One product of the Reformation in Europe was the Anabaptists, a coalition of Protestant groups bound together by a belief in adult baptism and literal interpretation of the Bible. This innovation began in 1525 and spread initially from two centres – the Swiss Brethren based in Switzerland, and the Mennonites based in the Netherlands.

The cohesiveness of the Anabaptist movement was regularly tested by minor points of religious dogma. Fragmentation was common, resulting in the breakaway of sub-groups that then became independent. One such reformist group was led by Jakob Amman between 1693 and 1697, which emerged within the Mennonites and evolved into the more conservative Amish branch of Anabaptism. Amman and his followers were concerned, amongst other things, to restore and strictly observe the lapsed Mennonite practice of *Meidung* (the shunning of excommunicated members).

In the early years of the Amish movement, congregations grew and spread largely through conversion by contagious expansion diffusion. Relocation diffusion, involving the migration of Amish into new areas, was limited during the early years. An interesting sign of how the dynamics of religious diffusion change through time, even within a given religious group, comes from the fact that relocation diffusion was soon to become the main vehicle for growth amongst Amish and the Amish today do not engage in conversion of non-believers. This initially progressive and expansionist group has turned into a conservative and static body.

Diffusion in Europe

Locational stability for the growing Amish communities was to be short-lived, and during the eighteenth and early nineteenth centuries religious persecution was to cause them to move a number of times. This triggered a phase of relocation diffusion, initially within Europe but subsequently across the Atlantic.

From about 1710 onwards, when the movement was about 15 years old, Amish filtered out from core areas in Bern Canton (Switzerland), Alsace (in north-east France) and the Palatinate (in south-west Germany). They took their religion with them as they settled in many areas across central and western Europe. The first large migration wave of Amish exiles from Bern Canton passed down the Rhine and settled in the Netherlands in 1711. By 1721 they had established separate Amish congregations amongst Dutch Mennonites in Groningen and Kampen. There were several major movements of Amish from Alsace between 1710 and 1825, to new locations in France, Luxemburg, Germany, Austria and Bavaria. Some migrants from the Palatinate joined their fellow Amish in Austria and Bavaria, while others moved to the Netherlands and other destinations.

Two things are striking about these waves of migrating Amish – one is that many of the transfers were over great distances; the other is that the persecution-driven movements spread Amish communities over a wide area within Europe. A number of factors appeared to have influenced the choice of where to relocate to. Some destinations were chosen because they guaranteed the Amish freedom to worship as they pleased, whilst others reflected the eagerness of some noblemen – who were well aware of the reputation of the Amish as hard workers and good stewards of resources – to employ them as farmers.

Faced with continuing religious persecution, the Amish adopted a number of strategies. Many moved to North America in the eighteenth and early nineteenth centuries. Some of those who stayed behind eventually relaxed the strict Amish code of conduct, which prohibited contact with outsiders, and associated with local non-Amish and even inter-married with them. In Switzerland only three Amish congregations existed after 1750, there were only two by 1810 and by 1850 all the Swiss Amish had rejoined the Mennonites. Amish congregations in the Netherlands remained independent for nearly 200 years and then merged into the general Dutch Mennonite body in the nineteenth century. The remaining European Amish, confronted with the real prospect of religious extinction, decided instead to rejoin their Mennonite brethren in 1937.

Diffusion in North America

Two main waves of Amish immigrants arrived in North America, seeking refuge from incessant religious persecution in Europe. The first wave lasted from about 1717 to 1750. It involved about 500 people, mainly from the Palatinate, who settled in Pennsylvania largely because of attractive land offers from William Penn's agents. Around 1,500 Amish, almost entirely from Alsace and Lorraine, arrived in the second wave between 1817 and 1861. They settled in Canada and the United States, but most chose Ohio, Illinois, Iowa and southern Ontario where land was cheaper and more easily available than in Pennsylvania. Both migration waves gave rise to new Amish settlements in the United States, and diffusion within the United States continued between the waves while few immigrants were arriving. Crowley (1978) identifies five distinct phases of Amish diffusion and settlement.

During the 'First Wave' (1717–1816) Amish settlements were established by newly arrived immigrant groups in south-eastern Pennsylvania, away from existing Mennonite communities (because of religious incompatibility). New colonies were soon established further west, mostly still in Pennsylvania. The Amish had a reputation as hardy pioneer settlers, and this is borne out by the fact that eleven out of the first fifteen Amish settlements eventually failed.

The 'Second Wave' (1817–1861) saw the arrival of the second group of immigrants. Those from Alsace founded colonies in western Ohio, central

Illinois and south-eastern Iowa, though few of them were to survive beyond 1870 without assistance from more progressive Mennonites. Many of the 'First Wave' settlements continued to expand during this phase, and new settlements were started in Ohio and north-eastern Indiana.

Phase three (1862–1899) brought to a close the 'Westward Advance' as the supply of virgin frontier land started to run out. The dynamics of diffusion altered radically, because any new Amish colonies would thus have to replace existing landowners, and non-Amish farmers were adopting the innovation of mechanisation. The Amish were also troubled by internal religious tensions, which resulted in the splitting-off of the conservative Old Order from the main body. Notwithstanding the more restricted opportunities for continued growth, the Amish continued to expand westwards into North Dakota, Nebraska, Kansas, Oklahoma and Colorado (the Great Plains) and into more southern states (Maryland, Missouri, Mississippi, Arkansas, Tennessee and Virginia). Few of these new colonies survived more than twenty years.

The establishment of new Amish settlements continued at a similar rate during the 'early modern era' (1900–1944). But patterns of expansion changed significantly. Little if any colonisation occurred within the core area, and most new settlements were located in Great Plains and Southern states surrounding the core. As in earlier phases, many (21 out of 32) Amish settlements failed. By now the Amish were more visibly different from their neighbours, and their traditional lifestyle (with its distinctive clothing, overall plainness, strong religious beliefs, primitive farming techniques and the use of horses and buggies for transport) was looking distinctly old-fashioned if not anachronistic to outsiders.

The impetus to start new Amish settlements has survived into the 'modern era' (1945–1970s). In fact, 42 per cent of all recorded Amish settlements attempted in the United States were founded between 1945 and 1975. Most new colonies were founded in states that already had an Amish presence, particularly Pennsylvania where 23 new communities were started over that thirty-year period. Crowley (1978: 261) comments that 'their high rate of natural increase has made the Old Order Amish one of the fastest-growing (percentage wise) religious denominations in the United States'. Reconstructions of estimated total Amish populations suggest an average annual growth rate of over 2.5 per cent, causing a rise in the total Amish population from around 9,100 in 1905 to 67,100 in 1976.

The diffusion of the Amish, initially within Europe and subsequently within the United States, illustrates some properties that do not exist in the other case studies. One is the continued relocation diffusion that has caused Amish groups to migrate a number of times both within and between countries. A second is that the evolution of the pattern of Amish settlements, particularly in the United States, has not been uni-directional nor linear. Settlements have been established and abandoned, and the distribution at any one point in time might bear little resemblance to the pattern either before or after it. A third property of the Amish diffusion is the way in which its objectives have changed through time, from an initial enthusiasm for growth via conversion to a more long-term commitment to survive and grow through natural increase.

CONCLUSION

The spread of religious beliefs and practices provides a classic illustration of innovation diffusion, which has occurred at a variety of spatial and temporal scales. The movement of people who transport their religion into new areas has been a major catalyst for such diffusion throughout history. The major world religions have spread out from initial source areas in northern India (Hinduism, Buddhism and Sikhism) and in the Middle East (Judaism, Christianity and Islam) and they now have essentially global distributions, though each major religion has its strongholds and areas of rapid growth.

Hanneman's detailed reconstruction of the diffusion of the Reformation in Europe shows how fast religious innovation can disperse, and highlights some of the key processes involved. Like all cultural innovations, the spread of religious innovation is not uniform through time but peaks after an early majority adopt it. Thereafter the rate at which the innovation spreads generally slows down, but it can continue diffusing over a long time.

A number of detailed case studies of the diffusion of new religious groups have been published. These reveal interesting processes and patterns, and suggest that further geographical study of such phenomena could yield rich returns. Most of the published case studies focus on North America.

5

DYNAMICS
How religions change

There is only one religion, though there are a hundred versions of it.
George Bernard Shaw, *Plays Pleasant and Unpleasant* (1898)

INTRODUCTION

Although the central theme in Chapter 4 was how religions diffuse through space and time, many of the case studies and illustrations there highlighted just how dynamic things are. Shaw's belief that there is only one religion is misplaced, because it overlooks the fundamental differences between religions and falsely assumes that they are all variations on an underlying common theme.

Few religions remain absolutely constant through time. Splits occur within individual religions, often promoted by differences of interpretation or practice which reform movements seek to restore to original forms. Good examples include the split in Christianity brought by the Reformation (Chapter 4, pp. 111–17), the split of Islam into Shiite and Sunni branches (Chapter 4, pp. 109–11), and the split of the Amish from the Mennonites (Chapter 4, pp. 123–7). Even where religions are not torn apart by schism, beliefs and practices can change through time particularly when a religion is planted in a new area. Cultural assimilation can include the absorption of indigenous beliefs and practices into the newly arrived (usually universalist) religion, which can thus evolve regional variants reflecting their unique cultural settings. The Black Christian Methodist Episcopal Church in North America (Chapter 4, pp. 117–18), for example, retains vestiges of its African genesis and its black slave heritage, although it still has great appeal in contemporary black American society.

In this chapter we examine some important dimensions of the dynamics of religions, in seeking to better understand how and why religions change. It is convenient to start by exploring patterns and processes of religious change.

PATTERNS AND PROCESSES OF RELIGIOUS CHANGE

Religious change involves both quantitative and qualitative changes, and in this section we will look at some of the empirical evidence of rates and patterns

128

of change through both time and space. The observed patterns must be interpreted in terms of the underlying processes, which explain both why and how religions expand and contract numerically and geographically. Religious innovation gives rise to new religious movements, and these alter the nature of competition within the religious market-place and provide a further dynamic for change.

Ebb and flow

Barrett (1982) argues that the continued numerical growth of all major world religions is the most significant trend in religious change over the twentieth century. Between 1900 and 1985 the total world population increased nearly threefold, although the relative importance of many religions changed through time (Table 5.1).

Christianity has maintained its position at number one in terms of global membership, with around a third of the total. Islam increased its relative strength by nearly half between 1900 and 1985, and further expansion is expected by the turn of the century. Hinduism slightly increased its relative position. One hallmark of the twentieth century has been the rise in the

Table 5.1 Changes in the relative strengths of the major world religions during the twentieth century

Religion	1900	Mid-1970	Mid-1985	2000
Christians	34.4	33.7	32.4	32.3
Muslims	12.4	15.3	17.1	19.2
Non-religious	0.2	15	16.9	17.1
Hindus	12.5	12.8	13.5	13.7
Buddhists	7.8	6.4	6.2	5.7
Chinese folk religion	23.5	5.9	3.9	2.5
Atheists	0	4.6	4.4	4.2
Tribal religion	6.6	2.4	1.9	1.6
New-religionists	0.4	2.1	2.2	2.2
Shamanists	0.7	0.4	0.4	0.3
Jews	0.8	0.4	0.4	0.3
Sikhs	0.2	0.3	0.3	0.4
Confucians	0	0.1	0.1	0.1
Shintoists	0.4	0.1	0.1	0.1
Baha'is	0	0.1	0.1	0.1
Jains	0.1	0.1	0.1	0.1
Total world population (millions)	1,620	3,610	4,781	6,260

Source: Summarised from Barrett (1982).
Note: The figures show the percentage of total world population that follow (or is predicted to follow) the major religions, ranked in decreasing order in mid-1985.

number of people with no religious beliefs, and this is borne out in the data for the non-religious and atheists who carved out a significant combined market share of around 20 per cent.

Some religions experienced relative decline through the century. Buddhism declined a little, but tribal religion fell sharply relative to other religions. The position of the Jews, whose market share fell by half between 1900 and 1970, is easy to explain given the wholesale massacres they have endured this century in the Holocaust and other atrocities. One of the most dramatic changes during this century has been in Chinese folk religion, whose membership fell sharply during China's turbulent political transitions and revolutions.

Most of the minor religions, in terms of numerical strength, more or less maintained their relative positions over the century.

For many religions the increase in numbers has been coupled with geographical spread. By the early 1980s, for example, there were large Muslim populations in 162 countries. Hindus and Buddhists were recorded in 84 countries (not always the same ones), though in most they accounted for a small percentage of the total population. The total Jewish population is relatively small, but they are scattered among 112 different countries.

Natural change and conversion

The global figures on changing relative strength of different religions (Table 5.1) reveal just how dynamic the situation is. But the summary statistics can only be interpreted in general terms because they represent the outcome of a series of processes that occur at different speeds and directions in different places. The relative strength data are the outcome of a mass balance, the components of which need to be examined by disaggregating the global data. Barrett (1982: 7) underlines the complexity of the mass balance, which involves – each year – millions of people changing their religious profession, super-imposed on demographic changes that themselves vary from religion to religion. Established universal religions win many converts, largely by aggressive missionary strategies, and new religions gain from mass defections from stagnant or declining religions.

The two key components of numerical changes in religions are natural change (brought about by population increase or decrease amongst believers) and conversion (the number of believers won from or lost to other religions). Barrett (1982) provides global data on the contribution of each factor to religious changes between 1970 and 1985, which are summarised in Table 5.2. Amongst the declining religions – those with negative values in columns (c) and (d) – the most prominent is Chinese folk religion, which lost just over a million and a half believers over the fifteen year period. The net loss represents a balance between an input of 3.4 million new believers (by natural increase) and an output of 5.1 million existing believers (by conversion to other religions).

130

Table 5.2 Changes in strength of the major world religions between 1970 and 1985 caused by natural population increase and by religious conversion

Religion	(a) Natural	(b) Conversion	(c) Total	(d) Rate
Christians	21.4	0.2	21.6	1.64
Muslims	17.1	0.1	17.2	2.74
Non-religious	9.3	8.0	17.3	2.76
Hindus	12.1	− 0.2	11.9	2.03
Buddhists	5.1	− 0.9	4.2	1.67
Chinese folk religion	3.4	− 5.1	−1.6	− 0.08
Atheists	2.4	0.5	3.0	1.66
New-religionists	1.8	0.1	2.0	2.28
Tribal religion	2.4	− 2.2	0.2	0.21
Jews	0.2	− 0.01	0.2	1.09
Sikhs	0.3	0.03	0.4	2.94
Shamanists	0.3	− 0.6	− 0.3	− 0.41
Confucians	0.09	− 0.05	0.04	0.98
Baha'is	0.08	0.04	0.1	3.63
Shintoists	0.05	− 0.1	− 0.06	− 1.66
Jains	0.07	− 0.01	0.06	2.00
Total world population	76.4	76.4	0	1.93

Source: Summarised from Barrett (1982.
Note: The figures show the average annual changes (in millions of people) between 1970 and 1985 attributable to population increase (a) and to religious conversion (b), along with the annual rate of change (d). The religions are ranked on the basis of total number of followers in mid-1985.

Christianity had the biggest numerical increase between 1970 and 1985, of more than 21 million people. Almost all of the growth was natural (Table 5.2). Islam increased by over 17 million people, again largely by natural growth. Hinduism grew at just over half the speed of Christianity, but had a net loss through conversion (i.e. more were converted out than in).

The drift away from formalised religion shows some interesting patterns. Nearly half of the 17 million new non-religious people between 1970 and 1985 were converters (Table 5.2), who abandoned their earlier religion and in effect abandoned religion entirely. However, the vast majority of the 2.4 million new atheists were existing atheists, not won from other religions. Tribal religion lost almost as many followers through conversion as it gained through natural population increase, so its total numbers changed little over the period.

The picture is even more interesting when we deconstruct the global change data for a single religion. Christianity, the largest and most stable of the world religions, provides a useful illustration. Barrett's (1982) figures of changes in the strength of Christianity by continent between 1970 and 1985 (Table 5.3)

Table 5.3 Changes in the strength of Christianity between 1970 and 1985 caused by natural population increase and by religious conversion

Continent	(a) Natural	(b) Conversion	(c) Total
Africa	4,586.648	1,466.149	6,052.797
East Asia	276.181	359.622	635.803
Europe	2,197.458	−1,150.645	1,046.813
Latin America	8,419.292	− 219.821	8,127.471
North America	2,008.880	− 669.881	1,338.999
Oceania	372.894	−128.200	244.694
South Asia	2,645.668	447.043	3,092.711
USSR	907.238	164.182	1,071.420
Total Christians	21,414.259	196.449	21,610.708
Total world population	76,388.313	0	76,388.313

Source: Summarised from Barrett (1982).
Note: The figures show the average annual changes (in thousands of people) attributable to population increase (a) and to religious conversion (b) between 1970 and 1985.

give a flavour of how variable the patterns are, and how different the explanations must be in different places. Almost all of the 21.6 million new Christians over the period were born into the faith rather than won by it – some 58,000 more Christians were born than died each day through the year. The small proportion (about 1 per cent) listed in column (b) under 'conversion' represents a net figure, the outcome of conversion in and transfer out. The 21 million new Christians were spread very unevenly between the continents, with most in Latin America (38 per cent) and Africa (28 per cent).

On a number of continents there were sizeable net increases in the number of Christians won through conversion. This was so particularly in Africa (which had a net increase of nearly a million and a half new Christians) and to a lesser extent in South Asia (nearly half a million). On other continents, however, there was a net loss of converts to other religions (or no religion). Europe suffered particularly heavy losses (over a million), while conversion losses were also high in North America (two-thirds of a million) and Latin America (over a quarter of a million).

Natural increase brought most new Christians on every continent except East Asia (Table 5.3). There it accounted for about 44 per cent of the three-quarters of a million new Christians between 1970 and 1985, and provided less than were won by conversion. Clearly the universalist tradition of Christianity survives and thrives in East Asia! Elsewhere, natural increase accounted for more than three-quarters of the total increase – 75 per cent in Africa, 85 per cent in South Asia and the former Soviet Union, and over 100 per cent in the other continents. Spatial variations in demographic variables (birth and death rates in particular) account for the broad differences in relative rates of natural increase between continents. More than 8 million more Christians were born than died in Latin America, 4.5 million in Africa, and more than 2 million in South Asia, Europe and North America over the fifteen-year period.

Migration

A potent force in changing the geographical distribution of religions is the mass movement of believers from one area, country or continent to another. This is different from the migration of small numbers of missionaries, which in itself causes minimal distributional change (although when they are successful in converting large numbers of native non-believers, their secondary impacts can be very significant).

The dispersion of the Jews from Palestine in the Diaspora and the ingathering of Jews from around the world into the modern state of Israel (see Chapter 6, pp. 191–5), provide graphic examples of the centrifugal (outward) and centripetal (inward) dimensions of migration.

Migration promoted by religion has also played a key part in the redistribution of Hindus, Sikhs and Muslims in post-partition India. On independence the Indian subcontinent was sub-divided along religious lines, with India being predominantly Hindu and Pakistan predominantly Muslim. Before and particularly after independence, there were sizeable migration flows of Hindus from Pakistan into India and of Muslims from India into Pakistan. Gosal's (1965) case study of religious population change in the Punjab between 1951 and 1961 picks up empirical traces of these changes. An excess of in-migration over out-migration allowed the Hindu population to increase by about 30 per cent over the decade. Sikhs increased by 19 per cent; this represents a net outflow by migration to other parts of India and other countries, partly offset by a natural increase of 28 per cent. In 1961 Hindus and Sikhs accounted for 97 per cent of the state's population. There have been pronounced regional redistributions of Hindus and Sikhs within Punjab, too. The more economically active Sikhs have been more mobile than Hindus, and large numbers have moved from one rural area to another. Hindus moved more within and between towns. As Gosal notes, these internal migration flows have encouraged the inter-mixing of Hindus and Sikhs and have been an important step towards national integration.

On a smaller scale, the migrations of Amish colonies, initially within Europe but later to and then across the United States (see Chapter 4, pp. 123–7), illustrate the importance of population migration to both the numerical status and distributional patterns of individual religions.

Involuntary conversion

Most of the converts won by the universalising religions adopt the new faith by choice and consciously, in a calculated step designed to improve their quality of life by transferring to a religion that appears to better meet their (spiritual, social and often practical) needs. But there are many instances where converts have been acquired in a more involuntary manner, particularly as the outcome of battles for religious supremacy. Thankfully, examples tend to be rare, but they also tend to be well known. Unfortunately these infrequent and context-specific incidents are often (wrongly) eagerly seized upon by iconoclasts as illustrations of the corruptness and unsuitability of all religions.

A classic example is the Crusades, the Christian holy wars against Islam between 1095 and the sixteenth century. These military expeditions from Europe, authorised by the Pope, were designed to recapture Christian lands (particularly Jerusalem, the Holy City) from the Muslims. The wisdom of the Crusades might be questioned, given that many lives were lost, vast sums of money wasted, and at the end Jerusalem was still under Muslim control. The spirit of the crusading movement was rekindled with the (Catholic) Spanish and Portuguese conquest of Latin America in the sixteenth century. Tyler (1990: 13) comments that 'if the conquests were in any sense a Christian mission, the conversion of the Indian tribes came a very poor second to stealing their gold and oppressing them'.

Experience of the Muslim holy war – the Islamic *jihad* – illustrates more successful military campaigns carried out in the name of religion. The Prophet Mohammed and his followers used such tactics in the early seventh century, and military might was used to expand Muslim territory westwards after the Prophet's death in 632 CE. By the end of the seventh century Islam had conquered the whole of the North African coast, and in less than two generations Arabs had won a vast empire stretching from Spain in the west to the River Indus in India (Tyler 1990). Although the main thrust of the *jihad* was to gain control of territory rather than to convert people, the newly acquired lands came under Muslim religious and legal rule.

The recent rise of Muslim fundamentalism in the Middle East has given new life to the Islamic *jihad*. Muslims are required by the Koran to give loyalty to the state only if the state is seen as loyal to Islam. If the state is not loyal to Islam, the mullahs (religious leaders) can declare it to be evil, and a holy war against the state can break out. To the fundamentalist Muslim the Koran may be interpreted as encouraging martyrdom as a high achievement in holy war against the enemies of Allah. Such factors were instrumental in bringing about the Islamic-inspired 1979 revolution in Iran that toppled the Shah and allowed the triumphant return from exile of Ayatolla Khomeini. They also promoted recent fundamentalist insurrections or separatist movements in Afghanistan, Iraq, Kuwait, Lebanon, Syria and the Philippines, and the violent emergence in 1988 of an alliance between Palestinian nationalism and Islamic fundamentalism within the Israeli-occupied territories of Gaza and the West Bank (Evans 1989b).

Religious innovation

The patterns and processes of religious change we have encountered above cannot be dealt with separately. Processes and patterns are not mutually exclusive; each reflects the other in reciprocal ways. In reflecting on the processes (such as natural change, migration and involuntary conversion) which underlie and shape the patterns, we have assumed up to now that the religious market-place has a fixed number of brands that simply compete with one another.

In reality, neither the number of religions nor the number of sects within religions is static. The major sub-groups within leading global religions such as Christianity, Islam and Buddhism (Table 5.4) often compete with each other for new believers. Differences in relative growth rates between different sects within a religion raise questions about intra- as well as inter-religious transfers of allegiance.

From time to time new sects develop, often reflecting tensions between existing religions and their surrounding socio-economic environment. Bainbridge and Stark (1980) studied a number of sects (including the Nazarenes, the Assemblies of God and the Seventh-Day Adventists) and found some common properties suggesting that they emerge and function as sub-cultures. These include different standards of behaviour expected of believers, antagonism between the sect and society that is manifested in mutual rejection, and the creation of separate social networks that can tend to make sects introverted.

Reformist sects sometimes emerge when there is a widely held view that existing religions have failed to move with the times or accommodate contemporary problems. Two new urban Buddhist cults have emerged in Thailand

Table 5.4 Changes in the strength of sects within major religions during the twentieth century

	Millions			
Religious group	*1900*	*1970*	*1985*	*2000*
Christians	558.1	1,216.6	1,548.6	2,020.0
Roman Catholic	272.0	668.0	884.2	1,169.4
Protestant	119.7	259.0	292.7	357.5
Orthodox	121.2	111.9	130.8	153.1
Muslims	200.1	550.9	817.1	1,200.7
Sunni	173.1	465.8	680.9	999.8
Shia	26.0	79.5	126.7	185.0
Schismatic	1.0	5.6	9.5	15.8
Buddhists	127.2	231.7	295.6	359.1
Mahayana	71.6	130.1	166.4	201.8
Theravada	48.1	87.7	111.6	135.9

Source: Summarised from Barrett (1982).

since the 1970s (the Thammakaai Religious Foundation and the smaller but more outspoken Santi Asok) in a vacuum left by the apparent failure of traditional institutions to create a new, relevant social order and moral framework (Taylor 1990).

Sectarian tension can give rise to new cults but it might not fully explain how a cult, which develops in a particular place at a particular point in time, can grow and spread into a religious movement. Detailed studies of the Pentecostal movement (a fundamentalist and charismatic group of Christian churches) in the United States, by Gerlach and Hine (1968), reveal some key factors that are important in the genesis of a movement (but are not essential to its spread). For a movement to emerge, it must have an appropriate organisational structure with clear networks. Pre-existing significant social relationships are also important, because they provide route ways for face-to-face recruitment of new members. Movements are more likely to emerge and persist if their ideology is oriented towards change, and if members perceive or experience opposition from non-believers. Commitment to such movements is generally strongest if it is marked by some 'rite of passage', such as a particular religious act or experience.

Many of these properties are exhibited in the origin and early expansion of the Disciples of Christ in the United States (Bigelow 1986). This sect originated in the Ohio River Valley in the early nineteenth century, from the merging together in 1832 of two groups who had earlier split from the Presbyterian Church in central Kentucky (the New Light group, led by Barton Stone, founded in 1804) and south-west Pennsylvania (a group led by Alexander Campbell, founded in 1808). The merger was fairly simple because both reformist groups had similar theologies and church organisations. Other groups were also brought together in the creation of the Disciples, including a restorationist church called the Old Christians, a group of Baptist reformers and the Independent Brethren. The new coalition appears to have had the right qualities to build on and survive – including good organisational networks, pre-existing social networks and a change-oriented ideology. By 1845 there were an estimated 7,000 Disciples in Indiana (equal in strength to the Baptists and the Presbyterians, both long-established churches), and between 1832 and 1860 the total number of Disciples in the United States rose from around 22,000 to an estimated 200,000.

The Shakers provide another interesting example of how a religious innovation is planted and grows in a climate of heightened sectarian tension with its surrounding socio-cultural environment. There were always far fewer Shakers than Disciples, and their geographical distribution was much more concentrated. What's more, the Shakers illustrate a complete cycle of growth and decay, and documentary sources allow a detailed reconstruction of the rise and fall of a distinct religious sect (Bainbridge 1982). Ann Lee, founder of the Shakers as an offshoot of the Quakers, established a group of eight followers in New York in 1774. The three most famous Shaker habits were celibacy, common ownership of property, and a tendency towards ecstatic shaking during worship (hence the name).

From the initial core of eight members, the group grew to several thousand members by the mid-nineteenth century – all by conversion (celibacy ruled out any natural increase!). Twenty-two Shaker communities were established between 1787 and 1898 (Figure 5.1). Census returns show that total numbers peaked at 3,608 in 1840 then fell to 3,489 in 1860, 1,849 in 1880 and 855 in 1900. The Shakers survived for two centuries, and – indeed – they prospered for half of that time in largely self-sufficient communities. Within the twentieth century the group has declined through defection and death; few new members have been admitted to the shrinking number of colonies. In 1906 there were 516 Shakers left, and this fell to 367 in 1916, 192 in 1926 and 92 in 1936. By the 1980s only a handful of Shakers survived at the Sabbathday Lake colony in Maine (Kay 1982). It is a curious testimony to the craftsmanship of the Shakers (and the nostalgia of contemporary America) that the group is doubtless best known for its distinctive hand-made furniture, which is much-admired, highly collectable and very profitable.

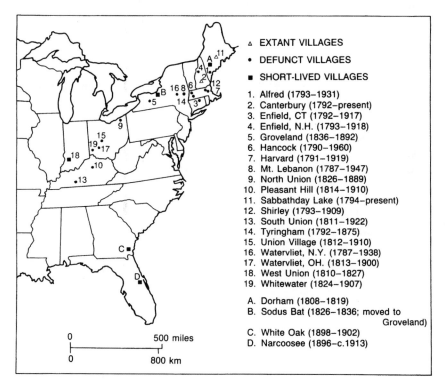

Δ EXTANT VILLAGES

• DEFUNCT VILLAGES

■ SHORT-LIVED VILLAGES

1. Alfred (1793–1931)
2. Canterbury (1792–present)
3. Enfield, CT (1792–1917)
4. Enfield, N.H. (1793–1918)
5. Groveland (1836–1892)
6. Hancock (1790–1960)
7. Harvard (1791–1919)
8. Mt. Lebanon (1787–1947)
9. North Union (1826–1889)
10. Pleasant Hill (1814–1910)
11. Sabbathday Lake (1794–present)
12. Shirley (1793–1909)
13. South Union (1811–1922)
14. Tyringham (1792–1875)
15. Union Village (1812–1910)
16. Watervliet, N.Y. (1787–1938)
17. Watervliet, OH. (1813–1900)
18. West Union (1810–1827)
19. Whitewater (1824–1907)

A. Dorham (1808–1819)
B. Sodus Bat (1826–1836; moved to Groveland)
C. White Oak (1898–1902)
D. Narcoosee (1896–c.1913)

0   500 miles

0   800 km

Figure 5.1 Distribution of Shaker communities within the United States, 1793–1980
Source: After Rooney, Zelinsky and Louder (1982).

In a somewhat different vein, but still illustrative of the interplay between beliefs and cultural setting, is the belief in witchcraft which was particularly strong in the colonies of New England during the seventeenth century and centred upon the town of Salem (Boyer and Nissenbaum 1974, Gregory 1988). Salem was founded in 1626 and by mid-1692 a number of local people were suffering from what were interpreted to be witchcraft afflictions. Suspected witches were arrested and tried; eighteen were hanged and more than a hundred more served jail sentences. The Salem witch trials left a sinister and enduring legacy on the spiritual landscape of New England!

DYNAMICS OF DIFFUSION

We looked at the general principles of religious diffusion in Chapter 4 (see pp. 99–101) and at some case studies of diffusion in the same chapter (see pp. 117–27). Here we will examine some aspects of the dynamics involved in the two main types of diffusion (relocation and expansion).

Relocation diffusion

Relocation diffusion involves the planting of a religion in a new area, which occurs mostly with deliberate intent. The three basic mechanisms involved are the movement of an entire religious group (like the Mormons *en route* to their chosen land in Utah), the dispersion of a group as individual members migrate, or the arrival of missionaries in an area with few if any followers of their particular religion.

Migration

The Pilgrims who settled in New England must be amongst the best known examples of migrant groups who take their religion with them and plant it in a new place. This infamous band of 74 men and 28 women – all English Puritans who had lived for some years in Holland to escape religious persecution at home – set sail aboard the Mayflower from Southampton on 15 August 1620 for America. Four months later they landed at Plymouth Rock, Massachusetts, and they subsequently founded the settlement of Plymouth. The Pilgrims are widely regarded as the pioneers of American colonisation, and they were also instrumental in introducing their fundamentalist brand of Christianity into North America. Puritanism was to remain a hallmark of New England in the ensuing centuries.

Nearly two decades before the Protestant Puritans arrived in the New World, French Catholics had established New France in south-west New Brunswick (Maclellan 1983). A commercial expedition led by de Monts, a Huguenot, chose the uninhabited St Croix Island as a suitable place for the initial settlement in June 1604. The dozen houses and a chapel they built there were designed as a bridgehead from which to befriend the Indians and convert them to Christianity. The settlement was abandoned the next summer (because the river froze in the harsh winter, blocking access for food and supplies), but the process of colonisation was under way and continued.

Many other countries – including South Africa, Australia and New Zealand – were to receive an injection of strong Protestant beliefs, values and practices from colonial groups of immigrants from Britain, the Netherlands and other countries in north-west Europe (Broek and Webb 1973). Other examples of religious diffusion through population migration include the planting of Roman Catholicism in both South America and parts of North America by European immigrants. Jewish immigrants from Europe took their faith and practices with them, and embedded them within the urban fabric of the United States.

An interesting micro-scale example of migration that affects the diffusion of religion is the movement of novice monks and preachers to Mexico City in the seventeenth and eighteenth centuries. Malvido's (1990) study of the migration patterns of novices of the Order of San Francisco between 1649 and 1749 highlights the attraction of an ecclesiastical education, which attracted an average of ten novices a year from among Mexican Creoles (Spaniards born in the city of Mexico), provincial Creoles (Spaniards born within New Spain) and the Spanish. The clerics would have dispersed widely after training, and served the Catholic Church over a wide area. The migration pattern would thus have been two-way – initially to Mexico City and subsequently away from it.

Additional dimensions of the dynamics of relocation diffusion are illustrated in the migration of large groups of Quakers in the nineteenth century within the United States. Rose (1986) describes the high concentration of Quakers in Indiana in 1850 (the state's 89 churches accounted for 12 per cent of the national Quaker population) as a consequence of wider migration flows during the early 1800s. By September 1807 the upper Whitewater Valley area contained 84 Quakers who formed the nucleus for several rapidly growing settlements that were to attract nearly 900 Quakers by 1815 and over 2,000 by 1820. Southern Quakers had freed their slaves by 1800 and were actively opposed to the 'peculiar institution' of slavery. Many free blacks lived near them, and when economic and political conditions in North Carolina became intolerable for the Quakers and blacks, large numbers of both groups moved together to Indiana. Although some blacks migrated separately and settled in different areas, many of them chose to locate close to the Quakers because of their anti-slavery and humanitarian philosophy. The net result was a close association in Indiana among natives of North Carolina, Quakers and blacks by 1850.

Missionaries

Migration has clearly been an important vehicle in the relocation diffusion of religion at different scales. Many migrating groups have been driven by religious persecution to abandon their original locations and seek new ones, and for them the preservation and strengthening of religion has usually been a high priority. Others have seized the opportunity to better their socio-economic prospects by migrating to new places, and for them religion is often merely part of the cultural baggage to be transported and transposed. A third type of migrant are missionaries, who deliberately set out to plant or reinforce their religion in new territory.

Missionaries have been key players in religious diffusion throughout history. But their role has inevitably not been confined to dispensing religion, because they have often represented the modern world in undeveloped nations. Christian missions played a prominent part in shaping the initial development of many parts of Africa during the late nineteenth century, before colonial occupation. Henkel (1989) has examined some of the impacts of these missions on education, health care, settlement structures, development projects and economic development (particularly in Zambia) as they propagated 'the spirit of capitalism'.

Many Christian missions were established in Africa during the nineteenth century as part of the white man's penetration into Africa. Johnson highlights the tensions between missionaries and locals:

> with their chapels, residences, dormitories, schools, dispensaries, gardens, utility buildings, water-supply systems, and good access roads [mission stations] stand in great contrast with their immediate surroundings. In the confrontation of Europeans with African ways of life these stations have been for the missionaries a refuge, a symbol of achievement and a home; for the Africans they have been strongholds of alien ways from religion to agriculture, an intrusion but also a promise of help, of learning and of a better life.
>
> (Johnson 1967: 168)

His paper describes how suitable sites were chosen for new missions, and it seems that factors such as climatic and scenic attractiveness for Europeans figured as highly as density of local populations of unbelievers. The relationships between the distribution of missions and local populations was two-way, however, because 'missions affected local distributions of population through the founding of villages for freed slaves, the settlement of converts on mission land, the opportunities for work, and the attraction of educational and medical services' (ibid.: 172).

The trend continues with contemporary mission initiatives in the Orinoco delta and the Gran Savana of Venezuela, which are described by Delavaud (1981). Missions there bring marginal populations into contact with the modern world and their far-reaching influences include education, agriculture and livestock-rearing, timber industries, and the marketing of Indian products. The mythology of missionary work (that missionaries always damage and eventually destroy indigenous cultures) does not apply to the Venezuela example, where the positive contributions introduced by missionaries include the consolidation of Indian settlements, the introduction of better hygiene and healthcare, attempts to preserve traditional ways of life, and provision of assistance in overcoming cultural problems.

Missionary endeavours in South America continue apace, and the competition between Christian denominations is as strong today as it was in the nineteenth century. Protestant groups, for example, are making great inroads in Bolivia, Brazil and Peru mainly through active proselytism. Many of their converts are being won from the Roman Catholic Church as well as from other religions, although an upsurge in activity by charismatic Catholics promises to keep the battle for souls alive (Brinkerhoff and Bibby 1985).

Despite their pivotal role in the diffusion of religion over the centuries, relatively few case studies of missionaries have been published. One of the most detailed is a study by Jolles (1989) of Protestant conversion among the Sivuqaghhmiit on St Lawrence Island, Alaska. This Yup'ik Eskimo community is located 38 miles from the Soviet shore, and through history the Sivuqaq people shared a socio-religious heritage shared with their Soviet Yup'ik relatives. Christianity entered St Lawrence Island in 1894 with the arrival of Presbyterian missionaries, and over the following century most locals were converted. Conversion meant much more for the Sivuqaq than simply attending church, because it also brought a new cosmology and a new language. But whilst Christianity did replace their earlier belief system it was more of a convergence than an outright take-over and from it emerged a uniquely Yup'ik Christian world view.

Missionary endeavours are often most successful when they involve enculturation (the setting of a new religion into its proper cultural context) rather than the dogmatic transplant of a belief system from a source area to a new destination. Such forces partly explain the success of The Canadian Presbyterian Mission in Trinidad, which is described by Prorok (1987). The Mission was founded by John Morton, and was much more successful than any other attempt to evangelise overseas Indian communities. One reason was that Morton and his missionaries used Hindi to communicate with the East Indians (many of whom had been imported into Trinidad as indented labour on sugar estates between 1845 and 1917). They also made a Hindi hymn book and observed Indian religious life in order to converse more freely with Indians, and provided them with much-needed medical, legal and educational services. The spread of their influence was greatly assisted by the country's expanding railroad system and road network.

Missionaries also played a key role in shaping the religious geography of the Philippines before the close of the nineteenth century (Doeppers 1976). Islam had ruled unchallenged in many parts of the archipelago from the thirteenth century, but the arrival of Roman Catholic missionaries 300 years later was to bring radical change. The Spanish settled first in the port village of Cebu in 1565, and in 1571 a small permanent capital was established in Manila. By

1574 six Augustinian missions were scattered in the main areas of interest to colonists, and they were later joined by Franciscans, Dominicans, Augustinian Recollects, and Jesuits. Initial mission sites were established in the larger villages, and missionary priests used the regional languages as a medium for carrying their message. Missionary efforts were concentrated on the children of village leaders and of other influential leaders, through whom in time most locals came to accept Christianity. Christianity was introduced in this way over many coastal and inland areas, though it did not fully displace Islam (particularly in western Minanao and Sulu) or the ethnic religious systems of the highlands.

Merchants

Relocation diffusion, as we have seen, involves a variety of processes and routes by which people transplant religion from one place to another. Sometimes the transplanting is done deliberately, and sometimes it is incidental to the main purpose of the move. Relocation can be a temporary as well as a permanent process, and people who move around regularly in the course of their day-to-day existence can also serve as very effective channels in the diffusion of innovations. This is true particularly of merchants and traders who, throughout history, have assisted in the spread of ideas, values and beliefs within their commercial networks.

Many major trading centres have been important cultural and religious crossroads. One example is Timbuktu in the kingdom of Songhai (now central Mali) in West Africa, a terminus of one of the trans-Saharan caravan routes and a thriving trading centre. Between the fourteenth and sixteenth centuries it was a major seat of Islamic learning (Lass 1987) and an important nucleus from which Islamic ideas and values filtered out over a wide area, in and beyond West Africa. In East Africa, many Muslim coastal settlements traded extensively throughout the Indian Ocean and the Mediterranean world in high-value commodities ranging from gold, rock crystal and ivory to slaves and timber. Archaeological excavations of the earliest Muslim mosques along the coast date the arrival of Islam to the beginning of the ninth century (Horton 1987), since when Muslims were attracted to the area and won some converts amongst local people.

Expansion diffusion

Expansion diffusion involves the spread of an innovation *within* an area, as opposed to the transfer of innovations *between* areas (relocation diffusion). The two primary mechanisms of expansion diffusion are hierarchical (top–down) and contagious (contact) expansion (see Chapter 4, pp. 99–101). We looked at a number of case studies of national-scale expansion diffusion in Chapter 4 (see pp. 117–27). Most studies focus on broad patterns of diffusion, and there are far fewer empirical descriptions of how the process works.

One such study by Meyer (1975) analysed the numerical and spatial expansion of the Lutheran Church–Missouri Synod (LCMS) in the United States, over a period of 125 years, from an original cluster of several thousand midwestern German settlers. Although natural increase and internal migration contributed to the expansion, most can be attributed to the effectiveness of organised evangelism. The first evangelists were leaders of the several immigrant factions that formed the synod, and their specific targets were German Lutherans. Non-German areas were not evangelised until later. The Missouri Synod's emphasis on congregational ministry severely limited geographical expansion through evangelism, and contact conversion led to a relatively compact spatial pattern, with occasional outliers. Spatial patterns of LCMS membership remained stable between 1870 and 1970 despite migration and conversion, largely because of its ethnic character and conservative theology.

More indirect forms of expansion diffusion are illustrated by monastic initiatives. Although most are designed to provide havens of tranquillity within which monks can live a life of prayer and reflection, monasteries often interact with their surrounding area in a variety of mostly practical ways (such as introducing new land management techniques, producing produce for sale in local markets, and supporting if not providing local schooling). In some senses this represents a type of hierarchical expansion diffusion, in which the monasteries provide the nuclei from which religions trickle down through an area. Thus monasteries can provide a persistent and effective agency of religious diffusion.

Inevitably, open communities (in which the religious group interacts regularly with its surrounding population) play a much stronger role in religious diffusion than closed communities (where the religious group is shut off from its surrounding population). Sometimes monastic foundations placed great emphasis on charitable and pastoral work amongst their local populations, and this sort of contact would provide channels for spreading the faith. Monastic houses in medieval Ireland, for example, often provided local health-care facilities – such as the Hospital of St John in Dublin (Hennessey 1988).

RELIGIOUS ADHERENCE

The most critical single factor in religious change, whether it be expansion or decline, is the adherence of individuals to the faith. Consequently it is important to examine how religious adherence varies spatially, and to search for causal factors to explain observed patterns.

Definition and measurement

It is quite easy to offer a definition of religious adherence, but it is much more difficult to arrive at an unambiguous way of measuring it. To adhere is 'to be devoted to . . . to follow closely or exactly', and an adherent is 'a supporter or follower' (*Collins English Dictionary* 1979). Religious adherence, therefore, refers to the support of religion and it reflects the fact that individuals are devoted to and followers of the religion of their choice.

The opposite of adherence is neglect, and in some ways it is probably easier to identify religious neglect (or decline) than adherence. Gilbert (1980) describes the making of what he terms 'Post-Christian Britain', which he defines not as a

> society from which Christianity has departed, but one in which it has been marginalised. It is a society where to be irreligious is to be normal, where to think and act in secular terms is to be conventional, where neither status nor respectability depend on the practice or profession of religious faith.
>
> (Gilbert 1980: ix)

These are the hallmarks of secularisation (see Chapter 2, pp. 48–52).

Measuring adherence is problematic because it has two dimensions – the *number* of adherents, and the *strength* of their adherence. The latter is similar to but not identical with 'religious observance', which means the degree of strictness with which individuals or groups follow the rules and practices of a religion.

If the number of adherents drops a religion goes into decline (at least numerically), and if the number increases the religion expands. Strength of adherence is much more difficult to measure than number of adherents, because even tangible expressions such as frequency of church attendance, participation in religious rituals or amount of time devoted to religious matters might not fully reflect a person's inner feelings and religious convictions. In the absence of more meaningful (but still readily available) measures of strength of adherence, however, most studies have to resort to quantifiable measures.

The simplest measure of religious adherence is level of church attendance, which can be counted with ease and accuracy. Attendance records can be used to examine variations in this aspect of adherence through space and time. A good example is Muller's (1990) study of religious observance in Lower Normandy in western France, which compared conditions in the mid-1950s and late 1980s. He found that traditional religious observance had declined throughout the area, but the decline was faster in the Pays d'Auge (Lisieux) and the Bocage Normand than in Cotentin, Bessin, the Plaine de Caen or Perche.

International comparisons of religious adherence using church attendance data are open to problems of variable data quality and incompatible definitions, because different churches and different countries collect the data in different ways. An alternative approach is to use an indirect measure of religious commitment for which at least compatible cross-national data are available. Wuthnow (1976) used United Nations data on the number of religious books published in 43 countries over the past quarter century.

Controls, evolution and impacts

The context within which religious adherence operates is vitally important, and this includes identifying what factors control it, how it changes through time, and how in turn it influences other aspects of behaviour amongst believers.

Controls

If measuring religious adherence is problematic, then seeking explanations of observed patterns raises even more difficulties. Studies have attempted to do this at a variety of scales, and inevitably the relevance of specific factors changes as we move from one scale to another.

The smallest scale is the individual believer. Naturally, a person's strength of religious belief will be influenced by a range of different determinants (conscious and unconscious), and not all of them can be isolated and measured unambiguously. One interesting attempt to unravel the maze of variables, by Hilty, Morgan and Burns (1984), used factor analysis to identify what influenced religious involvement in a sample of members of a Mennonite Church in the United States. The seven factors that emerged were personal faith, intolerance of ambiguity, orthodoxy, social conscience, knowledge of religious history, life purpose, and church involvement. No overlap on the seven factors was found. These sorts of results reinforce the conclusion that at the personal scale, highly personal factors determine religious adherence.

At a broader scale, such as an area or region, studies have shown that adherence varies spatially, largely reflecting historical and socio-economic factors. In Lower Normandy, for example, Muller (1990) found that levels of church-going were highest in the rural areas, and much higher among women than men. Religious observance in traditionally Catholic rural north-west Portugal has declined since the 1960s, as an increase in commuting to major cities has brought changes in local class structure and social diversification (de Almeida 1986). Other measures of strength of adherence to Catholicism in Portugal – including regional variations in the incidence of religious practices such as Catholic marriage and attendance at Sunday mass, as well as the proportion of the population that is Catholic (Figure 5.2) – also suggest that local social, economic, political and cultural factors can exert strong control over individual and collective practices (Andre and Patricio 1988).

Religious persecution is a powerful determinant of adherence, too, which can function in two opposing directions. In some situations persecution from outside binds members of a religious group together, and this cohesive force promotes greater adherence. This certainly seems to have been the case with Jewish communities throughout history. In other situations, however, persecution can promote a dwindling of adherence as harassed individuals gradually give up the fight to keep their collective religion intact. Such seems to have been the fate of the sheltered group of Bahais from the Iranian oasis of

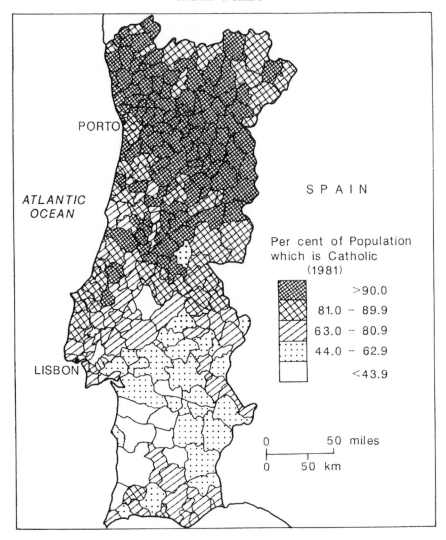

Figure 5.2 Distribution of Roman Catholic adherence in Portugal in 1981
Source: After Andre and Patricio (1988).

Nadjaf-Abad, who have long been persecuted by surrounding Muslims
(Fecharaki 1977). Most of the young have abandoned the community and
migrated to the relative anonymity of large cities, and the Bahai numbers at
Nadjaf-Abad decreased from 3,000 in 1956 to less than half that number in
1977.

At the international scale, Wuthnow (1976) found some interesting
correlations between religious book publishing (as a measure of religiosity) and

146

various indicators of modernity for 43 countries. Significant (95 per cent) negative correlations were established with variables such as level of literacy, university enrolment, proportion of population in professional and technical jobs, and gross domestic product. The results indicated an inverse relationship between religiosity and modernity; more modern countries were less religious, and vice versa.

Evolution

Few studies have looked at how religious adherence evolves through time within particular groups. Jeans and Kofman (1972) examined the links between religious adherence and population mobility in nineteenth-century New South Wales. Their analysis of census data between 1841 and 1901 revealed religious adherence (particularly amongst Anglicans, Catholics, Presbyterians, Methodists) to be relatively stable through time. Active proselytism was very limited, and most children appeared to adopt their parents' religion. Marked regional variations in the distribution of Protestants and Catholics were largely the outcome of distinct migration paths, rather than a reflection of conversion *in situ*.

Evolution of religious adherence extends to changes in practice within particular religions. Hey's (1973) study of the spread of nonconformism in Yorkshire (England), offers glimpses of the dynamism of such change. During the seventeenth and eighteenth centuries the so-called Old Dissent (the early Nonconformists) had grown strong particularly in remoter areas with weak parish structures, in the outlying settlements of huge moorland and marshland parishes, and also in the developing industrial centre of Sheffield. Dissent was easily suppressed in small and nucleated settlements where the farming system involved the whole community and where the squire and rector were in immediate supervision of their charges. The Evangelical Revival that came in the wake of the population boom and industrial development of South Yorkshire was eagerly embraced by the original dissenting strongholds, who were receptive to the various forms of the new preaching. Methodism was particularly successful in attracting converts of all classes and in most areas, and by 1851 the Methodists had become the chief alternative to the Church of England. Autonomous Methodist chapels appealed particularly to labourers and farm workers, who ignored the Anglican churches patronised by their squires. Most working-class people remained largely unaffected by organised religion, but those who did attend worship were increasingly to be found in the Methodist chapels.

The reverse side of the religious coin, the acceptance of Anglican ritualism, is illustrated in Yates's (1983) study of 'bells and smells' in southern English churches in the late nineteenth and early twentieth centuries. Some towns attracted ritualistic clergy and some did not. Many factors influenced the distribution of ritualistic parishes, including reliance on wealthy benefactors to

147

provide buildings and endowments, the role of aristocratic and ecclesiastical patrons, and the influence of national ritualist societies (particularly the Oxford Movement that was opposed to liberalisation of the church). The pace of change was often dictated by the congregation of a church, and many high church clergy were encouraged to introduce a more elaborate ritual (including the use of lighted candles) by popular demand.

Impacts

Religious adherence influences many aspects of human behaviour, and it is difficult to draw universal generalisations because different religions expect different behaviours of their followers. Islam, for example, fixes acceptable codes of conduct in almost every aspect of life, with tough sanctions for those who violate them, whereas many other religions rely on voluntary compliance. Some visible manifestations of the impact of religion on human geography were outlined in Chapter 1 (see pp. 2–7) and it will suffice here to look at two case studies of the impacts of religious adherence on behaviour. They are illustrative, not exhaustive!

Religion can strongly influence the behaviour of even non-believers in religious strongholds, by constraining what sorts of activities are regarded as socially acceptable. One such stronghold is the Isle of Lewis, the largest of the Western Isles (Outer Hebrides) off the north-west coast of Scotland. This strongly Calvinistic and isolated rural community, dominated by crofts and small townships, is renowned for its radical Protestantism; nine out of ten Lewis people still claim to belong to the Free Church of Scotland. Mewett (1985) describes how islanders who are 'called by God' are invited to take communion, and the communicant elect see themselves as guardians of the true faith.

> The secular mode of religious behaviour involves regular church attendance, daily bible readings and family prayers in the home, abstinence from alcohol, and a negative attitude towards music and dancing. Keeping up this mode of behaviour is essential for the very religious and deviation from it may prevent a person from taking communion ... on Sundays religion becomes dominant on the island as a whole and markedly affects the behaviour of everyone, irrespective of their professed religiosity. And the Sabbath is actively 'policed'. The biblical injunction that no work should be done on the seventh day is taken literally and is actively enforced. This restriction extends even into the home ... All trade ceases; every business closes down for the day. Pubs and hotel bars are closed to the public. The ferry and air services between the island and the mainland are curtailed ... Doctors and nurses are permitted to continue their labours, but it is impossible to buy a bottle of aspirins on the Sabbath.
>
> (Mewett 1985: 188–9)

On Lewis the religious observers claim a moral authority to legitimate policing of the Sabbath, often through pointed verbal attacks. It doesn't always work, but it certainly can be a potent factor in influencing how non-believers behave.

A second example is the impact which people's religious adherence can have on their mobility, particularly in terms of willingness to migrate. Kan and Kim (1981) looked at whether migration intentions in non-metropolitan Utah differed significantly between Mormons and non-Mormons. Although their multiple regression analysis failed to support the hypothesis that religion exerts significant influence on migration intention (after controlling the effects of other relevant variables), it did show that

> after proper control of all other variables, religion . . . [is] significant in influencing the residents' migration intentions. Specifically, membership in the Church of Jesus Christ of Latter-Day Saints retards the intention to move [away from the area]. The other significant variables in the regression analysis are length of residence, community satisfaction, and home ownership.

> (Kan and Kim 1981: 683)

They concluded that if the pattern of in-migration remains basically the same in the future (which is likely, given the Mormon's 'Gathering Doctrine' that defines Utah as their homeland), non-metropolitan Utah will probably maintain its cultural uniqueness via social forces such as differential migration. Religious adherence clearly plays a key part in this demographic system, particularly in driving positive feedback (reinforcement) which differentially favours or promotes Mormonism.

RELIGIOUS PERSISTENCE

Religious adherence is an important ingredient of the religious geography of different places, and geographical and historical variations in adherence show that it is by no means constant through either space or time. This raises profoundly significant questions about why some religions persist, whilst others ebb and flow. This is another dimension of the dynamics of religious change that has wide-ranging geographical implications, and helps to explain the distribution of religions at different scales (see Chapter 3).

There is abundant evidence that some religions, and some particular religious groups, are remarkably stable and persistent through time (and thus, by definition, through space). Great stability in denominational patterns within Nova Scotia between 1871 and 1941 was described by Clark (1960), even though groups from different origins migrated into the area. The stability was accounted for in terms of the lack of active proselytism, and enduring attachment to cultural affinities.

Some forms of religious persistence reflect lack of challenge from new religious innovations, caused by geographical inaccessibility. Other instances

149

of persistence reflect deliberate policies of separatism when confronted with competition from surrounding religions. Useful pointers to why some religions persist are also offered in examples of religions that have made conscious efforts to integrate into the wider religious scene.

Inaccessibility

There is no doubt that particular forms of religion have survived in some places because of lack of contact with outside agents of change. Topographic features can form effective barriers to the diffusion of many innovations, because lack of easy access prevents the ready transfer of new ideas of practices by carriers like travellers, missionaries or traders.

High mountain regions provide amongst the best illustrations of topographic barriers to the spread of religious innovation. Bhutan, in the Himalayas, has until recently remained relatively cut off from the modern world and indigenous Bhutanese forms of Buddhism have developed there over the centuries (Wilhelmy 1990). But such persistence might not last much longer, because this land of fortified castle-like Buddhist monasteries has started to open up its frontiers, particularly with the arrival of new roads like the Indo-Bhutan Highway. These new access routes will allow the penetration of external influences (including religion), and threaten the hitherto unchallenged survival of Buddhism in Bhutan.

Similarly remote and inaccessible conditions have persisted in parts of the Karakoram Mountains in Kashmir, along sections of the old Central Asian Caravan Route between Chinese Turkestan and India. Noack (1979) describes the Buddhist peoples of the previously forbidden land of Nubra, where access to outsiders has been tightly restricted and requires a permit from Indian officials. This land, previously part of Western Tibet and still inhabited largely by Tibetans, has many Lamaseries in which Buddhism has been preserved and which continue to have significant impact on both the landscape and culture of the area.

Separatism

Lack of contact with outsiders, caused by remoteness and inaccessibility, has greatly assisted the persistence of particular forms of religion in some areas. In other situations, religious groups have tried to maintain their own separation from outside influences, with mixed results. The survival of distinctive religious groups in the United States (like the Shakers, the Amish and the Mennonites) owes a great deal to their separatist ideologies and behaviour.

The survival of Arab refugee groups within modern Israel is another graphic example of persistence through separatism. This minority group of involuntary migrants and refugees, displaced by force, are very dynamic and have been displaced a number of times. Al-Haj (1988) identifies three phases in the displacement of Arab internal refugees within Israel – the search for asylum (1948–51), waiting and expectations (1952–6) and the period of stable settlement and resettlement (since 1957). These Arab refugees represent a 'minority within the minority', as refugees in the local Arab communities and as Arabs in a dominantly Jewish society.

The Ethiopian Falashas illustrate much more long-lived separatism and religious persistence. This black Jewish group (the word Falasha means 'exile' or 'stranger'), believed to be descendants of Jews who lived in Egypt and migrated to the land of Kush (now Ethiopia) over 2,000 years ago, until recently remained separated from all other groups of Jews around the world (Epstein 1992). Through two and a half millennia the Falashas kept themselves separate from the outside world (including other groups within Ethiopia), and their traditional lifestyle altered very little. Like all Jews they remained committed to return to their biblical homeland, but when the modern state of Israel was established in 1948 they were not accepted as truly Jewish and thus did not qualify for the Law of Return (which confers automatic rights of citizenship and settlement in Israel). In 1975 the Jewish Rabbinate reversed that ruling, and the Falashas were officially accepted into the Jewish community. A small-scale clandestine operation to smuggle the Falasha out of Ethiopia, between 1979 and 1982, prepared the way for the evacuation of large numbers of Falasha to Israel on 24 May 1991. There appears to be no precise record of the numbers involved – Epstein (1992) mentions 145,000 people, but Karadawi (1991) puts it at 17,000. The massive airlift, code-named Operation Solomon and involving over 40 planes, is widely believed to have been planned and resourced by Israel, the United States government (including the CIA) and private organisations (Karadawi 1991).

Integration

There is no shortage of studies of how the newly arrived and pre-existent religions interact during the early phase of absorption of migrant groups into new areas. Sometimes an immigrant group seeks to integrate into the mainstream society it has joined, but this does not always require them to sacrifice their (minority) religious beliefs and practices and adopt (majority) mainstream ones in their place. Integration does not necessarily mean extinction of religious diversity, as the following examples show.

The 12 million Indians in present-day Mexico comprise up to 15 per cent of the country's total population. Descended from the Aztecs these people have achieved a high degree of assimilation into the modern Europeanised world. None the less a large number of traditionalists have survived, and recent years have witnessed a revival of some old cultures and religious practices, often as a reaction to problems of unemployment, poverty and discrimination (Sanger 1991).

Over 1.5 million Muslims migrated from three North African countries to France in the last generation. Their children (popularly known in France as 'Beurs') are French nationals but ethnic minorities, who face considerable difficulties assimilating into mainstream French society (Begag 1990). Sweden has also received large numbers of Muslim immigrants from different countries, and until the late 1980s there were few signs that the Muslims were becoming integrated. Soderberg (1991) describes the symptoms of a change in attitude, using Malmo as a case study. Since they started to bury their dead in the city's Muslim burial ground, rather than returning them to the old country, Malmo Muslims have started to regard the city as their home town not a temporary place of exile. The building of mosques and other Muslim institutions within Sweden is further evidence of assimilation. Both symptoms reinforce rather than extinguish Muslim identity and cohesiveness, and indicate religious persistence through integration.

Integration of a different kind, this time involving loss of identity and cohesiveness, is illustrated in the fate of the Syrian–Lebanese Christians in Sydney (Australia). McKay (1985) examined three groups of Syrian–Lebanese immigrants – pre-Second World War (when most migrated from Christian hamlets in what is now Lebanon), 1950–1975 (mainly comprising Christian migrants at a time when Australia relaxed its immigration laws), and recent arrivals fleeing from the civil strife (post-1975). He found that, unlike some other ethno-religious populations in Australia, the different generations of immigrant Christians were unable to reconcile their religious differences for the sake of preserving ethnic solidarity. Members of the second and third generation of descendants from the immigrants do not feel Lebanese because of any linguistic, religious or nationalistic ties to Lebanon, and many of them have joined and become active in Australian churches. As a result, the survival of Syrian–Lebanese Christians in Australia as a distinct ethnic group is questionable.

Language is often a powerful cultural barrier to integration, and Jones (1976) used evidence of the dissemination of English-speaking Baptist chapels through the developing coal field of South Wales in the late nineteenth century as an indication of cultural transition in the area. Nonconformist chapels were foci of the emerging industrial culture, and the Baptists had a universal appeal to both Welsh- and English-speaking populations. Before 1860 the coal field was uniformly a Welsh cultural area, and the Welsh language was used throughout (including in chapel). Between 1860 and 1890, English culture and language were taking hold in the area, and this is reflected in the opening of new English-speaking chapels. After 1890, Welsh, as the single or even dominant language, was in decline (except in the western section of the coal field), bilingualism was increasing fast, and English as the main language was becoming the norm through most of the area.

MIGRATION, ETHNICITY AND RELIGION

The example of the Syrian–Lebanese Christians in Australia (above) raises interesting questions concerning the role of population migration in changing the pattern of religion within an area. Religious and ethnic identities are usually closely intertwined, and often one cannot be readily separated from the other. In this section we will look at some of the interactions between migration, ethnicity and religion using examples drawn from a variety of religions and contexts. Many illustrations relate to the United States, because most detailed studies have focused on North American case studies.

Migration and religious commitment

Some interesting studies have focused on the impact of migration on the degree of religious commitment, particularly amongst the migrants. Changes in the population into which they move can be equally pronounced. These two mechanisms represent important dimensions of the dynamics of religious diffusion. We concentrate here on changes amongst migrants, and will examine a number of hypotheses and models that figure prominently in the literature.

One model that has been tested in several United States studies is the so-called 'regional convergence' model. The model makes two basic assumptions – that there is great population migration across regions within a country, and that migrants retain their religious affiliation when they settle in the new area. It predicts increasing religious mixing and decreasing geographical concentration through time, thus increasing religious homogeneity and removing distinctive religious patterns. Newman and Halvorson (1984) tested the model using data for 35 religious denominations for two periods: 1952–71 and 1971–80. They found no evidence of greater religious dispersion through time, and concluded that (despite large movements of people around the country) patterns of religious membership had not converged among the regions in the United States.

Two other conceptual models have been proposed to describe changes in religious commitment among regional migrants – the 'dislocation' and 'adaptation' models. The 'dislocation' model predicts that all migrants undergo a decline in commitment after being uprooted from familiar surroundings. It is illustrated by De Vaus's (1982) study of migrating adolescents in Australia.

The 'adaptation' model predicts that commitment among migrants will rise or fall depending on the norms of religious behaviour in their new location. Stump (1984b) tested this model using data on patterns of church attendance and strengths of adherence within national samples of white and black Baptists, and white Lutherans, Methodists and Catholics in the United States. The results confirm the model's prediction that religious commitment rises among migrants to regions of higher native commitment (such as the South) and falls along migrants to regions of lower commitment (such as the West). The study showed that – contrary to expectation but reinforcing Newman and Halvorson's (1984) findings – migration may contribute to the persistence of regional religious differences in the United States.

More detailed studies of the impact of migration on church attendance throw some light on the social processes underlying observed patterns. Welch and Baltzell (1984) developed a path model based on population surveys from three states (Iowa, New Jersey and Oregon) to examine the effects of geographic mobility and social integration on church attendance. Their results suggest that geographic mobility had only limited effects on church-going. The effect was partly negative, by disrupting an individual's network of social ties and 'disrupting the stable web of communal attachments that promote conventional behaviours such as churchgoing' (ibid.: 76) (this is called the 'integration–disturbance' hypothesis). Mobility also inhibited the formation of new ties in their new community of residence. But people who had migrated frequently in the past were found to be motivated to attend church services frequently in their new communities of residence, partly because of their desire to establish new stable social ties within that community (this is the 'social linkage' hypothesis).

Ethnicity and identity

Meyer (1975: 197) notes that 'many current religious distributions in the United States may reflect ethnic origins traceable to immigration processes that began in the colonial period'. The link between religion and ethnicity can be viewed the other way too, because attempts to map ethnicity in North America often use religion as a key discriminating variable. As Raitz discovered,

definition of [ethnic] group is a serious problem in Canada and the United States because ethnicity has a number of forms and is based on a variety of cultural and historical components. Consequently, some groups are classified by national origin (first- or second-generation European immigrants), language (French-speaking Canadians), religion (Jews and possibly Mormons), race (American Indians and blacks), minority status (Mexicans and Puerto Ricans), the so-called new ethnics (fourth-generation European-Americans such as the Polish), or even region (Appalachian whites).

(Raitz, 1978: 349)

Religion and ethnicity

Williams notes how

religion provides a powerful means of shaping and preserving ethnic identity through the transmission of cultural patterns and the attribution

154

of sacral status to elements of those patterns. The formation and preservation of ethnic identity are negotiated between the immigrant community and the host society and between the first and second generation.

(Williams 1987: 25)

His case study, dealing with Gujarati immigrants to the United States, focused on a range of religious institutions that help to preserve ethnic identity for Asian Indians. These include home-based groups for newly arrived immigrants, local groups to study religious texts, and all-India religious groups that support a general Asian Indian identity. Gujarati immigrants also join with immigrants from other countries in some religious groups (such as Sunni and Shiite mosques, which serve immigrants from many Asian and African countries), or with major American religious groups with which they can share worship but retain their own distinct identity (such as a Tamil Christian congregation as part of the Reformed Church of America). In these ways religion is 'significant in the formation and preservation of ethnic identity and in the cultural negotiations of Gujarati immigrants with the host society' (Williams 1987: 27).

Many immigrant groups to the United States – including Romanians, Ukrainians, Hungarians, Poles, Greeks, Italians, Portuguese and Serbs – see religion and ethnicity as inseparable and mutually reinforcing. But some groups, including the Dutch in Canada – described by Palmer and Palmer (1982) – see them as conflicting. Three waves of migrants from the Netherlands to Canada during the twentieth century have largely assimilated into Canadian society in terms of language and traditions, although the two Dutch churches (the Reformed Church in America and the Christian Reformed Church) have acted as a strong cohesive force within the group. Paradoxically,

> religion has proven to be a significantly stronger force for group cohesion than ethnic identity. The deeply held religious convictions of many Dutch immigrants has prompted them to retain a viable and cohesive group identity, with a wide range of interlocking institutions and activities and minimal intermarriage outside the group, while simultaneously denying or minimising their ethnic affiliation.

(Palmer and Palmer 1982: 261)

North America has certainly been an important cultural melting-pot in which many different religions have come into contact with one another. This has sometimes produced patterns of religious dominance in an area, where one religion or denomination emerges as more important than others (numerically, in terms of social and political influence, or both). Religious culture regions (see pp. 158–67) reflect this sort of dominance in an inherently pluralistic continent. Sometimes, however, the mixing produces genuine diversity in which the spatial mosaic of religious variations is finely-grained and highly variable.

155

Other areas show similar diversity, reflecting cultural mixing and adjustment. The Mediterranean basin, for example, has throughout history been a cradle for the development of civilisation and a junction-point for different cultures and religions. Little wonder, therefore, that it houses 55 different ethnic groups and that thirteen of the eighteen Mediterranean states are multiethnic (Kliot 1989). Ethnic self-expression within these states is reflected partly through religious diversity. Preserving the identity of ethnic groups in multiethnic states often requires territorial policies that accord legal rights to inhabitants of culturally defined sub-state regions, and these also can be defined using religious criteria. Murphy (1989), for example, describes the creation of a Jewish Autonomous Oblast in the Soviet Union, although it was in an area that held no special meaning for Jews and was thousands of miles from the nearest significant concentration of Jews.

Ethnicity and the Roman Catholic Church

The Roman Catholic Church in the United States has long played an active part in helping immigrants to assimilate into American society. Between 1974 and 1989 over half a million immigrants and refugees entered the US each year, legally or illegally, for permanent residence legally or illegally, and the Church was a vital link in the chain for many of them.

As Keely notes,

> ethnic neighbourhoods, foreign language press and home town associations, along with churches, provide a place of security from which people could integrate. Parochial schools and ethnic parishes were meant to enable people to remain Catholic, even while as immigrants they were becoming Americans. . . . Catholic schools and national parishes were a peculiarly American approach. They attained two goals of immigrant communities simultaneously; to remain Catholic and to become American.
>
> (Keely 1989: 31)

He adds that 'people tend to become American when they can be true to their past while acquiring the skills to participate in American life' (ibid.: 33), and emphasises the role played by national parishes in this vital assimilation process.

One Christian group with a strong allegiance to the Catholic Church is the Maronites, who fled religious persecution in Lebanon and emigrated to the United States beginning in the second half of the nineteenth century. Labaki (1989) reports that by 1924 there were nearly 90,000 Maronites in the USA and the Church played a prominent role in helping to safeguard their identity. Initially they attended Italian and Irish churches, but were treated as outsiders. Through time the Catholic bishops approved the setting up of Maronite parishes, and by 1962 there were 33 Maronite parishes (looking after 80,000 Maronites) in the United States, spread over 34 Roman Catholic dioceses. By

1989 this had grown to nearly 60 Maronite churches with a community totalling almost half a million.

Some Catholic national parishes (particularly those for Vietnamese and Koreans) have been very successful in meeting the needs of their immigrant groups, but not all major groups have their own national parishes. Keely (1989) identifies gaps for Columbians, Mexicans, Puerto Ricans, Dominicans, Ecuadorians and other Hispanic groups, but there are doubtless other nationalities that remain unrepresented too.

Patterns in the survival of Catholic National Parishes (German, Italian and Polish) in the United States between 1940 and 1980 were studied by Stump (1986b), who found much faster decline in the Middle West than in the North-east. Specific local factors that influenced the survival of national parishes include the attitudes of local priests towards national parishes, the ability and desire of the ethnic community to maintain the parish, and the spatial distribution of the parish and its parishioners within the city. Alba (1981) argues that the Roman Catholic Church might, in fact, be doing itself out of business as an ethnic church because it has been so successful in helping immigrant groups to assimilate into American society. This seems to be the case particularly among American Catholics of European ancestry, who entered the United States in the nineteenth and early twentieth centuries as voluntary immigrants and thus 'had greater freedom than others to determine where they settle and what occupations they pursue' (ibid.: 96).

Ethnicity and social mobility

Another important dimension of the close relationship between religion and ethnicity, which can strongly influence both religious persistence and cultural assimilation, is the impact it has on the integration of immigrants into their new society. This is reflected in a variety of ways, including social integration, social mobility (particularly upwards), and social stability.

Social integration is widely seen as one of the most important aspects of assimilation because it involves acceptance of the immigrants into social cliques, clubs and institutions of the host society. Religion is one of the institutional areas of a society in which indigenous and immigrant populations can mix and integrate. Veglery (1988) examined patterns and processes of social integration among first generation Greeks in New York, who had arrived since 1965 and were interviewed in 1982. He discovered that church membership and attendance played stronger roles in integrating newcomers into mainstream American life than did schooling and fluency in English. But the integration was, at least initially, ethnically closed because most immigrants preferred Greek Orthodox Churches to the more mainstream American and Russian Orthodox Churches. Church, in this sense, serves as a cultural bridgehead through which ethnic immigrant groups develop social networks in their new environment. At least in the early stages, integration proceeds naturally through religious and social networks with which the newcomers have some natural affinity.

Integration is a critical step in the process of enculturation of ethnic groups, but some groups take the process much further and seek upward social mobility. Jews have been upwardly mobile in Western society over the last 150 years, and case studies of Jewish social mobility in New York City and Pittsburgh (Pennsylvania) illustrate the theme. Between 1880 and 1914 some 23 million European immigrants entered America, 17 million of them through New York City. Although they differed greatly from their Irish and German predecessors, the Italians and Jews readily joined in what Kessner (1977) calls 'the American religion of upward mobility'. The Jewish immigrants in particular used New York not only as an entry port to America, but also as a stepping stone to Americanisation. Silverman (1978) studied three generations of Conservative Jewish families of eastern European origin in Pittsburgh, and found that closely knit family networks and ethnic behaviour change but are not necessarily lost as families become upwardly mobile. He concluded (ibid.: 39) that 'each generation and social class reassesses, reinterprets, and reintegrates its symbols and meanings for ethnicity, substituting new symbols for older, less meaningful ones'.

Ethnicity can also have a strong impact on social stability, and Coburn's (1988) study of the women of Block, Kansas, between 1868 and 1940 highlights the significance of social networks. This separatist community in south-central Miami County, Kansas, was established by German Lutheran immigrants in the 1860s and its ethnic heritage and religious doctrines have survived relatively intact. Coburn used the group as a case study of the relationships between ethnicity, religion and gender. Four networks (church, school, family and the outside world) were found to play roles in the transmission of beliefs, values and culture, and the church network was most stable through time. Largely resistant to the passage of time and the process of Americanisation, 'for more than seventy years, its formal networks, structure of authority, and character of activities changed only by degree' (ibid.: 229)

RELIGIOUS CULTURE REGIONS

Shortridge has commented on the fact that many phenomena that are usually taught in cultural geography,

> including such seemingly diverse topics as domestication, food prejudices, house types, and the origin of cities, are linked with religion. These connections exist because religion not only involves theology, but also implies other values and attitudes. In fact, the role of religion in the creation and maintenance of cultural traits and regions is difficult to overstate. It is thus not surprising to note how religious regions resemble the more general culture areas of the United States.
>
> (Shortridge 1978: 56)

We have already noted the common spatial coincidence between ethnicity and religion (see pp. 154–8), and this association is brought alive perhaps most clearly in religious culture regions. These are geographical areas in which a particular religion or denomination is particularly strong and influences many aspects of the areas' lives and activities, which can be delimited on maps (some more definitively than others), and which serve as 'type areas' for that religion or denomination. Obvious examples in the United States include the Bible Belt and the South (see Chapter 3, pp. 87–9), and the Mormon realm (see below).

Cultural geographers like Jordan and Rowntree (1990) recognise two main types of culture region – formal and functional. A *formal culture region* is 'a uniform area inhabited by people who have one or more cultural traits in common' (ibid.: 7), although they 'are the geographer's somewhat arbitrary creations' (ibid.: 9). A *functional culture region*, in contrast, is an area that is related to surrounding areas by some functional or structural relationship. In terms of religious functional culture regions, the individual church congregation is an example of the smallest individual units, and the functional sub-division of most countries into networks of parishes and large church administrative units illustrates the broader scale. Sacred sites and religious places (see Chapter 8, pp. 249–58) are expressions of the functional designation of space, and they too can be important components of such functional culture regions.

We will return to the topic of functional culture regions in Chapter 8, where we look at themes such as the religious sub-division of territory, the designation of sacred space, and the dynamics of religious pilgrimage. Here we focus on formal culture regions, which we can illustrate with some examples from North America.

Formal culture regions based on religion can be delimited at various scales. The world map of major religions (Figure 3.1) offers a broad-brush subdivision of geographical space into major formal regions, because areas dominated by Christianity, for example, will usually have different cultural environments than areas dominated by Islam, Judaism or Hinduism. At the national scale, the spatial pattern of dominance by leading Christian denominations in North America (Figure 3.7) gives clues about the nation's cultural and ethnic mosaic. At the regional scale, definition is more problematic because boundaries of formal regions are rarely sharp and such regions are never totally homogeneous (there are always people of different religions living within the same area).

The Mormon culture region

Perhaps the best studied culture region in the USA is the Mormon one that has proved remarkably persistent through time. Zelinsky (1961: 193) notes that 'the Mormon Region is the most easily mapped and described', and it is one of 'only two or possibly three cases of regions whose religious distinctiveness is immediately apparent to the casual observer and is generally apprehended by their inhabitants'. The evolution of this Mormon culture region is described in detail in a classic paper by Meinig (1965).

Evolution

Joseph Smith (born in Vermont) was living in western New York when he had his early visions, produced his sacred text (the Book of Mormons) and in 1830 formally organised the Church of Jesus Christ of Latter-Day Saints. The first Mormon temple was built in New York, and Smith tried to establish the first full Mormon society at Kirtland in the Western Reserve in Ohio. Friction between the first Mormons and neighbouring communities in both New York and Ohio led Smith to start the search for a more suitable place to develop the ideal Mormon society. Here ground would be consecrated for the building of the Kingdom of God (Zion) by Smith and his followers

Smith visited Independence, Missouri in 1831. He liked it and proclaimed it the new Zion, and the Mormons moved there. Confrontation with Gentiles (non-Mormons) soon forced Smith to move on, initially into north-central Missouri and then to the north-east across the Missouri to Nauvoo in Illinois. Religious conflict continued, and Smith was murdered by a mob in 1844. By now the Mormons had established that they could not survive within the frontier zone, so they prepared to move well beyond it. A number of possible locations were considered (including Oregon, California and the Great Basin), and the Great Basin was chosen only after very careful study and evaluation.

Brigham Young (now leader) declared 'This is the place' when he first set eyes on the Valley of the Great Salt Lake on 24 July 1847. This marked the start of five decades of Mormon colonisation in the Far West. The Mormons' new city (New Jerusalem) was soon under construction, based on a strict gridiron of spacious blocks and streets, not dissimilar to the New England town formalised by the Biblical foursquare. Young declared that the area be called Deseret (a word from the Book of Mormon meaning 'honeybee'). Congress was asked to recognise the State of Deseret but refused, and chose instead to call it Utah Territory.

Once the Mormon core had been established, attempts were made to expand it outwards. Some failed, like the proposed new capital at Fillmore (150 miles south of Salt Lake City), because settlements were too scattered and access was limited. Others succeeded, like the Mormon agricultural community in San Bernardino in the Los Angeles Basin, which offered an important gateway to the Pacific coast. Within a decade the Mormons had transformed Deseret from a visionary idea into the physical basis of a Mormon culture region. Initial colonisation was mostly confined to low-lying and accessible valleys, and the Mormons had planted roots throughout the area. But – faced with the threat of attack by a federal army – they had to retreat from Carson Valley, San Bernardino and Las Vegas to the core area at Salt Lake City in 1857.

Former Mormon properties in the abandoned areas were bought or seized by Gentiles.

A second phase of Mormon expansion, after 1857, was based on colonisation spreading outwards from the core along successive valleys. During the 1860s settlement was pushed further southwards into the south-east corner of Nevada, and outwards into more distant valleys. Expansion to the east was constrained by shortage of land, and to the west by the inhospitable Great Basin. By the late 1860s the Mormons had occupied most of the suitable available land and were hemmed in by a ring of difficult terrain.

Phase three of Mormon colonisation started in the 1870s. This involved secondary colonisation (land was bought not just occupied) in Gentile ranching country to the south, in areas like Little Colorado, Salt River and Upper Gila. Mormon enclaves, often comprising small clusters or strips of farms and villages adjacent to a small river, were enclosed within Gentile rangelands and mining country. There was also limited contagious expansion eastwards, westwards and southwards, and some major expansion northwards during the 1880s along most of the tributaries of the upper Snake River.

Expansion of Mormon settlements has been limited since 1890, when colonisation was at its fullest. Since then the core of the Mormon culture region (Figure 5.3) has survived remarkably well, as a cultural island surrounded by Gentile expansion and development.

General model

Using the pattern and evolution of the Mormon culture region as a blueprint, Meinig (1965) proposed a general model of how culture areas develop. The model is based on ideal conditions that requires relative isolation over a long period, freedom from interference by others, and the opportunity for unhindered territorial expansion for a culture group. When such ideal conditions prevail, the model predicts the emergence of three concentric zones – the core, the domain and the sphere.

> A *core* area . . . is taken to mean a centralised zone of concentration, displaying the greatest density of occupance, intensity of organisation, strength and homogeneity of the particular features characteristic of the culture under study. It is the most vital centre, the seat of power, the focus of circulation.
>
> The *domain* refers to those areas in which the particular culture under study is dominant, but with markedly less intensity and complexity of development than in the core, where the bonds of the connection are fewer and more tenuous and where regional peculiarities are clearly evident.
>
> The *sphere* of a culture may be defined as the zone of outer influence and, often, peripheral acculturation, wherein that culture is represented

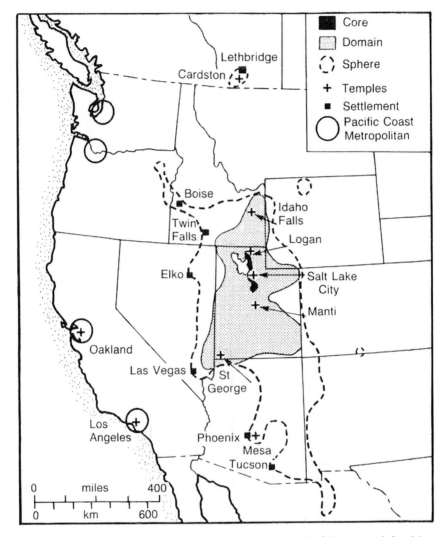

Figure 5.3 The Mormon culture region in the western United States, as defined by Meinig (1965)

only by certain of its elements or where its peoples reside as minorities among those of a different culture.

(Meinig 1965: 213–17)

Ideal conditions did exist in the Mormon culture region, and the three zones can be mapped meaningfully (Figure 5.3). Definition of the zones is not based just on the distribution of Mormons, but on a host of cultural factors including landscape (particularly the farm–village pattern), economy (primarily

162

commercial farming with relatively little tenancy) and population (a unique combination of a high average standard of living, a very high birth-rate, and large families).

Not everywhere is blessed with such ideal conditions, however, and Zelinsky (1973: 116) points out that although the Meinig model works fine for the Mormon and Texas culture regions, it only loosely applies to New England, Midland or French Canadian cases and 'is almost useless in describing or explaining other North American culture areas'.

Texas German culture region

The Texas culture region, which Zelinsky (1973) commented on, is different from the Mormon culture region in both evolution and manifestation. In Texas the dominant influence has been the survival of some characteristic ethnic groups and settlements, particularly of German origin.

Definition and delimitation of the Texas culture region owe much to the work of cultural geographer Terry Jordan. The initial designation, in an ethnic map of Texas compiled by Jordan (1970), outlined the extent of many German settlements. To be defined as an ethnic community, 'the population in question had to have the feeling of belonging to a particular group and live in close proximity to other members of the group'. The main data source for the map was 1910 and 1930 federal census material, complemented by information from church groups (including handbooks from churches with predominantly Germanic-American membership, such as the American Lutherans, United Lutherans and Church of the Brethren). The draft map was verified using field data on cemetery and mailbox name counts, interviews and letters to editors of county newspapers.

Jordan (1976) later concentrated on the rural religious cultures of North Texas, which are reflected in many ways including folk architecture, dialect, political party allegiance, livelihood, and diet. He noted how

> the widely divergent origins of rural North Texans are expressed by equally contrasted religious faiths, including fundamentalist Protestantism, introduced by old-stock Americans from the Upper South; Roman Catholicism, implanted by immigrant farmers from the German lands and Bohemia; northern Methodism, brought to this southern state by Midwesterners; and a scattering of smaller sects, such as the Seventh-Day Adventist Church.
>
> (Jordan 1976: 135)

Churches appeared to serve as partial shields against assimilation, by providing cultural rallying points for the immigrant ethnic minorities. But it was in the physical landscape that Jordan found most traces of the ethnic history of North Texas, because the different religions carried with them contrasting material cultures, particularly in terms of the architecture of church buildings and to a lesser extent in the design and layout of graveyards and cemeteries.

A detailed case study of the Hill Country Germans in Texas provided Jordan (1980) with the opportunity to examine how religious acculturation progressed and what processes encouraged the survival of ethnic religious groups in a non-German area. Between 1844 and the Civil War many peasant farmers, artisans and university-educated political refugees from western and central Germany emigrated to the sub-humid Hill Country of south-central Texas. Church affiliation emerged as a significant factor in the preservation of ethnic identity (many Germans were Lutherans, and others were Methodists, Roman Catholics or non-religious Freethinkers). Residential segregation by sect was also important, and he found evidence of considerable regionalisation of religious groups within the German Hill Country. Religious spatial organisation played its part, too, because those Germans who converted to Methodism quickly discarded the village-church tradition and adopted an Anglo-American Protestant spatial system. Those Germans who remained true to Lutheranism and Catholicism hung on for decades to a quasi-European ecclesiastical pattern, despite its impracticability, and they never fully adopted the Anglo-Protestant concept of the small rural chapel. An important tangible sign of the German heritage of the area is church architecture, because three-quarters of the German structures are ornamental Gothic or Gothic-inspired and they contrast sharply with neighbouring simple white-frame Anglo-American chapels. Burial practices in the Texas Hill Country demonstrate acculturation rather than persistence, because family cemeteries replaced the European custom of burial on sanctified ground.

Other religious culture regions in the United States

Such other religious culture regions in the United States that exist are much smaller and less distinct than the Mormon and Texas ones. A brief study of several such areas highlights the diversity of this lesser group.

The Dutch Reformed in Michigan

The ethnic island of Dutch Reformed immigrants in Michigan described by Bjorklund (1964) illustrates how ethnic persistence has been translated into visible landscape components, which in turn provide a useful means of delimiting the religious region.

The Dutch Calvinists established a series of communities in south-west Michigan in the 1840s. Their belief systems were based on the doctrines of 'unconditional election' and 'irresistible grace' originally set forth by John Calvin in 1556. This gave them a set of guiding principles by which to live, including the need to recognise and obey religious rules, to perform both physical and spiritual works, and to reject the intrusion of conflicting rules of conduct (which would threaten the salvation of their souls).

Their strict religious adherence and obedience to the Bible fostered an austere lifestyle in which Sunday trading, drinking in taverns, and socialising in movie theatres were banned outright. They saw their Calvinist Church as the only true and proper Church, and would not tolerate the setting up of any non-Calvinist churches in their area. Mapping the distribution of such components in the contemporary landscape provides a means of delimiting the boundaries of the Dutch Reformed culture region (Figure 5.4). The core area contains no taverns and very few churches other than Dutch Reformed ones, and no Sunday business is allowed within it.

Interestingly, Bjorklund (1964: 241) concludes that most of the tangible signs of the Dutch Reformed culture were in fact not imported but created *in situ*. They 'evolved during their occupancy of Michigan and were not known ways or practices followed in the Netherlands. Instead they are innovations designed to support and propagate their ideology.'

Roman Catholics in Louisiana

The Cajuns of Louisiana represent an ethnic group that, since 1945 in particular, has started to assimilate into mainstream American society and is thus losing its ethnic identity. It is questionable whether the term 'culture region' can properly be used in this case. The Cajuns are Catholics, descended from Acadian exiles from Nova Scotia who first settled in south-west Louisiana in 1755. They maintained a distinct way of life there until well into the twentieth century, strongly reinforced by their French culture and language. Evidence of changing patterns of religious affiliation in the post-war period, analysed by Clarke (1985), is compatible with the regional convergence model (see pp. 153–4) and shows that at least in religious terms the regional culture of the Catholic Cajun community is on the decline.

The Roman Catholic Church has traditionally played an important role in French Louisiana and has significantly assisted French ethnic persistence there. But extensive fieldwork in 35 French Louisiana communities led Trepanier (1986) to conclude that the Louisiana Catholic Church has failed to maintain its position as an ethnic institution, and has lost contact with the cultural character of the population it has served. He argues that it has ceased to be a paternalistic colonial Church, adopted American values (including racial segregation and English language supremacy) and so helped to undermine the position of French in French Louisiana society.

Amish/Mennonite ethnic areas in the USA

Like the Catholics in Louisiana, Amish and Mennonite areas are not strictly culture regions *per se* but they have such distinctive characteristics that they are

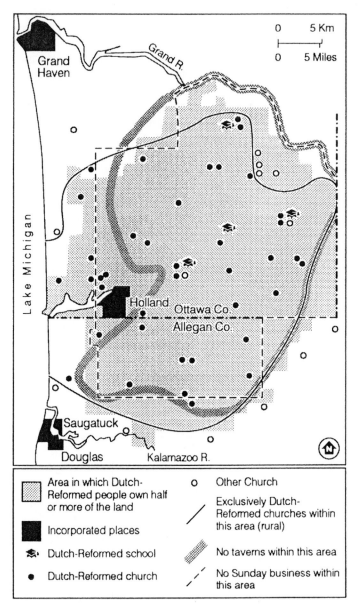

Figure 5.4 The Dutch Reformed (Calvinist) area in south-western Michigan, about
1960
Source: After Bjorklund (1964).

worthy of mention here. Perhaps more than any other religious groups in the United States, they not only show remarkable religious persistence but their world view and religious beliefs shape every aspect of their existence.

One of the most cohesive of these groups is the Old Order Amish who continue to live in traditional ways, unlike the Amish Mennonites who have adopted many modern practices. Hostetler emphasises the binding significance of small-scale activities for the Old Order Amish who settled on the Great Plains, who

> work constantly to keep scale. Their *Ordnung* (community rules) functions to keep the physical environment limited, modelled after the Garden of Eden and the valleys of Switzerland. All their social structures proclaim that 'small is godly'. Their limitations on transportation (automobile, truck, aeroplane), their limitations on the telephone, their limitations on farm equipment and electricity, make large-scale operations dysfunctional. Their economic structure (labour-intensive family farms) and their social structure (married couples, children, and inter-generational family) maladapts them for large-scale agriculture. Their work socialisation trains them for small-scale, family-sized, diversified, intensive farming on varied soils and terrain. Supportive of the small scale are a common faith, dialect, simplicity, neighbourliness, and consumptive austerity.
>
> (Hostetler 1980: 93)

Amish–Mennonite settlements in the area are unique components in the rural landscape, and they offer vital clues to the distribution and dynamics of these minority groups within American society. Kent and Neugebauer (1990) evaluated a number of ways of identifying Amish–Mennonite settlements in Ohio, including county-based population data on religious affiliation, the location of Amish church districts and Mennonite churches, topographic maps, and surnames and cadastral maps. The latter three data sources proved the most useful. On the ground, the characteristic features of Amish landscapes (see Chapter 7, pp. 232–5) are reliable indicators of the spread of Amish influences within a particular area.

CONCLUSION

In this chapter we have explored some important dimensions of how religions change through space and time, picking up from Chapter 4 the theme of diffusion and focusing on the role of factors such as natural change, conversion and migration. The evidence suggests that different factors are important in different places at different times, and that there are fundamental differences between the universal and the ethnic religions. Religious adherence is a relatively unexplored property of the cultural landscape, but it is difficult to measure unambiguously. Geographical factors are doubtless very important influences on religious persistence, particularly in remote and inaccessible regions, and broader cultural factors underlie separatist or integrationist movements in different places. Ethnicity, which might or might not be associated with religious heritage, is an important ingredient in the socio-cultural landscape and it assumes great prominence and significance when people migrate to new areas. One of the most tangible expressions of religious adherence and persistence is the religious culture region, where prevailing religious values and norms are reflected in the physical appearance of the area. The Mormon culture region is best described, but it is not the only religious culture region in the United States.

6

RELIGION AND POPULATION

It is well said, in every sense, that a man's religion is the chief fact with regard to him.

Thomas Carlyle, *Heroes and Hero-Worship*

INTRODUCTION

The exploration of the dynamics of religious change in Chapter 5 emphasised a range of important factors that affect diffusion, adherence and persistence. Underlying many of those factors and themes is the broader issue of how people affect religion, because religion in itself can only be transferred and have an impact via people who believe in it. Religion is not an entity in its own right; it has importance only in so far as it involves people. Yet the relationship between people and religion is reciprocal, with each one both influencing and being influenced by the other. This reciprocity is illustrated, for example, in the emergence of new religious movements (see Chapter 2, pp. 44–8) and the diffusion of existing ones (see Chapter 4, pp. 99–101).

In this chapter we focus on some important dimensions of the impact of religion on people. The three focal points – demography, development and politics – are chosen to illustrate ways in which religious factors can exert powerful influences on how people behave, and in turn give rise to some quite pronounced geographical patterns and features. There are many other ways in which religion influences behaviour, and such dynamics are also evident in terms of landscapes (Chapter 7) and pilgrimage (Chapter 8).

RELIGION AND DEMOGRAPHY

Religious beliefs can be a powerful determinant of personal decisions which most individuals and families face, including family size and mobility. In reality, however, the relationship is two-way because demographic factors also influence the diffusion, stability and persistence of religious groups.

It is often difficult to isolate the impact of religion as an influence on demography from the broad matrix of socio-cultural factors in which it is embedded.

169

Brah's (1987) study of the reality of Asian women's lives in Britain, for example, discovered many shared experiences and frustrations amongst this group that represents different religions, languages, cultures and origin nationalities. Race, class and gender – as well as religion – constrain their prospects within mainstream British society. Another example of complexity is the pattern of social inequality, particularly related to health care, in Malta (Agius 1990). The traditional Catholic way of life of the Maltese, in which the Church played a significant role, is giving way to a more secular lifestyle dominated by politics and economics. Such factors determine the distribution of resources including health resources, and so indirectly influence the pattern and persistence of health problems such as infectious diseases.

It is also often difficult to isolate particular components within demographic change, which reflects the interplay of many variables including fertility, birth and death rates, and migration. We have already looked (in Chapter 5, pp. 138–9) at some examples of population migration either caused by or associated with religion. One of the most graphic illustrations of such migration is the return of many European Jews to Israel, their holy ancestral homeland.

Population dynamics

Population distributions and patterns of religious adherence are often closely correlated. In Europe, for example, religious vitality is strongest in rural areas and organised religion is declining fastest in major cities (Barrett 1982). In many countries Christianity – Catholic, Protestant and Orthodox – is strongest and most active in rural areas (Jordan 1973: 164).

These are symptoms of the influence of population on religion, but reciprocity is shown in the influence of religion on population. Islam provides an interesting case study because most Muslim populations have profiles characteristic of the early stages of the classic demographic transition (Clarke 1985). Like a number of major religions, Islam has particular attitudes to family planning and this has strongly influenced rates and patterns of population growth in countries like Bangladesh, Indonesia, Malaysia, Iran, Egypt, North Africa and Bahrain (Rowley 1979).

Weeks (1988) provides a comprehensive account of the demography of Islamic nations, the world's fastest growing population group. In 1988 there were 47 nations in which a majority of the population is Muslim, particularly in sub-Saharan Africa and South Asia (Table 6.1). In some, such as Iran and Saudi Arabia, Islamic law is the only law of the land. In others, including Algeria and Pakistan, Islam is the declared official state religion although some secular aspects of state law complement the religious rules.

At current rates of growth the 1988 estimated population of 980 million Muslims is likely to nearly double to 1.9 thousand million before 2020, accounting for 23 per cent of the then world total population. The growth of

Table 6.1 Islamic nations by geographic region

Region	Number of countries	1988 total (millions)	Per cent of Islamic total
Middle East			
Northern Africa	7	138.4	13.9
Western Africa	15	170.5	17.2
Sub-Saharan Africa	17	255.3	25.7
South Asia	4	231.7	23.3
Southeast Asia	3	194.7	19.6
Europe	1	3.1	0.3
Total	47	993.7	100.0

Source: Weeks (1988).

Islam in the modern world owes more to natural increase (the excess of births over deaths) than to the conversion of non-Muslims to the Islamic faith, as we saw in Chapter 5 (see pp. 130–3).

The Islamic nations, both individually and as a group, are characterised by higher-than-average fertility, higher-than-average mortality, and rapid rates of population growth (Table 6.2). The average crude birth rate in Islamic nations is a quarter higher than the average for other developing nations, nearly three times as high as in nations with centrally planned economies, and more than three times that in developed nations. This high birth rate reflects very high fertility. The average woman in a Muslim nation has six births, compared with under five in other developing countries, just over two in nations with centrally planned economies and below the replacement level of two in developed nations.

Whilst Muslim populations reproduce faster than most others they also have a greater turn-over because life expectancies are relatively short and death rates relatively high (Table 6.2). The average crude death rate in Islamic nations (13.8 per 1,000 population) is a third higher than in other developing countries, a fifth higher than in countries with centrally planned economies and nearly half as high again as developed nations. These death rates reflect average life expectancies at birth. An average life expectancy of 55 years in Islamic nations compares with 62 in other developing nations, 70 in the nations with centrally planned economies and 75 in developed nations. Differences in infant mortality are particularly noteworthy, because a new-born baby in the average Islamic nation has a ten times greater chance of dying in infancy than one born in a developed society.

Despite the high mortality, the very high birth rate in Islamic countries produces the world's most rapidly growing population group. Islamic nations are growing at an average of 2.8 per cent per year, which is a population-doubling-time of 25 years. This is a quarter faster than the other developing

Table 6.2 Demographic characteristics of Islamic nations compared with other nations of the world, 1988

Demographic characteristic	Islamic nations	Other developing nations	Centrally planned economies	Developed nations
Number of nations	47	94	7	25
Average crude birth rate per 1,000 population	42.1	33.6	15.0	13.1
Average crude death rate per 1,000 population	13.8	10.1	11.6	9.4
Average rate of natural increase (per cent)	2.8	2.3	0.4	0.3
Average total fertility rate	6.0	4.5	2.1	1.7
Average life expectancy (years)	55	62	70	75
Average infant mortality rate per 1,000 births	104	63	18	10
Average per cent of population under age 15	43	39	23	20

Source: Weeks (1988).

countries and nearly nine times as fast as developed nations. One direct result of the high rates of fertility in Islamic nations is a very young population – an average of 43 per cent of the population is below the age of 15.

The high fertility of Islamic nations owes much more to socio-economic factors than to religious belief. It is a historic trend, reflecting the adaptation of Islamic society to high mortality by encouraging youthful marriages and an early start to childbearing within marriage. Similarly, high rates of mortality have little if anything to do with Islam. They are a function of economic disadvantage and reflect poverty, underdevelopment and a lack of sufficient resources to improve the health profile and living conditions faster.

Whatever the causes, the rapid rate of population growth in Islamic nations is putting national infrastructures under great strain. Weeks notes how

the young age structure that is produced by higher-than-average fertility combined with declining mortality not only pushes a nation's resources to the limits right now in attempting to educate, clothe, feed, house, and find jobs for a greater number of people each year; it also has a built-in momentum for growth that will continue to produce ever greater numbers of babies each year even though the overall levels of reproduction may be falling.

(Weeks 1988: 48)

An important variable in population dynamics is marriage pattern, both in terms of the number of married partners an individual can have under religious rules and the breadth of the population pool from which those partners are normally selected. Monogamy (marriage to one partner) is the normal marriage pattern in many cultures and religions, and it was typical of ancient Sparta, Rome, early Christianity and Western Europe during the early Middle Ages. MacDonald (1990) shows that there was great diversity in the origin and function of monogamous mating arrangements, and suggests that the Christianisation of barbarian Europe greatly assisted the top-down diffusion of monogamy from the aristocracy to ordinary folk.

Patterns of inter-group marriage (particularly between different religious groups) throw some light on the perceived strength of group boundaries, might have an impact on population growth rates (via differential fertility) and can interfere with the transmission of religious beliefs from one generation to the next. The population of Switzerland is roughly half Protestant and half Catholic, although most cantons are dominated by one group or the other. Studies by Schoen and Thomas (1990) show that most Swiss marry within their own religious group, but that inter-religious marriage is common and becoming more so. Between 1969 and 1972 some 24 per cent of Protestant and Catholic marriages were to a person of the other faith, and between 1979 and 1982 that figure had risen to 30 per cent.

Religious missionaries have had marked effects on population dynamics amongst indigenous groups in many places. A common result of the adoption of missionary values is changing native fertility, brought about mainly by the abandonment of traditional pursuits such as polygamy and wife exchange and the encouragement of unrestricted reproduction. Scheffel (1983) also uncovered evidence of other demographic impacts of nineteenth-century Christian missionary activities amongst the Inuit Indians of northern Labrador. One of the most pronounced changes was a large increase in mortality triggered by the introduction and spread of European epidemic diseases. This occurred at the same time as the missionaries displaced traditional Inuit health care systems with an evangelistic 'theology of death' designed to encourage natives to convert to Christianity, thus compounding the problem.

Other studies have also suggested indirect links between mortality and religion. Infant mortality rates in Cameroon, Ghana and Kenya (in sub-Saharan Africa) are highly variable, and surveys show that rates are lower for children of Christian mothers (Akoto 1990). This is not to imply that God's favours are given more sparingly to non-Christians, however, because the correlation between religion and mortality was found to be indirect. Membership of a religious group was associated with other health and socio-economic indicators such as place of delivery, literacy, and socio-occupational group. When these factors were taken into account with religion, the correlation almost disappeared and the impact of religion on infant mortality rates was negligible.

Fertility and family planning

Perhaps the most easily detectable links between religion and demography are those which reflect attitudes towards family planning. To some world religions any form of artificial family planning is unacceptable, whereas other religions have more liberal teachings and practices.

Much controversy surrounds the morality of birth control, and the debate is nowhere more heated than when dealing with abortion. Here conflicting interpretations about the status of the foetus and about a woman's rights to make choices regularly bring different religious and secular arguments into sharp focus. Whatever the morality of the situation, Jacobson (1990) emphasises the tangible benefits achieved over the last three decades during which many laws governing access to family planning have been liberalised and 36 countries have legalised abortion. The relative number of unwanted pregnancies has decreased dramatically in countries where abortion is legal, and the number of deaths associated with illegal abortion have also fallen sharply.

Striking contrasts in fertility between Islamic and other nations have already been noted (Table 6.2), and a number of studies have focused on differential fertility between different religions. At the small scale, for example, Rele and Kanithar (1977) have shown that within a single city – Bombay in India – Muslims have much higher fertility than Christians and Hindus.

Some of the most detailed research has explored the impacts of Catholicism on fertility. Roman Catholic areas (where birth control is against Church teaching) tend to have higher birth rates than other areas. This is so even within the same country, as spatial variations in fertility within Northern Ireland confirm (Compton and Coward 1989). Murphy (1981) traces the roots of Catholic perspectives on population control back to the early Christian Church in the second century. Traditional Catholic doctrine emphasises the meaning of human sexuality, love and marriage and it defines the right and duty of parents to determine the number of children they should have. Although Pope Pius XII declared in 1958 that the pill was not an acceptable form of birth control, a growing percentage of Catholic women in the United States use modern forms of contraception (Blake 1984).

Catholic opposition to organised birth control extends beyond the Church's condemnation of particular contraceptive methods. Other factors in America include class and ethnic hostility between Catholics and the promoters of family planning, jurisdictional disputes over what institutions would guide family life, and the perceived threat that government support of family planning was to the authority of the Church and the Church's understanding of sex and women (Donaldson 1988).

A number of case studies of Catholic fertility reveal the importance of indirect factors as well as Church teaching. Changing levels of Catholic fertility in Australia and New Zealand between 1911 and 1936, for example, appear to reflect occupational structure and rural residence, rather than Catholicity *per se* (Bourke 1986). Fertility has declined much more slowly in the Netherlands since the 1880s than in most other European countries, and van Poppel (1985) attributes this mainly to the continued high fertility of the Catholic population. Catholic priests maintained their active opposition to the use of contraception, but they also had an indirect impact by encouraging attitudes that preserve the traditional system of family norms and kept Catholic family fertility high.

A convergence through time of fertility levels between different religions in particular places has been documented in many studies of religion and fertility. In pre-independence India, for example, the highest rates of population growth were recorded among Christians, although since independence birth rates in this group have fallen closer to other religious groups (Breton 1988). Gutmann's (1990) study of religion and fertility in Gillespie County, Texas, between 1850 and 1910 also shows convergence through time. In the mid-nineteenth century, Protestants (Lutherans and Methodists) in the area had significantly lower fertility than Catholics. Within four generations, however, both groups were limiting their fertility in significant ways and the convergence of Protestant and Catholic fertility began much earlier here than in most of the United States.

One of the undisputed generalisations in demography has been that Catholics have higher fertility than Protestants and Jews, but Williams and Zimmer (1990) point out that this is no longer necessarily true. During the 1970s, in particular, American Catholic birth-control practices were increasingly secularised and as a result Catholic fertility declined and converged with non-Catholic fertility. Declining fertility is not uniform between all Catholics, however, and survey results show that fertility has varied among Catholics who differ in their religious practices (as measured by frequency of attendance at mass and frequency of communion).

Between the mid-1950s and the mid-1970s, contraceptive patterns converged between white Protestant, Catholic and Jewish couples in the United States, and there was a considerable narrowing of differences by religious affiliation in whether contraception was used at all (Mosher and Goldscheider 1984). Between 1955 and 1982 all the major religious groups had experienced downward changes in expected family size and all used effective contraceptive methods, including sterilisation, the pill and the IUD. Patterns of contraceptive use converged among the three religious groups. Observed changes include a threefold increase in the proportion sterilised (from 7 to 33 per cent overall, with the fastest increase among Catholics), a sharp fall in the proportion not practising any contraception for both Protestants and Catholics (from 22 to 4 per cent), and a pronounced contraceptive shift for both Protestants and Catholics towards the pill and IUD (Goldscheider and Mosher 1988). A variety of factors help to explain this convergence, including peer pressure and social norms, differential sex roles, male-female communication patterns, and the differential use of physician-based versus other sources of contraceptives.

RELIGION AND DEVELOPMENT

A second important dimension of the link between religion and population is the way in which religious views, values and practices have impacts – both direct and indirect – on socio-economic development.

The foundations for this theme were set by German sociologist Max Weber in *The Protestant Ethic and the Spirit of Capitalism* (1930). In that book he explored the effect of Protestant ethics on the development of capitalism in north-west Europe. Weber argued that religion produces a distinct attitude towards life, which is in turn reflected in development. Broek and Webb (1973) point out that Weber was reacting to the Marxist idea that the social super-structure within society (including religion) is determined primarily by the methods of material production. In reality, they add, the situation is much more complex because culture and lifestyle are products of the interaction betwe‍ religion, social, political and economic factors.

Rer‍ t theoretical and empirical work has focused on the significance of religious values to social and economic development. Wilber and Jameson (1980), for example, suggest that religion affects development in a variety of ways. Religious values can significantly shape the priorities and strategies of people involved in development initiatives, and religious institutions can be major international agents of development and change. Moreover, religious values and objectives can be both catalysts and resistors to development in different contexts. Empirical surveys by Morris and Adelman (1980), based on a sample of 55 countries, have established a positive relationship between religion and socio-economic development that is reflected in a range of political and social indicators.

Expressions of relationship

Religion and development are associated in a variety of different ways. Some-times a religious group acts directly as an agent of development, particularly in terms of providing an infrastructure for aid and welfare support and providing a widely trusted vehicle for implementing socio-economic change. The independent Christian churches in many parts of Black Africa have traditionally served in this capacity (Turner 1980), generally with some success. Sometimes the very process of development – which can disorientate individuals and weaken their sense of purpose, belonging and attachment to community – encourages people to continue religious traditions and practices. A survey of the behaviour of male heads of household in Dakar, Senegal, by Creevey (1980), found that they were equally or more likely to perform Muslim rituals as development occurs.

Religion sometimes hinders development by providing constraints or distortions. Many development initiatives in modern India have failed to be of real benefit to Hindu women, for example, because of the traditional

176

male-centred religious ideology that expects Hindu women to remain submissive and dependent. Dhruvarajan (1990) suggests that this lack of gender equality will inevitably persist while there is a lack of political will to change the oppressive religious ideology and allow women to be the beneficiaries of development.

Another dimension to the religion-development link is the contribution which particular religions can make to the identification of moral principles on which sustainable and acceptable development might best be founded. A good example is McKee's (1991) critique of Christian commentaries on debt problems in less developed countries (LDCs). McKee argues that the world's richer countries have an obligation in social justice and charity to lighten the burden weighing on LDCs, especially the poorest, and that this duty extends to national governments, electorates, and private and international institutions. Beyond offering general moral criteria, however, he suggests that 'a Christian approach has no special insights to offer concerning causes or solutions . . . it is precisely moral goodwill and generosity from the heart that are needed to make technical plans workable'.

The impact of Catholicism on development has been explored in a number of papers. This impact often arises in an indirect way, via the attitudes of Catholics to development. Houtart (1980) found that these attitudes and impacts vary between priests and lay people in his study of 429 development projects started by the Catholic Church in Sri Lanka. The Catholic Church itself has a key role to play in assisting the poor to participate in social change. In Latin America, however, it seems that for both historical and structural reasons the national churches lack influence and are poorly equipped to relate to the poor. Bruneau (1980) advocates a strengthening and empowering of Basic Christian Communities to reach these largely disenfranchised groups. In the Caribbean, in contrast, the Church played a prominent role in establishing a development agency (CADEC) to pioneer its own development programmes and to develop a type of development that embraced equity, justice, liberation and self-reliance (Davis 1980).

The Catholic Church has traditionally shaped many aspects of life and livelihood in many Western European countries. In Ireland, for example, the Church has long held great importance in national life. Whelan's (1988) analysis of the regional impact of Catholicism in Ireland in the eighteenth and nineteenth centuries reveals close spatial correlations between the strength of Catholicism and the pace of development. Studies of Church revenue and expenditure in nineteenth century Ireland (Kennedy 1978) also show that the Catholic Church made a positive if modest contribution to the development of the national economy.

Different branches of Buddhism also project different attitudes towards development. Ling (1980) examined the association between Buddhist values and development problems in Sri Lanka, and concluded that traditional Sinhalese Buddhism bears some responsibility for retarding economic development through merit-making practices, non-rational attitudes to life and population increase. Buddhist monks also play a part in development decision-making in some countries, particularly by promoting the ideals of a righteous and economically progressive society (Ariyaratne 1980).

Islam and development

Whilst most major religions have some detectable impact on the pace and pattern of development in many countries, Islam offers an interesting case study because throughout the Muslim world all aspects of politics and law are firmly founded on religious principles. There is no shortage of illustrations. Islam has significantly influenced the development of Pakistan, for example. Indeed, the state was established on religious grounds and the implementation of the Islamic Laws in Pakistan in 1979 heralded an important milestone in the establishment of an Islamic state (Qureshi 1980). The recent rise of Islamic fundamentalism in many countries in South-East Asia (including Burma, Thailand, Indonesia and Malaysia) threatens significantly to shift the axis of development efforts there (van der Mehden 1980).

One way in which Islam influences development directly is via its banking system. Muslim financial practice reflects a religious belief in interest-free banking, which is common in Islamic countries such as Iran and Pakistan. Dual banking systems, embracing both Islamic and secular banking systems, are emerging in Kuwait, Turkey, Egypt, Sudan and Jordan (Wilson 1990). Naughton and Shanmugam (1990) suggest that although Islamic interest-free banking exists in Malaysia it is likely to remain limited there, particularly given the continued growth of non-Islamic banks there and the country's proximity to the major international banking centre of Singapore.

The Islamic Development Bank plays an important role in providing financial assistance to less developed countries in accordance with permitted Muslim financial practice. It is the only aid institution that functions in accordance with the Sharia (Islamic religious laws) and its policies reflect the need to redistribute wealth between rich and poor Muslim countries. Wilson (1989) describes some of the schemes supported by the Bank, including providing emergency aid to drought-stricken Muslim states in North Africa, funding the provision of vaccines and animal feed for livestock and clean drinking water for people and animals. Although the Bank's resources could more profitably be invested in the richer Muslim countries, Islamic financial practice favours narrowing rather than widening the gap between rich and poor.

RELIGION AND POLITICS

Many commentators on religion emphasise that it is a mixed blessing, which affects society in both positive and negative ways. De Blij and Muller capture this perspective in writing that

> religion has been a major force in the improvement of social ills, the sustenance and protection of the poor, in the furtherance of the arts, in education of the deprived, and in medicine. However, religion has also thwarted scientific work, encouraged oppression of dissidents, supported colonialism and exploitation, and condemned women to inferior status in society. In common with other bureaucracies and establishments, large-scale organised religion has all too often been unable to adjust to the needs of the times.
>
> (de Blij and Muller 1986: 199)

This blend of positive and negative dimensions of religion is also reflected on the political landscape. One way in which religion is an important positive force is through unifying believers into a religious community with shared values and beliefs. The worlds of Christianity, Islam or Buddhism, for example, embrace people from all parts of the world and from all walks of life. Religion can transcend other dimensions of human variability (including race, colour, language, and nationality) and serve to unite people of diverse origins.

The positive benefits of religion tend not to be widely commented upon, unlike the negative aspects or problems commonly associated with religion. Non-religious people often criticise religion as divisive and antagonistic, and point to major areas of conflict between religious groups to support their case. There is no escaping the fact that conflict within and between religions has occurred in the past and continues today, sometimes with disastrous results (such as the Crusades). Deep-rooted religious conflicts have given rise to distinct patterns of religion at various scales. At the national scale, for example, Catholicism is common in the south of Germany and Protestant Lutheranism is common in the north. At the local scale, Catholics dominate the Falls Road district in Belfast and Protestants dominate the Shankill Road district.

Some of the most heated conflicts are associated with the rise of religious fundamentalism, founded on the belief that there should be no separation between government and religion. The rise of Islam in Iran since 1979, for example, has been accompanied by mass street demonstrations and opposition to economic development and modernisation that has created social tension within the country (Nash 1980). Similar radical religious movements are evident throughout the Muslim world, from Morocco in North Africa to Indonesia in the Far East.

Fischer (1957) argued that whilst a population's religious affiliation does not necessarily determine its political attitudes, in most countries which contain a significant religious minority this factor has some political significance. In some cases, he notes (ibid.: 408), 'such religious minority status has hindered the assimilation of national groups: Armenians, Jews, French Canadians, Irish Catholics and many other groups have preserved their separate existence primarily because of religious differences'.

179

Other forms of conflict arise as a result of splits within religious groups. Many major religions have been affected by schisms, including the split between Catholics and Protestants within Christianity and the split in Islam between Shia and Sunni Muslims.

Religion and political parties

Religious minorities sometimes form separate political parties or support the political party that is most disposed towards them. In Germany, for example, Poles tended to support the Catholic Centre Party while Lutherans support the German National Party. Golde's (1982) detailed study of voting patterns in south-west Germany highlights the importance of social context and religious affiliation as determinants of voter behaviour.

Religion can strongly influence voting patterns and political party membership. In Germany, Italy and the Netherlands some political parties have long been established on religious affiliations. The links between religion and party support are not always immediately obvious, but in some cases the political party's name reflects its religious constituency. In the Netherlands, for example, the Catholic People's party is particularly strong in the Roman Catholic south of the country. There are Catholic parties in Italy and Austria, too. Some political parties even carry the name 'Christian Democrat' to reinforce their association with the Church (Jordan 1973).

Gilbert's (1991) analysis of voting patterns in South Bend (Indiana) during the 1984 American Presidential campaign focused on how denominational allegiance affected the way people voted. He found that church affiliation was a major determinant of voting behaviour, even after controlling for other socio-economic variables. Neighbourhood environment was a significant factor too, but it varied in impact from one denomination to another. The results from this empirical study strengthen the argument for considering churches and other religious institutions as significant units of political influence.

Persecution of religious minorities

Religious intolerance and persecution also lead to political and ethnic conflict, as is evident in countries like Ireland, Israel, Lebanon and Israel during the twentieth century. Persecution has often acted as a catalyst for migration. Historical examples include the emigration of persecuted religious groups from England to New England in the New World, and the Mormon movement west to settle in the remote state of Utah. Modern examples include the ingathering in Israel of large numbers of Jews from around the world and the migration population between India and Pakistan at the time of partition.

There is no shortage of examples from around the world of religious minorities that face political persecution. One such group is the Copts of Egypt, a Christian ethnic minority whose 10 million members account for about 23 per cent of the total population and whose church (the Coptic Orthodox Church of Egypt) has survived over 1,300 years of severe religious persecution and intolerant rule. Gregorius (1982) emphasises what a significant impact the early Coptic Church and the respected theological school of Alexandria had on the Christian world. Coptic language and music have survived within the liturgy of the Church, as reminders of its important past.

Ibrahim (1982) points out that at least 90 per cent of the Egyptian Muslims who have persecuted the Coptic Church are of Coptic origin themselves. The Copts are fully integrated into the Egyptian population, although during times of religious persecution they retreated to Coptic strongholds in villages of Upper Egypt in the provinces of El Minia and Assiut. Many Copts have through history been pressurised or terrorised into declaring their acceptance of Islam, and many others escaped constant Muslim threats in villages to live more anonymously in cities. Even today many key positions are barred to the Copts, including province governors, senior positions in universities and colleges, vice-presidents of the Republic, police directors, heads of town councils, heads and members of the higher councils of the Republic, Attorney-Generals, and most directors of the nationalised banks. As Ibrahim (1982: 66) notes, 'this creates in the Copts a feeling of bitterness, inferiority or isolation in a country in which they live, for which they work and to which they belong'.

Other groups that have endured long periods of religious persecution include the Muslims in the former Soviet Union (Hassal et al. 1990) and Christians in Turkey (Birchall 1983).

Jews have suffered from perhaps the most persistent and vindictive persecution of any religious group. Anti-Semitism – acts of persecution against the Jews, whose ancestors were members of the Semitic branch of the Caucasian race – has occurred throughout history, but the greatest genocidal atrocities took place in Nazi Europe under Adolph Hitler during World War II. In 1939 about 10 million of the estimated 16 million Jews in the world lived in Europe, but by 1945 almost 6 million had been killed (most of them in the nineteen main concentration camps). Only about 4,000 of pre-war Czechoslovakia's 281,000 Jews survived, along with about 5,000 of pre-war Austria's 70,000 Jews. An estimated 4.6 million Jews were killed in Poland and in those areas of the Soviet Union seized and occupied by the Germans (Nyrop 1979). Even during the First World War Jewish minorities faced much persecution and oppression, in Britain and other countries (Cesarani 1989).

Religion and residential segregation

Social and political instability commonly arises when incompatible religious groups occupy the same area, particularly when they compete for recognition and resources. Residential segregation on religious grounds often compounds already sensitive problems of integration and representation, by superimposing territorial claims on to socio-economic ones (Hershkowitz 1987).

Martin (1989) notes that most communities are subject to constant change – physically, socially and in terms of the values and ideologies of their residents. As a result many neighbourhoods change character through time, and neighbourhood boundaries shift in response to this dynamic evolution. He observes that most literature on neighbourhood transition ignores religious change, and most writing on religion ignores spatial patterns and differences. His study of three neighbourhoods in inner-city Vancouver tried to redress the balance by searching for possible connections between social transition and religious values.

Residential segregation is often most pronounced when it involves ethnic groups, which might or might not also reflect religious differences. Jewish communities are found in many of the world's larger cities, and within them they are normally concentrated into clearly defined urban ghettos. Jews in Britain largely conform to this pattern (Newman 1985). Studies of residential patterns amongst Jews in London, for example, show remarkable spatial cohesion amongst Jewish communities and hence pronounced stability of Jewish territoriality. Waterman and Kosmin (1988) describe cohesive Jewish communities in the Greater London boroughs of Hackney, Redbridge and Barnet and suggest that commonly used measures such as Indices of Dissimilarity, Segregation and Isolation should be replaced by new measures (such as an Ethnic Intensity score) which emphasise residential 'congregation' rather than 'separation'. Waterman (1989) examined the situation in Barnet in greater detail, and stressed the analytical difficulties of applying the results of Jewish residential segregation to other middle-class ethnic groups in Britain.

There are many other contexts in which religious differences have given rise to residential segregation. We shall examine two case studies which throw some light on the controls and products of such segregation, dealing with Catholics and Protestants in Irish cities and with Jews and Arabs in Israel.

Catholics and Protestants in Belfast

Residential segregation within the larger towns of Northern Ireland, particularly in Belfast, has been examined in a number of studies (including Boal 1969, 1972, 1976). Boal and Buchanan (1969) describe the plural society in Ireland, where sanctions against mixed marriages between Catholics and Protestants ensure that religious differences are strictly maintained. This plurality is reflected in schooling (there are separate education systems for Roman Catholic and Protestant children up to university level in Northern Ireland) and in housing.

Residential segregation arises in various ways, including voluntary segregation, administered segregation, and inter-group tension.

> Voluntary segregation occurs where people express a preference for a house in an area that is predominantly of their own religious persuasion. ...

182

Administered segregation covers [cases where] houses on particular local authority estates are specifically allocated either to Catholic or to Protestant families. .... Administered segregation also operates through the education system. .... Since there are very strong group pressures as well as strong family preferences operating towards sending children to the 'right' kind of school, and since proximity to school is a major factor in house selection, the segregated school systems accentuate and perpetuate the segregated residential areas.

(Boal and Buchanan 1969: 336)

The most detailed study of residential segregation in Belfast was carried out by Poole and Boal (1973) early in 1969, just before the present 'troubles' began. They found a high degree of segregation by religion, which echoed patterns observed in the late 1950s (Figure 6.1). In almost half of the streets in the city, 97 per cent or more of the inhabitants were of the same religious denomination, whilst two-thirds of the total households in the city lived in streets that had a minority of less than 10 per cent. Protestants appeared to be less willing to live in Catholic streets than vice versa, because less than 2 per cent of the total Protestant households lived in streets with a Catholic majority, whereas approximately a quarter of the total Catholic households lived in streets with a Protestant majority. Very few streets had a mixed religious population and the distribution overall was very bimodal (either very Catholic or very Protestant). The predominantly Catholic streets formed six clusters or ghettos, although one of them (centred on the Falls Road) was much larger than the others and it contained 70 per cent of all Catholic households in the city.

Boal (1969) studied a small area of the Shankill-Falls divide and found a high degree of territoriality. Protestant areas were marked by loyalist marches and bunting, particularly in the summer 'marching season'. Semi-permanent markings of Protestant territories include gable-end wall paintings, slogans relating favourably to the Queen and unfavourably to the Pope, graffiti relating to paramilitary organisations, and kerbstones painted red, white and blue. Catholic areas were similarly marked by territorial identifiers, including wall slogans opposite to those in Protestant areas. Territoriality was also evident in people's movement patterns, because people on both sides of the Shankill-Falls divide generally avoided entering the other group's territory and would prefer to take detours to avoid passing through unfriendly and thus potentially risky areas.

Residential segregation intensified after the outbreak of 'the troubles' in 1969 as minority households on both sides of the religious divide moved to areas where they would form part of the local majority (Poole and Boal 1973). Most moves were inspired by a perceived threat of violence or by direct intimidation. More than 2,000 households (representing about 1 per cent of the total population of the city) moved in a three-week period following the introduction of internment in August 1971. By 1972 the proportion of

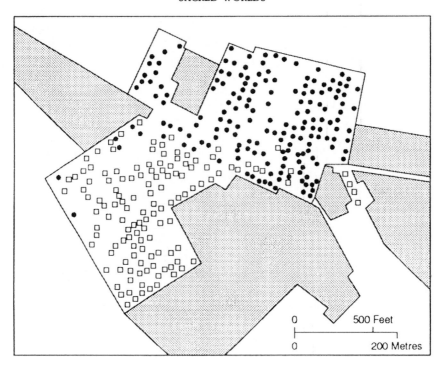

Figure 6.1 Sectarian residential segregation in part of Belfast in 1958
Source: After Boal (1969).

households living in streets with less than a 10 per cent minority had increased from 66 per cent to 76 per cent.

Jews and Arabs in Israeli cities

There are a number of mixed cities in Israel where Jews and Arabs have been forced to co-exist since Israel's War of Independence in 1948, in which new Jewish immigrants occupied abandoned Arab homes. The mixed city is the only place where Jews and Arabs live together in a distinctive geographical territory, with the Arab enclave surviving as an island within a sea of Jewish residences. Kipnis and Schnell (1978) describe the Arab enclave as a unique type of ghetto – a minority ghetto occupied by long-term residents in their own city (as opposed to rural or foreign immigrants) and created and sustained by a combination of voluntary and discriminatory factors.

The problems of religious segregation are particularly well illustrated in Jerusalem. Although the Israelis have made it their national capital, Jerusalem

184

is the Holy City of three world religions – Judaism, Christianity and Islam – each of which claims the right to influence what happens within its urban space (Rowley 1984). Many ingredients of the contemporary residential patterns within Jerusalem owe their origins to the nineteenth-century growth and development of the city (Ben-Arieh 1975, 1976, 1978), including the spatial segregation between religious groups (Hershkowitz 1987).

Baly (1979: 9) outlines the problems of resolving territorial claims on Jerusalem, which 'no thoughtful Jew, or Christian, or Muslim, can approach . . . without the most profound emotion. It is bound up with everything that makes him what he is, and apart from this city, in his understanding, he has no identity'. He suggests that a *status quo* policy be continued as a means of minimising conflict among the three different claimant groups because unification is unlikely to ever be more than a form of unity imposed by one group of people upon another. His proposal (ibid.: 14) is to divide most of Jerusalem up between the main claimant groups, but keep the Old City within the city walls as a separate, shared area that should be 'set apart, to be administered by a mixed body, which should include certainly the local people, Jewish, Muslim and Christian, but also representatives of those smaller nations whom no one could accuse of imperial pretensions'.

Religion and national identity

According to the *United Nations Declaration of Human Rights* (1948) every individual has the fundamental right to say to what religion s/he belongs and to have this accepted by state and society. Article 18 states that

> Everyone has the right to freedom of thought, conscience and religion; this right includes freedom to change his religion or belief, either alone or in community with others, and in public or private, to manifest his religion or belief in teaching, practise, worship and observance.

This phrase has since been incorporated into the state constitution of many countries around the world.

Despite the declared recognition of this fundamental right by all civilised societies, it is not universally implemented or respected. Religious freedom in a country may be quite different in practice to what the state professes about it. Records show that in 1980 half of the world's population – some 2.2 thousand million people in 79 countries – was living under restrictions on their religious freedom (Barrett 1982).

Most countries can be placed somewhere along a spectrum, from complete acceptance of religion at one end to complete rejection of it at the other. Figures published by Barrett (1982) show that in 1900 there were 145 religious countries (which regard themselves as officially religious), 78 secular countries and no officially atheistic ones. With increasing secularisation the number of religious countries fell to 114 by 1970 and to 101 by 1980. By 1980 there were

92 secular countries and 30 atheistic ones. Total world population in 1980 fell roughly equally into the three categories – 1,307 million people in religious countries, 1,579 million under secular regimes, and 1,488 million under atheistic regimes.

In some countries the Church is actively involved in governing a country or region. In such theocracies the head of the Church is often also the head of state. Examples include Vatican City (an independent state within Rome) ruled by the Pope, and the Mormon culture region in Utah, where in the nineteenth century Brigham Young was both leader of the Mormon Church and state governor (Jordan and Rowntree 1990: 216). In some other countries a state church is formally recognised by law as the only church in the state, and the government controls both church and state. Examples include the Anglican Church in England and the Lutheran Church in Norway.

Although many countries are dominated by single religions, there are many multi-religious countries in which a number of religions compete for presence and power. Oomen (1990) points out that South Asia contains many such multi-religious nation-states, in which 'irrespective of the official status accorded to religion in their constitutions, the state policy of religion is bound to be substantially moulded by the norms, values and lifestyles of the dominant religious collectivity'. All three Muslim majority states in South Asia – Pakistan, Bangladesh and the Maldives – have declared Islam as the state religion, although Bangladesh initially attempted to pursue a policy of secularism. The Buddhist (Bhutan, Burma and Sri Lanka) and Hindu (India and Nepal) majority nation-states seem to be able to accommodate more varied patterns of state-religion relationship. Although countries accord differing constitutional status to religion, Oomen found relatively few differences in national policies and priorities.

Perhaps the clearest illustrations of state-religion relationships are those where religion has provided either a reason or a catalyst for the existence of independent nations or states. Well-documented examples include India, Ireland and Israel.

India

India has long been a religious melting-pot as well as the cradle of Hinduism (see Chapter 4, pp. 101–2). Brush (1949) describes the historic cleavage of Hindus and Muslims in India, originating in a medieval conflict between native Hindus and new adherents to Islam. The conflict was eclipsed with the arrival of the British and the decline of Muslim dynasties, but it was to provide a foretaste of more recent Muslim nationalism on the Indian subcontinent.

The 1941 Census revealed that the total population of the British Indian Empire was 387 million, 92 million (24 per cent) of whom were Muslims. The Muslims were not distributed evenly around the country. The western border territory was an Islamic stronghold, there was a pronounced Islamic gradient within Punjab (decreasing from over 90 per cent in the west to 30 per cent in the east), and there was a further Muslim concentration in eastern Bengal (accounting for three-quarters of the population).

The movement for the partition of India into Hindu and Muslim territories developed over many years, and the disintegration of internal law and order came to a head by 1947. As Chapman notes,

> it was in this atmosphere that the last Viceroy, Mountbatten, reached an agreement with Nehru [the Indian leader] and Jinnah [the Muslim leader] for the partition of India, something which was acceptable to all only in so far as all could see each other equally miserable and disappointed by the conclusion.

> (Chapman 1990)

Spate (1943, 1948a, 1948b) outlines the background and geographical dimensions of the partition debate, including the difficulties of where to locate boundaries.

When the former British Commonwealth colony of India was granted independence in 1947, the territory was split into a Hindu state (India) and a Muslim state (Pakistan). At the time of partition West Pakistan housed some 23 million Muslims and East Pakistan had about 30 million Muslims. It was impossible to draw the boundaries to allow Hindus and Muslims to occupy mutually exclusive space, and around 30 million Muslims remained in India. Similarly, not all Hindus from the former India were contained in the post-partition India, and there were an estimated 20 million in the newly created Pakistan.

Partition provoked large migrations of people in and out of the newly created Pakistan. Many Muslims relocated in an orderly manner, but there were also waves of immigrants fleeing from riots and violent clashes in parts of India. Some estimates put the number of migrants as high as 16 million, but most agree that at least 12 million Muslims moved as a result of partition. By the early 1970s the population of West Pakistan was almost entirely Muslim. East Pakistan declared its independence from Pakistan in 1971, and adopted the name Bangladesh (meaning 'land of the Bengalis').

The partition of India was largely designed to create a separate territory in which Muslims could establish their own government. The decision to partition, and the choice of where to draw the boundary lines, were directly inspired by religious factors. But, as Brush comments,

> the recognition of religion as a basis for the establishment of a state and demarcation of its boundary posed a problem. . . . Conflict of loyalties has been created in every town and village where Muslims live who believe that their welfare depends upon uniting with Pakistan. Events have demonstrated the failure of partition alone to solve the political problem. Exchange of minorities has altered the distribution of Muslims, Hindus,

and Sikhs to conform with the boundary only in the Punjab and adjacent areas. Elsewhere the adherents of the conflicting groups remain widely intermingled.

(Brush 1949: 81)

Ireland

Ireland provides an equally graphic example of territorial partition along religious lines. Pringle (1990) has outlined the background and history of sectarian division within Ireland, which provides a context for the recent 'troubles'.

By the beginning of the eighteenth century the population of Ireland comprised three main groups. The 'native Irish' were indigenous, descended from the Celts, and by then almost exclusively Roman Catholic. The 'new English', in contrast, were colonial immigrants and mostly Anglican (Church of Ireland). There were far fewer English than Irish, but they were politically dominant and controlled most of the land (which had been confiscated from the Irish after the Cromwellian wars in the 1640s and the Williamite wars in the 1690s and given to the new English landlords). The third group were the 'new Scots', who were mostly nonconformist Presbyterians and were mainly confined to Ulster.

Political conflicts among these three groups were common in Ireland during the eighteenth century, and the defeat of the United Irish rebellion in 1798 brought home to the British government the need for radical political reform. This took the form of the integration of Ireland within the United Kingdom, under the Act of Union (1800). Enforced union antagonised the Irish population, and fuelled the rise of Irish nationalism and separatism from the mid-nineteenth century onwards. Two strong political camps emerged which closely reflected the religious split in Ireland between Catholics (nationalists) and Protestants (unionists). In this way modern Ireland's political problems are intimately tied to its religious composition.

Unrest continued in Ireland through the first decades of the twentieth century, between nationalists and unionists and between nationalists and the British authorities. Separation of the nationalist southern part of Ireland from the United Kingdom was the only way forward, and it was introduced by the Government of Ireland Act (1920) and the Anglo-Irish Treaty (1921). These granted independence to the 26 southern counties which created the Irish Free State (which was officially renamed the Republic of Ireland in 1948). The six dominantly unionist counties in the north were separated from the rest of Ireland to form Northern Ireland, which remained part of the United Kingdom but was granted partial autonomy through its own local Parliament.

The boundary between Northern Ireland and the Republic of Ireland quite closely corresponds to the major watershed between the Catholic (southern) and Protestant (northern) populations (Figure 6.2). The Republic of Ireland is

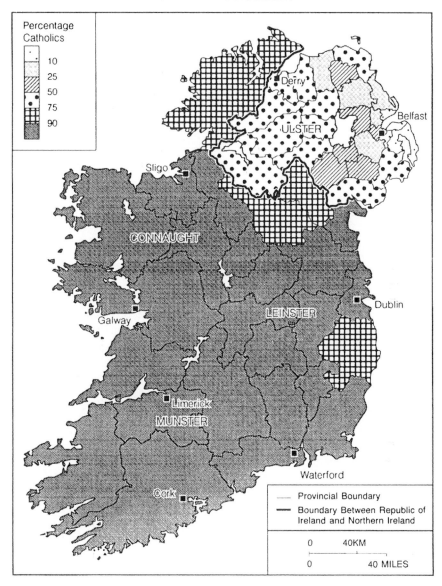

Figure 6.2 The distribution of Catholics and Protestants in Ireland in 1981
Source: After Pringle (1990).

overwhelmingly Catholic, and within Northern Ireland as a whole Protestants outnumber Catholics roughly two to one (Table 6.3). What appears, at least on a small map, as a clear solution to the sectarian tensions in Ireland is in reality a fragile and volatile holding operation in which partition and

Table 6.3 Religious affiliation in Ireland, 1911–81

| | Northern Ireland | | Republic of Ireland | |
| | Roman | Others | Roman | Others |
Year	Catholic (%)	(%)	Catholic (%)	(%)
1911	34.4	65.6	89.6	10.4
1926	33.5	66.5	92.6	7.4
1936			93.4	6.6
1937	33.5	66.5		
1946			94.3	5.7
1951	34.4	65.6		
1961	34.9	65.1	94.9	5.1
1971	36.8	63.2	93.9	6.1
1981	38.5	61.5	93.1	6.9

Source: Summarised from Pringle (1990).

segregation are vital ingredients but perhaps not permanent answers (Jones 1960).

As a result of religious partition, both parts of former united Ireland have radically different political complexions (Johnson 1962). Within Northern Ireland, political allegiance closely matches the religious distribution, so there is a core area of strong Protestantism and fiercely unionist politics, sandwiched between peripheral zones of stronger Catholicism and fierce nationalism (Figure 6.2). Protestants are strong in the east, reflecting a history of Scottish settlement in Antrim and English settlement in north Down and north Armagh and especially in Belfast. Catholics dominate much of the rural west (Jones and Eyles 1977).

Analysis of census records for 1971 and 1981 shows that the pattern of spatial segregation on religious groups became more intense during the 1970s. The Catholic presence declined in predominantly Protestant areas, and vice versa. Compton and Power (1986: 101) conclude that 'the increase in the geographical polarisation of Protestants and Roman Catholics . . . is clearly a result of the political tensions existing in Northern Ireland since 1969. As long as these persist one may expect this polarisation to intensify'.

Whilst spatial segregation between Protestants and Catholics is quite clear throughout much of Northern Ireland, the border zone with the Republic has a strong Catholic majority (Figure 6.2). Areas like south Armagh, south and west Fermanagh, and Derry west of the Foyle show greater similarity to the south than to the north, at least in terms of religious composition and nationalist politics. Little wonder, therefore, that these areas have witnessed some of the most recurrent sectarian violence since 1969.

Many of the political problems of Northern Ireland reflect socio-economic differences between the Catholic minority and the Protestant majority, which are deep-rooted and historical in origin and in themselves have little if anything to do with religious differences (Dillon 1990). Pringle (1990) comments on some of the more critical and persistent problems of social justice, particularly for the minority Catholic population. Catholics (collectively) have been widely regarded by the Protestant majority as an internal threat. As a result they have been excluded from positions of authority within Northern Ireland and discriminated against with regard to employment opportunities and many other forms of benefit controlled by the state. Links between patterns of unemployment and religious discrimination are strong and persistent, and they further fuel the sectarian tensions (Osborne and Cormack 1986, Miller 1989).

In response, many Catholics regard Northern Ireland as unreformable, so they look towards a re-united Ireland and the removal of their minority status as the only possible way of achieving social justice. There is no excuse for the sort of violence and terrorism which nationalist groups have inflicted on both Northern Ireland and mainland Britain since 1969, but there are signs that the geographical distribution of violence might be partly correlated with patterns of Catholic grievances and nationalism (Hewitt 1981).

Sectarian tensions are heightened further by the differential demographic patterns of Catholics and Protestants in Northern Ireland. Analysis of census data shows that the Catholic birth rate in Northern Ireland is significantly higher than the Protestant birth rate (Compton 1985). This will inevitably alter the religious distribution within Northern Ireland in favour of the Catholic minority. It has also given rise to Protestant fears that they will eventually be 'outbred' by Catholics, become the minority group and then be outvoted into a united Ireland.

Israel

The return of large numbers of the Jews to Palestine, from where they originated and had dispersed around the world, represents another significant illustration of the links between religion and national identity.

Throughout history Jews have been a persecuted people who have deliberately (and very successfully) resisted any temptation to assimilate into the new cultures to which they migrated. Reclaiming their former land in Palestine as a homeland for God's Chosen People has been both central to their religious beliefs and a prominent driving factor of their political ambitions. The traditional Jewish demand for a sanctuary and homeland was made even more poignant by Hitler's genocidal efforts in the Holocaust (Nyrop 1979).

Between 1918 and 1948 several proposals were made for the delimitation of a Jewish state within Palestine, and the matter was apparently finally resolved in 1947 when the United Nations voted in favour of partition along lines suggested by their Special Committee on Palestine. Israel was dissatisfied with the United Nations territorial proposals, and after the Arab–Israeli war of 1948 it occupied a much larger area that constituted the *de facto* state of Israel until June 1967.

Beaumont, Blake and Wagstaff (1976: 298) comment that 'although [it is] a relatively small addition to the political map, Israel is anathema to the Arabs, chiefly on account of injustices to the Palestinians, and the long-term implications of a Jewish state located at the heart of the Arab world'. Religion and state are intimately interlocked in a modern Israel that, as Weissbrod (1983) points out, still bears the imprint of Covenantal thinking (that Jewish nationhood derives from a Covenant with God). This is reflected in many ways, including the maintenance of religious symbols such as Jewish holidays, the revival of the Hebrew language, the use of the Star of David in the Israeli flag, and the use of the Menorah (the seven-branched candelabrum) as the symbol of the state.

By the end of 1948 the former British Mandate Palestine had been subdivided into three territories, each one ruled by different authorities. Israel occupied 20,700 sq. km (78 per cent), the Kingdom of Jordan occupied about 5,600 sq. km (20.7 per cent) in an area that became known as the West Bank, and the remaining 1.3 per cent – the Gaza Strip – was put under Egyptian administration (Saleh 1990).

Even before 1948 many Jews had settled in Palestine, upholding the Zionist belief that the creation of a Jewish state was the only means of ensuring the survival of the Jewish people. Up to 30,000 Jewish immigrants, mostly from Russia, Romania and Poland, had already settled by 1903. A further 40,000 Jews arrived from Russia between 1904 and 1914, and by 1931 they had been joined by another 35,000 immigrants, many of them from Poland. Between 1931 and 1939 an estimated 230,000 Jews entered Palestine to escape Nazi oppression in Germany and Austria (Beaumont, Blake and Wagstaff 1976). Preparation for the establishment of a Jewish homeland is also illustrated in the large amount of agricultural land in Palestine that was bought by Jews from Arabs between 1900 and 1914 (Katz and Neuman 1990).

Once the State of Israel was established, what had been a steady flow of Jews returning to their homeland in Palestine turned into a major tide. The so-called Ingathering of the Exiles was encouraged by law, because the Israeli parliament (the Knesset) passed a Law of Return in July 1950 that declared that 'Every Jew has the right to come to this country' (Nyrop 1979). The State of Israel was declared on 14 May 1948 and within six months the Jewish population had risen from just under a third to more than three-quarters, after up to 700,000 Arabs had fled or been expelled from the borders of the new state (Beaumont, Blake and Wagstaff 1976).

Israel's expansionist intentions and territorial objectives were made clear in the Arab–Israeli war of June 1967 that brought more changes in the map of Palestine (Smith 1968). Israel occupied the West Bank, Gaza Strip, Sinai (towards Egypt) and the Golan Heights (in southern Syria) (Figure 6.3). Sinai was returned to Egypt as a result of the Camp David peace settlement signed between Sadat and Begin; Golan Heights remains under Israeli occupation and the status of the West Bank and Gaza Strip has yet to be determined (although they are still occupied by Israel).

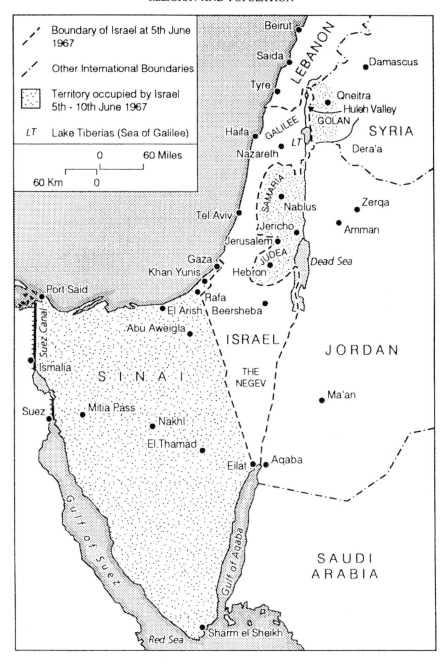

Figure 6.3 The State of Israel and territory occupied by Israel during the June 1967 war

Source: After Smith (1968).

Israel has established a number of Jewish settlements in the occupied territories since the war as part of its Zionist settlement process and to establish the basis for territorial claims (Portugali 1991). The Israeli authorities have also invested heavily in the demolition of Palestinian refugee camps in the Gaza Strip and the resettlement of refugees elsewhere (Dahlan 1990). Romann (1990: 381) emphasises that 'the territorial issue mainly concerns collective rights to the contested land, still mutually denied, and this remaining at the core of the Jewish–Arab national conflict'.

Since the 1967 war the demographic, settlement and economic structure of the West Bank region has been altered significantly. It has been argued that most changes reflect an unofficial Israeli aim to evacuate Arab natives from their homelands and to replace them by Jewish immigrants, as a means of expanding Eretz Yisrael (the Land of Israel) along the Jordan valley (Saleh 1990). The Jewish settlement strategy in the West Bank has evolved through time. The first structures were defence strongholds and military agricultural settlements dotted along the escarpment of the Jordan valley. These were later to be enlarged and given the status of permanent settlements. By September 1982 there were nearly 3,000 Jewish families living in 18 urban settlements in the West Bank, and a further 2,100 families in 65 smaller rural settlements.

Israeli developments plans for the West Bank (which have been prepared but not approved) include reserving extensive areas around Jerusalem for Jewish settlement, building a road network to better integrate Israel and the West Bank, sterilising development in large areas of the West Bank, and building hundreds of villages and municipalities (Coon 1990).

The Jewish settlers in the West Bank have significantly altered the pace and pattern of economic development of the area. Since occupation began in 1967 a dual economy has evolved, in which the native Arab population is heavily eclipsed by the Israeli incomers. Although the settlers represent about 3 per cent of the total population of the West Bank, their economic activity constitutes at least 35 per cent of the area's gross domestic product (Saleh 1990).

A central ingredient in the domestic politics of Israel is the 'Judaization' policy designed to enable Jewish control of territories within the state that contain predominantly Arab populations. Galilee, to the north of the West Bank (Figure 6.3), is typical of such areas. The Arab population in Galilee is numerically stronger than the Jewish population, and it dominates the cities. Israeli attempts to create a new and alternative 'Jewish core' in mountain Galilee have not been successful (Falah 1989). None the less the development of industrial zones in the Galilee region is a major element of Israel's national planning policy, and the design of the developments takes into account environmental as well as socio-economic factors (Amir 1990).

Studies have shown that the economic benefits of the industrial developments in Galilee are not shared evenly between Jews and Arabs. The

Arabs are significantly under-rewarded, which reinforces inter-ethnic economic gaps and minority dependence (Yiftachel 1991). Power struggles between Jews and Arabs are also evident in the unequal distribution of land, water and local budgets in Galilee, which indicates that central government's allocation policies are positively skewed in favour of the Jewish population (Falah 1990).

Whilst some aspects of Israeli settlement policy are clearly oriented towards the ideological objective of achieving Jewish population majorities in dominantly Arab territories, this is not always the case. Jewish colonisation of west Samaria, for example, was only begun in 1975 and it largely reflects a desire to acquire low-cost housing in areas accessible to Israel's urban core (Grossman 1991). The Nablus region in particular has attracted large numbers of Jewish settlers (many from the former Soviet Union) as a form of suburbanisation outwards from Israeli metropolitan space into the mainly rural Palestinian domain of the West Bank (Rowley 1990).

Many Arabs remain in what is now Israel, and attempts to integrate them socially into the economy and policies of the new Jewish State are beset with problems. Schnell (1990) highlights the critical problems of ethnic identity, social connections and territoriality and insists that through the 1970s separation between Arabs and Jews was more prominent than integration. Local labour market segregation in Israel also creates and perpetuates socio-economic inequality between Jews and Arabs in Israel, particularly for Arabs working in Jewish communities who suffer both occupational and income discrimination (Semyonov 1988).

CONCLUSION

This chapter has introduced a wide variety of themes relating to the interactions of religion with demography, development and politics. In each area of human activity it is clear that religious beliefs and practices can and do exert powerful influences, to such a degree that it is questionable whether the conventional human geography perspectives (which largely ignore religion) can really claim fully to account for observed patterns and processes.

Teaching and traditions within many of the major religions are powerful determinants of demographic variations, particularly fertility and family planning. To ignore the pervasive influence of religion on such fundamental dimensions of socio-economic life is to overlook a significant determinant of human behaviour. Religion can also profoundly affect rates and patterns of economic development within and between countries, yet conventional economic geography rarely concedes the importance of such factors.

It is in the political arena that the influence of religion is often most visible and tangible. The imprint is both physical (witness the strong residential segregation along religious lines in many major cities around the world) and socio-cultural (witness the strong persecution of religious minorities in many countries).

In these ways the impact of religion is not just confined to believers, because adjustments to religious variations and responses to religious tension often underpin the very fabric of society.

7

RELIGION AND LANDSCAPE

My country is the world, and my religion is to do good.
Tom Paine, *The Rights of Man* (1792)

INTRODUCTION

The landscape is a manuscript on which is written the cultural history of the area, although some traces of the past are more enduring than others. This manuscript can be read, and its messages uncoded. Whilst there are many types of landscape on which religion has left a dominant and indelible imprint, which we might think of as specifically religious landscapes, the religious influence does not stop there. Evidence of the impact of religious beliefs and adherence can be found in a wide variety of landscapes, although it is often preserved in quite minor details.

In this chapter we examine how religion has affected landscape in the past, and continues to do so today. Examples drawn from a range of religions, cultures and continents show how the religious imprint can be discovered at different scales and in different parts of the landscape. Key themes include the impact of religion on architecture (especially on building styles), on settlement forms and functions, on farming practices, and on the overall physical appearance of the landscape.

This theme has traditionally attracted much geographical attention, partly because of the discipline's enduring interest in landscape as a product of natural and cultural processes. Some would agree with Erich Isaac that the theme of religion and landscape is not simply a component of the geography of religion, it is the very core of the subject. Indeed, Isaac defines the geography of religion as

> the study of the part played by the religious motive in man's trans-
> formation of the landscape. It presumes the existence of a religious
> impulse in man which leads him to act upon his environment in a
> manner which responds secondarily, if at all, to any other need.
>
> (Isaac 1960: 14)

Landscape is a palimpsest, or a manuscript on which two or more successive texts have been written, each one being erased to make room for the next. Quite how much is erased varies from place to place, so that some landscapes preserve a great deal of evidence of their past whilst others are predominantly contemporary in appearance. This dynamic gives rise to the variable mosaic of landscape even within an area, and it provides opportunities for historic reconstruction based on visible landscape features. Jackson (1952: 5) insists that 'the cultural history of America is just as legible in the appearance of our landscape (for those who know how to read it) as it is in the monuments and institutions of our cities', and this is equally true in all countries.

Interpretation of landscape as a product of culture requires some understanding of how people translate values and beliefs into architectural forms, and how their values and beliefs inform their use of space, such as in the spatial organisation of settlements (Sitwell and Bilash 1986). Sitwell and Latham (1979) have called upon geographers to direct more effort to understanding the role in shaping landscape of explicit beliefs (expressed by what people say) and functional beliefs (expressed by how people behave). Sitwell explored the possibility of interpreting elements in the cultural landscape as being 'concrete' signifiers standing in place of verbal signifiers, using the metaphors of height, durability and central location.

Some landscapes preserve cultural artefacts from the past that are as enigmatic as they are informative. A vivid example is Easter Island in the Pacific, which is 22 miles long and dominated by over 17,000 archaeological sites witnessing to a remarkable past culture. Most striking are the 600 or so giant long-eared stone statues (*moai*) cut from local volcanic tuff that are dotted around the island on long stone ceremonial altars (*ahu*) to watch over tribal lands (Croad 1992). How they were moved remains a mystery; the largest is 30 ft high and weighs over 80 tons.

Such a landscape poses many interesting questions about form and function. It would certainly appeal to Isaac (1962: 12), who sees the task of a geography of religion as 'to separate the specifically religious from the social, economic and ethnic matrix in which it is embedded, and to determine its relative weight in relation to other forces in transforming the landscape'.

Isaac (1962) identifies a polarity between those religions that deliberately change landscapes and those which don't. Transformation is promoted in religions that see in the process of shaping the world a meaning of human existence. The Dogon of the Upper Volta Republic in West Africa, for example, deliberately seek to reproduce in the landscape what they understand to be the original cosmic pattern. Other religions conceive the meaning of existence as deriving not from the process of world creation as such but from a divine charter granted to them (such as Israel's covenant in Judaism and Christ's crucifixion in Christianity). The Budja, a Shona tribe from north-east Southern Rhodesia (now Zimbabwe) which sees no tie between religion and land, and therefore makes no effort to transform the land, illustrate this second group.

Deliberate transformation of landscape is, however, only one component of the link between religion and nature. The relationship is more reciprocal than one-way, because environment has doubtless had an influence on the shape of the major world religions (see Chapter 4, pp. 95–6). Nature also influences religion at a smaller scale, such as regional variations in some religious rituals. Sopher (1964) describes various inter-relationships between season, religion and landscape in India, such as the Hindu folk festival of Rathajatra that marks the transition from the drought season to the rain season.

Inevitably there is more than one way of reading a landscape, and – equally inevitably – the perspective adopted strongly influences what we see and how we interpret it. A technological perspective, for example, would emphasise physical properties that met particular human needs (such as space, light and access) and pay little attention to symbolism, imagery and religious attachment. Walter (1985) suggests that a functionalist perspective might focus on psychological impacts of landscape; a Marxist perspective might question who, why and for whom landscapes have been created; an ecological approach would stress unity and naturalness; a Christian perspective would emphasise the rightness of human beings ordering and giving meaning to the physical world.

The type of questions we ask about landscape, the evidence we regard as important, and the interpretations we place on that evidence all depend on our perspective. Those who look for signs of religion's influence on landscape are likely to find them, therefore, so we must be careful not to over-exaggerate the significance of reported evidence and the interpretations that have been attached to it.

LANDSCAPES OF WORSHIP

Landscape features which serve some function in worship, such as churches and temples, provide the most obvious visible signs of the imprint of religion on an area. Indeed, symbols of religious worship are woven into the very cultural fabric of many areas and give them a special and sometimes unique identity. Jordan (1973: 138) has commented that 'one cannot imagine a European cultural heritage devoid of the magnificent cathedrals, altarpieces, crucifixes and religious statuary'. Similarly, a Western visitor to Muslim countries or areas where Hinduism or Buddhism are strong is immediately struck by their religious buildings that contrast so sharply with the Western landscapes they are more familiar with. Through history, too, the architecture of worship centres has usually been striking, as perhaps befits places dedicated to the gods. Grand temples were among the first and finest monuments throughout the Roman world, and they occupied prominent places in the Roman city landscape in terms of size, appearance and social significance (Blagg 1986).

In this section we examine evidence of how religious worship is imprinted on landscape, in a variety of settings and cultures.

Domestic and roadside religious features

Some religious landscape elements are small scale and local, yet their cultural significance far outweighs their size and limited distribution. Typical examples include domestic altars, yard and roadside shrines, and roadside chapels.

Domestic altars

Hindu homes in many areas have altars that are used daily in worship. They might be rooms set aside specifically for the purpose, but quite commonly an altar is set up and maintained permanently in a living room or other room within the residence. Domestic altars figure in other religions too, including Buddhism.

Some of the most colourful domestic altars are to be found in the South-west United States, close to Mexico, where religious traditions (particularly within Catholicism) from both countries have blended together to create a rich and varied Mexican American culture (Rose 1992). Shops and bars often display discrete *nichos* (niche altars to patron saints) and many homes have *altarcitos* (home-made altars). The home altars are survivors of a pre-conquest Mexican heritage of keeping indoor shrines (Arreola 1988). They are dedicated to God, Christ, the Virgin Mary and patron saints, and are usually constructed and cared for by the women of the family who use the sacred space for prayer and meditation. Some are simple assemblies of sacred statues and family photographs, whilst others are elaborate edifices that include heirlooms, pop culture and souvenirs as well as the usual religious icons.

Yard and roadside shrines

Religions differ in their attitudes towards roadside shrines and structures. Visible symbols of faith are common in Roman Catholic and some Eastern Orthodox Christian areas, for example, where popular wayside structures include shrines, crucifixes and crosses. Jordan and Rowntree (1990) note that roadside religious structures are rare in Protestant areas.

Few geographical studies have focused on roadside religious landscapes, but two papers on Catholic yard shrines in different parts of the United States (Manyo 1983, Arreola 1988) reveal interesting aspects of this particular feature. Yard shrines resemble miniature churches, are usually open-fronted, and are permanently sited in the yard (garden) of a house. Designs vary but functions do not, because the yard shrines were usually built as personal responses to crises during which an individual promised to erect one as an offering in return for answered prayers.

Arreola (1988) found more than 200 yard shrines in the West Side of San Antonio (Texas), most of which were promised by women but built by men. Many were dedicated to the Virgin of San Juan de los Lagos (the patroness of journeys) who is popular among migrant workers and newly arrived immigrants. Some 400 yard shrines were counted in Tucson (Arizona), nearly a third of them built by Mexican immigrants.

Yard shrines are by no means confined to Mexican American areas; they are a fairly common feature in many Catholic strongholds. Manyo (1983) studied 15 Italian-American yard shrines in Kansas City (Missouri) and 12 in Archibald (Pennsylvania). The shrines were dedicated to many different religious figures, but most commonly to the Madonna (the Virgin Mary, found mainly in front yards) and Saint Francis (found only in back yards). All were built by homeowners of southern Italian descent, and many were referred to as grottoes (suggesting a possible connection with major European shrines like Lourdes and Fatima).

Roadside chapels

Small roadside chapels are a common feature of the religious landscape in many areas, particularly those associated with Catholicism. Arreloa (1988) suggests that mendicants (itinerant monks who lived by begging) introduced the Spanish hilltop hermitage into Mexico in the sixteenth century, which evolved into corner chapels on the forecourt of early open-air Catholic churches. From the close of the sixteenth century, the Church abandoned most such chapels but Indians continued to use them.

Different in origin and form are the Belgian roadside chapels of the Door Peninsula in Wisconsin, 24 of which Laatsch and Calkins (1986) studied in great detail. Like the yard shrines, these structures were built by devout Catholics as thanks for answers received to specific prayers of favour. Most date from the second half of the nineteenth century and were made by Belgian immigrants who arrived mainly between 1853 and 1860, lured by the availability of cheap government land. Nearly 70 per cent of the 3,800 foreign-born Belgians clustered in an area that was to evolve into the rural Belgian ethnic island, where the small roadside chapels were built on individual farmsteads by devout Roman Catholic Walloon-speaking Belgians. The wooden chapels are small, single-roomed, windowless and painted in pale colours. Inside is a small altar, containing a cross, crucifix or statue (commonly the Virgin Mary or an appropriate saint) and other symbolic elements including smaller statues and crosses. Many families whose ancestors built the chapels continue to look after them, and use them for private prayer and meditation.

Christian church architecture

The domestic and roadside religious structures are usually confined to specific areas where religious and cultural traditions gave rise to them and allowed them to survive. Much more ubiquitous components within the religious landscape in Christian areas are churches and other large structures used for worship.

Modern churches have evolved from primitive stock. The pace and pattern of change vary greatly from place to place, but the history of church building in Ireland – outlined by Whelan (1983) – illustrates a common sequence. The first Christian centres of worship were open-air gathering points. By 1731 in Ireland rudimentary chapels or mass houses (simple, small, thatched, undecorated buildings with mud walls and clay floors) had been built on the open-air sites. These penal chapels were increasingly replaced in the late eighteenth and early nineteenth centuries by the first modern chapels. The so-called 'barn chapels' were bigger and grander than the mass houses, with slate roofs, flagged floors, a gallery. Built by local craftsmen in the vernacular tradition, the barn chapels were still austere by modern standards (with no pews or internal decoration).

Jordan and Rowntee (1990) point out that churches vary a great deal in size, function, style of architecture, construction material and degree of ornateness, and they account for some of the variations in terms of significant denominational differences in interpretation of what church buildings actually mean. To Roman Catholics the church is literally the house of God, so their churches are typically large, elaborately decorated, and visually dominant. As Jordan (1973: 152) puts it, ' ie Roman church places great value on providing visible beauty for the faithful'. To many Protestants, in contrast, the church is a place to assemble for worship (God visits the church but does not live there), so their churches tend to be smaller, less grand, less ornate and more functional than many Catholic churches.

The pilgrimage churches built in the Holy Land during the twelfth and thirteenth centuries to cater for the needs of Christian pilgrims to Palestine during the Crusades transplanted European church architecture into the Middle East but there was no blueprint form or layout. Pringle describes how

> the planning of Crusader pilgrimage churches in the Holy Land varied considerably from site to site, taking account of the particular nature of the holy place being commemorated, the natural topography, the existence (or not) or earlier structures and the need to accommodate a resident religious community.
>
> (Pringle 1987: 358)

Variations in church architecture can provide important clues about ethnic origins in areas where large numbers of immigrants have settled. Two distinct waves of pioneer settlers landed in south-central Pennsylvania during the eighteenth century, for example, and each etched its beliefs into the structures of their church buildings (Milspaw 1980). Settlers from Scotland and Ireland, who arrived before 1720, built plain and simple churches with little decoration. German immigrants, who arrived after 1750, preferred their churches to be highly decorated and ornate. Surviving churches are helpful in reconstructing the original distribution of both groups.

The survival of 78 Russian Orthodox churches on the Aleutian Peninsula and coastal Alaska is in fact the only evidence that Russians once occupied Alaska (Straight 1989). Although the churches appear to be close copies of their Russian counterparts, with characteristic onion-dome styles, the harsh Alaskan climate required some important design modifications. The Alaskan churches have lower roofs (to survive stronger gale-force winds) and steeper roofs and steeples (to cope with heavy snowfalls).

In the last two examples the survival of original architectural forms has assisted the interpretation of the cultural history of particular places. But sometimes immigrant groups abandon their ethnic architecture as they assimilate into the new culture. This seems to have been the case with the Mennonites in many parts of North America. Heatwole (1989) has traced the evolution of their church architecture in the United States from plain buildings (typical of their ancestral homeland) soon after arriving, to a modern generic form (typical of most contemporary American churches). This shift reflects progressive acculturation and the adoption of new technologies and ideas, and means that it is difficult to distinguish recent Mennonite churches from other Christian churches.

Amongst the most persistent Christian architectural forms is the Gothic cathedral that is common in many European cities. The largest and one of the finest of the world's cathedrals is St Peter's in Rome, founded in 1450 on the site of an earlier basilica built in 306. The new cathedral, the basilica of the Vatican City, was built mainly between 1506 and 1615. Other fine examples include Notre Dame in Paris (built between 1163 and 1257), and the cathedrals of Cologne and Milan. St Paul's in London, built between 1675 and 1710 to replace an earlier cathedral destroyed during the Great Fire of London in 1666, is widely regarded as Christopher Wren's greatest masterpiece.

Hindu temple architecture

Hinduism also has distinctive architectural forms for its temples and other religious structures that are usually highly ornate and richly decorated, and this gives rise to unique and characteristically Hindu landscapes.

Hindus believe that building a temple, even a very modest one, endows the builder with spiritual merit and brings divine reward. A typical Hindu landscape has shrines of many sizes, and several factors guide the selection of suitable sites (de Blij and Muller 1986: 208). Sites should bring minimal disruption of the natural landscape, be in a comfortable location (shaded, for example), and face the village from a prominent position. Sites should also be close to water (which has a holy function in Hinduism, because many gods are believed to stay close to it), wherever possible.

Temples in cities vary a great deal in size, style and location. Temple form has probably evolved through time, reflecting adjustment to and accommodation of broader socio-economic changes. This certainly seems to have been the case in Calcutta, judging by Biswas's (1984) study of the evolution of Hindu temples over three centuries while Calcutta grew from an insignificant fishing village to a large urban complex.

Traditional temple architecture reflects the view that the temple is a sacred abode of God, which in turn reflects the traditional Hindu idea that mountains are dwelling places of the gods (lord Vishnu and lord Shiva). In all classical temple styles (Figure 7.1), the temple towers resemble mountain peaks, with tapering crowns. Small dark *cellas* inside (inner room housing the statue of the

Figure 7.1 Structure of a north-Indian-style Hindu temple

Note: The temple has a *cella* with one doorway and pointing finials at the top. It is aligned with the image of the deity, to signify the settling down of the divine spirit.

Source: After Biswas (1984).

deity) resemble caverns. Finials (ornaments) on top point to the sky and are aligned with the image inside, to signify the settling down of the divine spirit.

The earliest Hindu temple style in the Calcutta area is the thatched-hut Bengal-style (*chala*) temple modelled on village residential huts, commonly located at crossroads up to the eighteenth century. A variant is the larger *Ratna* temple, which has four miniature *chala* temples surrounding a large central five-pinnacled (*Pancha Ratna*) temple, located by a river or large pond. From the late eighteenth century onwards new temple forms appear, including the *Thakurbari* (family) temple built attached to or close to the house of its founder. This flat-topped and often richly decorated building, commonly two or three storeys high, had a central courtyard or assembly hall surrounded by a series of stately rooms. Two new temple forms have appeared in Calcutta during the twentieth century. Modern institutional temples (locally called *Maths*) show much stronger Western architectural influences than earlier temples. Community temples are small and neighbourhood-based. One colourful recent trend, most evident since 1960, has been the growth of numerous small shrines devoted to folk deities, which are usually owned and operated by traditional Brahmin priests. These shrines are widely distributed, and particularly common along major streets, in crowded market places and at busy road junctions.

Continuity and adaptation of Hindu temple form have also been described in Trinidad, by Prorok (1988, 1991). Her detailed field study of Hindu material culture revealed how religious structures had evolved from traditional temple forms, and showed that the contemporary landscape includes temples representing each stage of the evolutionary process. The main catalyst for the change appears to have been a shift in the manifest function of the temples from god-centred to community-centred. Between the 1860s and 1920, two temple forms were common in Trinidad. These were the *Simple Traditional* temple (the most common type before 1920, built mainly of bamboo poles and thatch material) and *Traditional* temples (stone or clay-brick temples that appeared in the 1880s). Both types are typical of village temples built in many parts of India. A new temple form (the *Koutia*) appeared between 1921 and 1940, largely by building extensions to existing temples to turn them into permanent community meeting places. Many traditional temple forms were built between 1945 and 1960, but by now they served a broader functional role within the community.

Sacred space in Hinduism extends beyond the temple to include the surrounding temple town. Bohle (1987) explains this in terms of traditional Brahmin cosmology, based on a circular central continent with the mountain of the gods (*meru*) in its centre, surrounded concentrically by six ring-continents and seven ring-oceans, and an outer ring of rocky mountains. Hindus view the temple as a representation of *meru* (equating it with the centre of the cosmos) and lay out the temple town to reflect the rest of the cosmos as closely as possible. A typical temple is square or rectangular in shape, oriented towards the cardinal points with the main temple gates pointing east, with up to four concentric temple walls. Up to three further street networks are concentrically arranged around the central temple enclosure, intersected by diagonal streets aligned with the cardinal points. The street leading out from the main temple entrance and the streets around the outer temple wall are particularly wide. The largest houses and plots of land are located in the centre of the temple town, and caste status decreases outward from the temple centre.

Buddhist temple architecture

Buddhist temples are often not so visually striking outside as many Hindu temples (which can be virtually encrusted with ornate carvings of people and creatures), but they make up for it inside. They often have quite distinct sights (a windowless room lit mainly by candles, with statues of the Buddha and varied gift offerings), sounds (gongs and chimes, people chanting quietly) and smells (the air is heavy with the odour and haze of burning incense) which are quite unlike the atmosphere in any other places of worship.

Buddhist countries commonly have many different temples and shrines dedicated to different deities. There are an estimated 5,000 temples and shrines in Taiwan, for example, and most of them are within the capital city Taipei. One of the most striking is the Lungshan temple, dedicated to Kuan Yin (the Buddhist goddess of mercy), which was substantially rebuilt in the 1950s.

A flavour of the atmosphere within traditional Buddhist temples is given in Noack's description of the Samtanling Lamasery in the high Karakoram Mountains:

> perched atop a cliff, high above the few stone and mud dwellings of the inhabitants, is the imposing Samtanling Lamasery, by far the largest, oldest and grandest sanctuary in the remote Nubra Valley of the yellow hat sect of Tibetan Buddhism ... [its] innermost sanctuary, chapels and altars, with intricate carvings on wooden beams, columns and wall panels portraying painted images of gods, demons and animals. Each and every true believer is urged to become thoroughly familiar with the bizarre creatures he is destined to meet as he wanders along the path of trans-migration, after death, towards attainment of Nirvana. The painted images on the walls, the mysterious darkness of the chapel with its shadows of devils and demons and the strong aroma of burning incense have a restraining influence on pious believers ambling through the sanctuary.
>
> (Noack 1979: 135)

Equally remote and visually striking are the Lama monasteries of the Yellow Church in Ladakh on the high mountain pass between India and Tibet, described by Pieper (1979). This inhospitable area, closed to outsiders until 1974, has a small and scattered population with primitive ways of life who live in small houses made from sun-dried bricks. Yet the landscape is punctuated with monasteries (*gompas*) which look like huge, lavishly decorated houses set in imposing, inaccessible sites. Their interiors, also richly decorated, contain temples, kitchens, courtyards and a labyrinth of chambers, caves and corridors. They act as repositories for art, paintings, sculptures and decorative crafts, partly paid for by proceeds from the sale of *mani* stones (prayer stones) and the provision of *lathos* (little tantric temples) on mountain peaks.

Like Christian churches and Hindu temples, Buddhist temple forms have evolved through time as a response to wider cultural and ideological changes. Tanaka (1984) has clearly shown how temples in the Kyoto–Nara region of Japan, which reflect four different schools of Buddhism, have evolved particularly in terms of location, internal layout and garden arrangements. The main structures within the temples include the *stupa* (pagoda) which symbolises the Buddha, the main hall (home of the Buddha) which contains the images and *sutra* scrolls, the lecture hall, the corridor (which surrounds the main structures and demarcates the sacred interior from the outside world), and the gate.

Mikkyo temples (dating from 794–1192) are located away from the city centre, often in hills and mountains and surrounded by forest. The preferred orientation is to the south, and these temples have no gardens. Most Jodokyo temples (founded between 1133–1212), in contrast, are located on the plain in accessible sites and are oriented towards the east. A typical Jodo garden has a large lotus pond spanned by a bridge, with the temple as the Buddhist Paradise on an island in the middle, mirrored in the pond. Zen temples (dating from 1141–1253) are found on level ground close to mountains, in quiet natural settings, and they have no obvious preferred orientation (except that north-facing sites were avoided). The small Zen garden that seeks the beauty of nothingness – devoid of bright colours, with rocks and moss-covered mounds representing mountains and cascades, sand or raked gravel symbolising rivers and oceans, and perhaps a solitary pine tree – is perhaps best known of all the Buddhist gardens.

CHURCH DISTRIBUTION AND DYNAMICS

Many of the geographical studies of religious structures have focused on Christian churches, and particularly on the themes of distribution and dynamics. Why are they located where they are, and what factors influenced choice of location? How and why have distributions and functions changed through time?

Choice of location

Despite the ubiquity of churches on the Christian landscape, it is surprising that little detailed research has been done on how sites are selected. The question is most relevant at two scales – how locations within an area are selected, and then how the precise spot on the ground is selected. The answer to the former lies partly in the practice adopted by major Christian denominations (particularly Catholics and Anglicans) of sub-dividing regions into smaller territorial units (parishes), each of which has their own church. The answer to the latter reflects mainly practical considerations, such as topography, aspect and access.

One of the few empirical studies of church site selection is Homan and Rowley's (1979) analysis of churches and chapels in nineteenth century Sheffield (England). Many new churches (particularly Anglican and Nonconformist ones) were built in the growing industrial city, and the rate of growth was fastest after 1840 when church provision grew faster than population growth. Threshold populations dictated where and when churches were built, and churches were established within the zone of building activity rather than beyond in green-field sites. It was quite common for population growth in part of the city to stimulate the establishment of a mission church or an independent church. They concluded (ibid.: 150) that 'in the main, religious sites were established, if not actually during the main phase of local house building then certainly before the area was completely taken up for building purposes'.

Religious considerations play a much smaller part in the selection of locations for Christian churches than they do for synagogues, which have special symbolic status for Jews. Shilhav (1983: 324) notes how 'the importance of the synagogue to the spiritual life of the Jewish observant community found its expression in the shape and size of the building itself, in its architecture, building materials, and especially in its location'. The Talmud lists the presence of a synagogue as one of ten vital functions of a town, and it states that a city whose roofs are higher than that of the synagogue is doomed to destruction. Synagogues are inevitably only built in settlements with Jewish communities, and in Palestine at the time of the Talmud they were ideally built taller than other buildings (but synagogues in modern cosmopolitan cities are usually not very high). Access is another important criterion, because all Jewish males aged 13 or over are expected to attend prayers in the synagogue regularly (the devout attend up to three times a day). Hence Jerusalem's new central synagogue, dedicated in 1982, was built on one of the city's main arterial roads.

Parishes

Parishes (the areas from which Christians came to a central place for communal worship) are amongst the oldest spatial divisions in English landscape history. Ravenhill (1985) has traced the evolution of the parish system as Christianity spread through the countryside, and landowners began to build churches on their villas and estates for local worship.

By the seventh century England had been divided into dioceses, within which the pastoral work of the Church was being carried out from mission stations (minsters) surrounded by their often large and ill-defined individual parishes. Stone or timber roadside crosses, at which locals would pray daily, were progressively replaced by permanent buildings for congregational worship (churches) overseen by a priest. The emerging church system was intimately tied with the feudal system because parishes often coincided with manor boundaries, although in Cornwall and other fringe areas many pre-medieval Celtic centres of worship were also incorporated into the parish system. Rural churches were being built in many areas by the start of the twelfth century, which was the golden age for the establishment of parish churches.

Parishes on Jersey had different origins to those in England, according to Myres (1978). Most of the island's churches were in existence by 1150, but the parish system was irregular. Here the Church rather than manorial landlords was the prime creator of the parishes, when archdeacons from Coutance in France drafted the lines of a regular parochial pattern out of the bits and pieces of Christian tradition, ruined buildings and surviving memories of local saints left behind by pagan Norsemen. The parish spatial units survive today, despite a mixed ecclesiastical history within Jersey.

Parishes provide the basic territorial church units, but their significance is not confined to spatial patterns. Their churches have often provided core nuclei around which settlements evolve, and this has made them important catalysts of broader cultural and socio-economic change. The development of the chapel-villages in Ireland, described by Whelan (1983), provides a graphic example.

The Reformation and colonial settlement in Ireland brought a wholesale redistribution of land, wealth and political power. One particularly significant development was the planting of Protestantism, which often involved the take-over of existing Catholic churches (and their parish structure). During the Counter Reformation (between 1600 and 1800) the Catholic Church responded by establishing a totally new parish structure and building new churches for its disenfranchised congregations.

The new parish framework better suited the demographic, social and economic conditions of contemporary Ireland, and turns out to have been a blessing in disguise. The pattern and pace of church building varied according to the attitudes of landlords, the wealth of the local Catholic community, and the sponsorship of prominent Catholic families. Chapel-villages grew within this reinvigorated parish and chapel network, strung along new roads which greatly improved access. Local priests helped to channel the agencies of the welfare state – such as schools – into their villages, and commercial establish-ments (such as forges, public houses, co-operative creameries and grocery shops) soon followed.

Historical changes

As with most other aspects of religion, churches are rarely static institutions. The size, activities and wider societal context of many churches has changed through time in response to shifting needs and opportunities.

One interesting dimension of church dynamics is the way in which congregations alter through time. For a variety of reasons, some related directly to local factors (such as the immigration or emigration of local people) and some related to wider change (such as broad demographic trends or religious revival), congregations can grow or shrink.

Growth and relocation

Numerical growth often requires physical expansion of the church building, which can be achieved by extending existing buildings, converting others from existing uses, or building new ones.

Historical records can be used to reconstruct patterns of church growth through time. An example is Proudfoot's (1983) study of the influence of population pressure on church size in Warwickshire (England) between about 1200 and 1535. Periods of local population growth are mirrored in phases of greater resource allocation to church building and in construction of larger aisles and naves in existing churches, paid for by parishioners.

More recent changes often reflect the dynamics of contemporary population change. Berdichevsky (1980), for example, has examined the changing distribution and character of churches in New Orleans between 1957 and 1977. The city's diverse cultural heritage (including Creole, Anglo, Black and European immigrants) has endowed it with many historic churches and synagogues, the most prominent of which is the St Louis Cathedral, facing the Mississippi River. Rapid suburban growth since 1960 had been a catalyst for significant changes in the religious landscape, including new church construction and the closure of 27 inner-city churches. The racial composition of many surviving congregations has changed markedly, and exotic denominations (including the Jehovah's Witnesses, Hare Krishna, and Black Muslims) have started churches in the area. Church architecture has changed, too, for largely practical reasons. Berdichevsky bemoans the fact that

> today's modern suburban churches are essentially functional, and geo-metric in design, constructed of standard materials, and often without any distinguishing denominational or ethnic characteristics. They are often completed in less than a year and require large parking lots. . . . Gone forever is the lavish care, patience, and craftsmanship which were the hallmarks of church construction in the past.
>
> (Berdichevsky 1980: 53)

Decline and recycling

Whilst numerical growth requires expansion of church plant, numerical decline creates surplus capacity and can ultimately lead to the closure of churches. When decline sets in church income also falls, so that declining churches often face real financial difficulties and resource shortages.

One solution to the financial problems posed by congregational decline is to establish yoked parishes, in which participating churches share resources (including personnel and programmes). Cantrell, Krile and Donahue (1982) examined how such a scheme worked in 131 rural congregations in Minnesota. They found (ibid.: 83) that 'the typical yoked church arrangements involved a primate church, larger and usually located within the limits of an incorporated community, and usually one or more smaller churches located either in the open country or in smaller neighbouring towns'. The formation of yoked parishes does help individual congregations to survive as identifiable units, but there are costs for both the church (the pastor spends less time in each church, provides less organisation) and the rural community (community activity is reduced because the pastor's role is constrained).

Closure is often the only option when church congregations fall below a threshold size. Some religious landscapes (particularly in rural areas) are largely relic, the churches having been abandoned and taken over by other functions. Jordan and Rowntree (1990: 218) note that 'along the United States–Canada border west of the Great Lakes, some former churches are now used as granaries, clinics, American Legion halls, garages, and apartments. In the Soviet Union, many have been converted into museums.'

Many redundant churches in the United States have been recycled for secular uses, and field studies by Foster (1981, 1983) in Minnesota and Manitoba (Figure 7.2) throw some light on how the process occurs. Large numbers of churches were built in rural areas to serve scattered populations and immigrant minority groups, particularly during the nineteenth century. Shrinking congregations have forced the closure of rural churches for regular worship services, and this trend has increased across the American Midwest and Canadian prairies since 1945. Many of the closed churches continue to be used periodically, for occasional marriage, burial or anniversary services, but maintenance can be costly and security problematic. Vandalism and neglect lead to the collapse of some abandoned buildings. An alternative fate for closed churches is sale and re-use for secular purposes. Field studies revealed a diversity of different subsequent uses, including residences, community centres, day-care centres, senior citizen centres, and museums. Some (in Manitoba) were even used as farm buildings, including chicken coops, hog barns, and granaries.

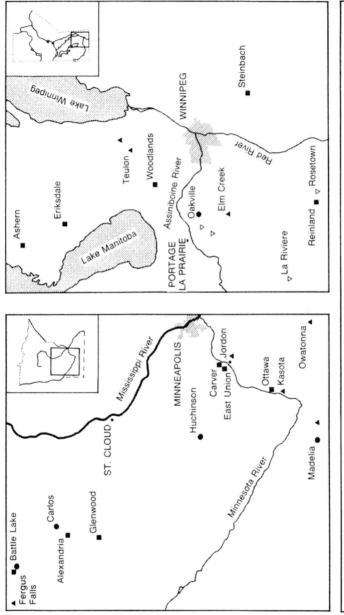

Figure 7.2 Distribution of recycled rural churches in parts of Minnesota and Manitoba in the early 1980s

Source: After Foster (1983).

LANDSCAPES OF DEATH

Landscapes of worship are clearly important and visually striking ingredients of the geography of religion within specific areas. Churches, temples and other religious structures not only dominate the appearance and character of many areas, they also serve as valuable indicators of both religious and cultural changes through time and space.

Much the same is true of the various landscapes of death, which reflect attitudes and practices concerning both the living and the dead. This might seem like a morbid or marginal area of study for geography, but it is a profoundly important dimension of the link between geography and religion because there are great variations between religions in the way they treat their dead.

Geographers have devoted more attention to landscapes of death than to landscapes of worship, which hints at how fertile an area of inquiry this promises to be. Indeed, the area has even attracted its own vocabulary, with Kniffen (1967: 427) and Jordan (1973: 153) writing about the so-called 'necrogeography' of an area, and Bhardwaj (1987b: 321) describing the meaning of 'necral space' (as 'space related to the deceased, whether an individual grave, a tomb, a cemetery or a cremation ground').

Most geographical attention has focused on burial practices in general, and on cemetery landscapes in particular. Before we explore that material, it is useful to reflect on attitudes towards death and whatever comes after it. This is, after all, perhaps the most fundamental question that any individual has to ponder!

The dead and the afterlife

Dying is the one thing that all humans have in common, other than being born. It is unavoidable if unwelcome, and although modern medicine can sometimes prolong life it cannot do so indefinitely. Cryogenic suspension and similar techno-fixes can only postpone the process of 'shuffling off this mortal coil', they cannot eliminate it.

Causes of death have little relevance to our discussion here which looks at how the deceased are disposed of and how the world of the dead is perceived.

Disposal of the dead

Different cultures and religions dispose of their dead in different ways. Cremation, involving the burning of the deceased, is an ancient custom which has been revived in modern times (particularly where land for burial is in short supply, or where hygienic considerations favour it). Hindus and Buddhists have traditionally cremated their dead. Traditional forms of cremation, in which funeral pyres are constructed where necessary and the ashes are scattered, leave

213

no obvious imprint on the landscape. Modern Western forms are based on crematoria, permanent structures in the landscape that might include chapels, gardens of remembrance, roads and other infrastructure.

Burial is more common than cremation, and it creates its very own landscapes or structures that dominate surrounding landscapes. Amongst the most spectacular architectural forms are above-ground tombs, usually built to house dead leaders and other important people. Famous examples include the pyramids of ancient Egypt, vast brick or stone edifices with inner chambers and subterranean entrances. The Pyramid of Cheops took twenty years and an estimated 100,000 men to construct; the Egyptian King who built it as his tomb clearly intended to be remembered! Lesser examples of above-ground tombs include numerous mausoleums, built to house the remains of members of distinguished families. Islam entombs the deceased and has created some extremely imposing landscapes of the dead. The best known Muslim tomb is the Taj Mahal, a white marble mausoleum built at Agra in India by Shah Jehan in memory of his favourite wife.

Christians, Jews and Muslims typically bury their dead in graves, on land set aside specifically for that purpose (cemeteries). Monuments (including headstones and statues) are erected to mark where the deceased are buried. Much of the rest of this section is devoted to the subject of cemetery location, design and architecture.

In some European cities, such as Rome and Paris, remains of the dead are entombed in catacombs (Jordan 1973). These are underground burial places consisting of tunnels with vaults or niches leading off them for tombs. An estimated 6 million skeletons are housed in the Catacombs, a dimly lit underworld beneath Paris's Left Bank. An alternative form of burial, particularly for the rich and famous, is in the walls and floors of church buildings.

Attitudes towards afterlife

Some of the observed variations in ways of disposing of the deceased reflect differences in attitudes towards whatever follows death. Simard (1984) suggests that many aspects of human activity during life are influenced if not shaped by religious imagery dealing with 'the next world', which in many religions is viewed as real territory. Bhardwaj (1987b) goes further in pointing out that although attitudes towards the 'hereafter' vary from one culture to another, this whole realm has received remarkably little comparative study.

Attitudes towards the afterlife are sometimes heavily influenced by mythology. In Greek mythology, for example, the souls of the dead were ferried by Charon (the ferryman) across the Rivers Styx and Acheron to Hades, the underworld where they would henceforth live. In Scandinavian (Viking) mythology, the souls of warriors who die as heroes were transported to Valhalla, the great hall of Odin (the supreme creator god) where they would dwell eternally.

Religion is the primary shaper of attitudes towards the afterlife amongst believers. Buddhists, certain Hindus, and some others believe in the trans-migration of souls, in which after death the soul of a person passes into the body of a new-born infant or an animal. The cycle of reincarnation continues, in which the soul survives but the body is just a temporary vehicle for it. In Christianity the souls of the departed are believed to pass to Heaven, the perfect spiritual realm of God, where they remain through eternity. Catholic theology recognises an intermediary place of torment or cleansing (purgatory), where evil deeds are punished before the soul is able to enter Heaven.

Ancestor worship of deceased family members, clan chiefs or ancient rulers is prominent in some religions (Meine 1957). The ancient Romans, for example, made ancestral deities into images, set them up in the home and appeased them with offerings. The ancient Egyptians embalmed the bodies of the dead with great care and held them in veneration. Ancestor worship was traditionally important in China and Japan, where departed souls were offered food, drink and prayers every day in the hope that they would aid the living.

Different attitudes towards the afterlife are illustrated in cultures that bury possessions with the deceased, most commonly to help the departed in their future existence. Grave-goods discovered in many Anglo-Saxon warrior graves in England, dating from the fifth to the eighth centuries, include dress ornaments, toilet implements, containers, tools and weapons (Harke 1990). Archaeological excavation of much earlier sites in North America – in north-western Vermont, radiocarbon dated at between 700 BCE and 100 CE – reveals bodies buried with lavish personal decoration including native copper, marine shell beads, and personal garments (Heckenberger, Petersen and Basa 1990).

Implications for landscape

The sort of differences we have seen above in religious attitudes towards the dead and the afterlife inevitably translates into landscape differences, some of which are quite obvious. The most fundamental and visually striking distinc-tion is between those funerary practices (such as Christian burial) which seek to preserve and localise the body, and those (such as Hindu cremation) which seek to commit it to the cosmos without leaving any locational trace behind. Bhardwaj (1987b) refers to this as the cosmicisation/localisation polarity, and gives some illuminating examples of each.

Hindu cremation practices leave no physical trace of the dead person behind in the world of the living. According to Hindu belief the individual no longer exists after death, the soul casts off its worn-out body and will enter a new body. Consequently, in India there are no monuments to the dead or tomb-stones to mark the site. Cremation sites are re-used many times, and the Hindu cremation ground is regarded as ritually impure.

This contrasts markedly with Christian practice, such as the American cemetery, with its permanent gravesites in which each occupant is clearly identified and the personal space protected in a well-maintained, well laid-out, consecrated memorial park. Christianity exerts strong influences over landscapes of death because, as de Blij and Muller (1986: 215) point out, 'no faith (other than perhaps those of China) uses land so liberally for the disposition of the departed'. The remainder of this section deals with the geography of cemeteries, which differ a great deal in size, situation, character and architecture.

Cemetery locations

Amongst the most obvious geographical studies are those which examine the location of cemeteries, and factors that control this. Archaeologists have long been interested in the evolution of cemetery landscapes and patterns, because they are widespread and reveal a lot about cultural attitudes and technical abilities. Cooney's (1990) examination of megalithic tomb cemeteries in Ireland illustrates this line of inquiry.

Site selection using practical criteria

Geographical studies of contemporary cemetery landscapes and patterns are less numerous and more context-specific. Darden's (1972) case study of the location of cemeteries in Pittsburgh is one of the most detailed yet published. On the basis of field observations, analysis of documentary sources and interviews with cemetery managers and funeral directors, it raises interesting questions about how and why spatial patterns of this cultural phenomenon change through time. Three main cultural factors influenced location and type of cemetery – proximity to the majority of potential users, ease of access, and cost of land – and they varied in significance for different types of cemetery. Cost was the most important factor in the location of corporation cemeteries, but the least important factor among family cemeteries. Proximity was the most important factor in sectarian cemeteries, and ease of access was important in both sectarian and non-sectarian cemeteries (most cemeteries in Pittsburgh were located on or near main lines of transportation).

The primary locational factors were topographic, because most Pittsburgh cemeteries were located on fertile land (suitable for grass and trees) characterised by steep slopes or relatively high elevation. High elevation appeared to be the most important factor, because of fear of flooding, fear of encroachment by other land uses, and tradition. Founders of cemeteries used similar site selection criteria, including an elevated and relatively secluded position, lack of pollution, accessibility and fair price.

Each community in Pittsburgh had its own cemetery, and the clustered pattern of cemeteries in 1850 closely mirrored the pattern of settlement. Population increase was matched by the establishment of more cemeteries, and some were relocated to make way for the city's expansion. By 1900 the cemetery pattern was much more dispersed, and cemeteries moved towards the periphery as other land uses penetrated the interior.

Site selection using cosmological criteria

The Chinese traditionally used a radically different approach to the selection of burial sites, based on geomancy. Sites were chosen not on practical or cultural criteria, but mystical or spiritual ones. Each burial site must display the *Feng Shui* (literally 'wind and water'), 'the perfect combination of tangible and intangible elements that would leave the dead in harmony with their surroundings. It is believed that if the Feng Shui of a grave site is wrong, the dead will be restless and their descendants will suffer' (Jordan and Rowntree 1990: 220). The ideal Chinese grave site, with good *Feng Shui*, would be neither featureless and flat, nor steep and rugged.

The ideal site would be properly surrounded by *Yin* and *Yang*, the active and passive energy forces of Chinese cosmology. *Yang* energy, representing positive, bright and masculine dimensions of the universe, is expressed topographically as a tall mountain range (called the Azure Dragon). *Yin* energy, representing the negative, dark, feminine dimensions of the universe, is expressed as a lower ridge (called the White Tiger). In Chinese cosmology, the interaction of *Yin* and *Yang* is thought to maintain the harmony of the universe and to influence everything within it. Consequently, the best *Feng Shui* topography is a secluded place where the two energies converge, interact and are maintained by surrounding mountains and streams.

Lai (1974: 506–7) notes that 'no traditional Chinese would choose a site for a tomb without first studying its *Feng Shui* . . . [because] the fortunes and misfortunes of succeeding generations depended on the Feng Shui of their forefathers' graves'. He tested the *Feng Shui* topographic model, using air photos, maps, archive material and land registry records, to pinpoint a site suitable for a Chinese cemetery in Victoria, Canada. The site (Figure 7.3) was not used because of objections by local residents, but its location clearly conformed to the cosmological model. Lai stresses that

> in traditional China, *Feng Shui* was a powerful determinant in the location of tomb sites for royal families, officials and commoners. It also played a role in the location and layout of individual houses, temples, palaces, hamlets, villages and cities.
>
> (Lai 1974: 512)

Cemetery landscape and architecture

Much more of the literature on landscapes of death is devoted to cemetery architecture and morphology than to cemetery location. The long archaeological tradition of studying burial sites for vital clues to the cultures that

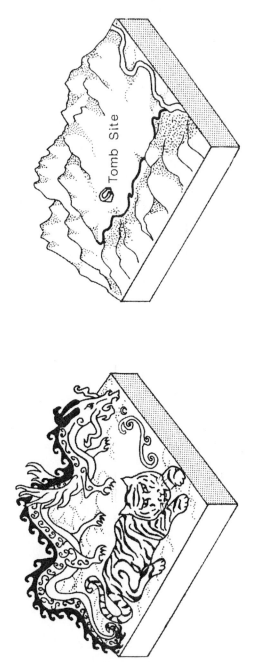

Figure 7.3 Selection of an ideal tomb site according to the Chinese principles of *Feng Shui*

Source: After Lai (1974).

created them continues, and it is illustrated in Whittle's (1990) description of pre-enclosure burials at the Neolithic Windmill Hill site in southern England. Cooney's (1990) analysis of megalithic tomb cemeteries in Ireland dating from 3800 to 2000 BCE shows clustered patterns and suggests organised development reflecting social interaction and stratification.

As well as providing invaluable time-capsules, cemeteries can be significant symbols on the landscape. Jordan notes how

> amongst the most striking landscapes in Europe are those devoted to the war dead, as for example the sea of identical small white crosses which mark American military cemeteries in parts of western Europe, the huge Soviet memorial at Treptow Park in (former) East Berlin, or the almost inevitable stone shafts in the commons of rural German villages commemorating under the iron cross the list of those who fell in world wars.
>
> (Jordan 1973: 154)

Amongst the more relevant themes which geographers have examined are the evolution of cemetery morphology, cemetery architecture, and residential segregation within cemeteries.

Evolution of cemetery morphology

Most studies of cemetery evolution – including those by Price (1966), Jackson (1967), Francaviglia (1971) and Howett (1977) – have focused on the United States. Price (1966) carried out detailed fieldwork on 214 rural cemeteries in south-eastern Illinois, and suggested a fourfold classification based on size and period of most active use. The earliest and smallest type, which he termed 'undifferentiated', was most common before 1860 and contained up to ten graves. 'Small family plots', with up to twenty graves, dominated between 1860 and 1880 and were replaced by the larger 'rural activity focus' cemeteries (with up to 250 graves) between 1880 and 1900. From 1900 to at least 1950 the 'population centre' cemeteries, with at least 250 graves, were dominant.

This pattern of evolution – from small, local family burials towards ever-larger community cemeteries – is reflected in many parts of the United States. The earliest years of settlement saw most families bury their dead in individual plots set aside on farms or close to homes. As New World towns grew and churches flourished, burial grounds moved to the churchyard or town commons.

Jackson (1967–8) has noted how many aspects of burial practice in Colonial New England resembled aspects in contemporary England (from where they had been imported). The most common form of burial was in the churchyard, but solitary graves and family burial grounds far from the villages were common too. Graveyards were usually located in the centre of the town or village and were available to all members of the community. They were functional rather than aesthetically attractive places – lawns were not well kept, there was no ornamental planting, and grave markers were simple and inform- ative (recording the name and sometimes the profession of the deceased) rather than grand and decorative.

Many American town and city churchyards quickly became overcrowded and unsanitary, and the need to establish new burial grounds on the city margins was widely recognised. New developments in public transportation, particularly along new roadways, made such relocation possible and attractive.

America led the world in this particular aspect of town planning, because the world's first properly planned and landscaped parkland cemetery was estab- lished in New Haven, Connecticut in 1796. The so-called New Burying Ground was designed and paid for by James Hillhouse, a wealthy local citizen. Situated in what was then a remote location, the large square plot was divided by smooth walks into small squares, which were further subdivided into smaller squares (often occupied by families). The design was regular and orderly. Many squares were bordered by Lombardy poplar trees, and some had weeping willows in the centre. Graves were of equal size, and aligned in the same direction. The ground was levelled, the grass was cut regularly, and rails around squares were painted black and white.

As Jackson (1967–8) points out, the layout devised by Hillhouse became standard in most American cemeteries and its real novelty was not its design but its 'non-public, almost domestic quality'. One prominent aspect of that quality was the elaborate markers used – including obelisks, draped urns, columns, statues – which were symbols of wealth and taste as well as inform- ative grave markers. This use of sculptural monuments in cemeteries was a sign of things to come, because new parkland cemeteries were soon to be built, modelled on Père Lachaise in Paris (a heavily ornamented wooded park that opened in 1804) and heavily influenced by English landscape gardening (Howett 1977).

The first American 'rural cemetery' – the Mount Auburn Cemetery in Cambridge, Massachusetts' – reflected this new parkland trend. Laying out the new cemetery began in 1932, and the early graves with simple markers were scattered through woods on the heavily undulating (glacial moraine) topography (Howett 1977). Extensive landscaping of the area began in 1835, and it involved draining of waterlogged areas, grading of hill slopes, filling of depressions, cropping hilltops, thinning of trees and extensive replanting. The Cemetery occupied part of the botanical garden founded by the Massachusetts Horticultural Society in 1825, which had long been a popular excursion venue for Boston residents (Jackson 1967–8). What became important now was the overall setting of the grave, not the grave itself, within a tranquil and grandly ordered landscape.

Many American cemeteries adopted the Hillhouse-type design as both practical and pleasing. But these monumental gardens represent one step in the evolution of modern cemeteries, which was to be replaced by the lawn cemetery. These are extensive grassed areas that are subdivided into individual sections by sculptures and small gardens (Howett 1977). They are noted mainly for their great diversity of monuments, sculptures and outdoor furniture.

More recently the memorial garden (or park) has become common. This type of 'entrepreneurial heaven-on-earth' (Howett 1977: 12) is run commercially and yields good rates of return, because graves are packed tightly together (up to 1,500 grave spaces per acre) on cheap land on the outskirts of large cities. Landscaping is minimal and functional, comprising a flat lawn with small bronze markers and containers for cut flowers sunk below turf level (for ease of grass cutting). Jackson (1967–8: 26) comments that 'often such memory gardens provide all the necessary funeral services in a central building – embalming, cremating, a funeral chapel; a supply of artificial flowers is always on hand, and there is someone to conduct the service; everything is taken care of'.

Cemetery architecture

Cemetery morphology varies through time and space, and so does the architecture of structures within cemeteries. The most common structures are the grave markers or headstones. Because headstones include information on when the person died, they offer useful prospects for absolute dating that have been explored in biogeography (such as lichen colonisation), geology (such as rates of natural weathering on different stones) and historical geography (such as reconstructing demographic change in an area). Headstones also provide important information about contemporary economy and technology, and about attitudes towards death. Nakagawa (1990b), for example, has commented on the utility of tombstones for reconstructing settlement history and attitude changes in rural Japan.

Graves in Britain were marked with simple wooden crosses or graveboards up to the late seventeenth century, when gravestones became fashionable. The early gravestones were made of local material, and reflected the rise of a prosperous middle class who could afford the cost of transporting the heavy material from a local quarry (Dove 1992). Gravestone design also changed through time, evolving from very plain designs to elaborate chest tombs and carved headstones that were common in the late eighteenth century. Simpler forms came back into fashion during the nineteenth century, and have remained so since.

A number of detailed studies of gravestone size and design have been published, which illustrate diffusion of taste as well as development of technology. Dethlefsen and Deetz (1965–66) examined the traditional designs inscribed on gravestones erected in eastern Massachusetts during the seventeenth and eighteenth centuries, and found evidence of the progressive break-up of strict Puritan ideas. Price (1966) found a sequence of tombstone styles in Illinois which mirrored period and size of cemetery type. Before 1840 the only tombstones used were of a durable local sandstone; most were simply crude rocks apparently carved by a single local craftsman. A plain marble marker appeared during the period 1840 and 1900, and an obelisk-shaped marker of marble or granite was prevalent between 1880 and 1900. By 1930 a lower and wider stone (usually granite) was dominant. Brass or bronze plates set flush with the surface are recent innovations that improve ease of ground maintenance.

The most detailed study of cemetery architecture, by Francaviglia (1971), looked at five cemeteries in a mixed Protestant area of Willamette Valley, Oregon, dating back to around 1865. He identified nine different types of tombstone (Table 7.1) which accounted for 95 per cent of the monuments in the study cemeteries.

Francaviglia (1971) detected four phases in the metamorphosis of tombstones in Oregon. During the pioneer period (1850–1879) cemetery monuments were mainly gothic, tablet and block forms, made of white marble with little ornamentation. The Victorian period (1880–1905) saw a radical change with the arrival in Oregon of ornate architectural styles and patterns. Obelisks and columns became popular in cemeteries, with ornate wrought iron enclosures for family plots. Simplicity was the hallmark of the conservative period (1906–1929), when geometric forms (rounded scroll, slant pulpit, block form) were popular. Much of this simplicity has survived into the modern period (1930–1970), when plaques flush with the ground surface have become more common than raised tombstones.

Residential segregation within cemeteries

In death, as in life, residential segregation according to wealth, religion or other socio-economic criteria groups together people with shared attributes and separates them from others (usually lesser mortals). Residential segregation in

Table 7.1 Types of tombstone in Oregon cemeteries

---

1 *Gothic*; pointed upright, usually marble, *c*. 90 cm high; pointed arch
2 *Obelisk*; vertical shaft of marble, *c*. 150 cm high, topped by pyramidal point or ball
3 *Cross-vault obelisk*; marble spire *c*. 150 cm high topped by a cross-vaulted roof
4 *Tablet*; usually marble, rounded top; average height 70 cm
5 *Pulpit*; marble or granite, *c*. 67 cm high, resembling a pulpit (perhaps with inscription on top)
6 *Scroll*; usually granite, resembles a horizontal broken column; *c*. 30 cm high
7 *Block*; tabular block of granite, 60 cm high by 60 cm wide by 15 cm deep
8 *Raised-top inscription*; usually granite, *c*. 15 cm high; top inscription horizontal
9 *Lawn type (plaque)*; plate, usually of granite or bronze, top flush with or below 2.5 cm above ground

---

Source: Francaviglia (1971).

cemeteries immortalises the inequalities inherent within most |
nities, and hints (wrongly) that such dimensions of humanhood
the afterlife.

Such segregation is sometimes based on colour. Pattison (195?) reported that
almost all cemeteries in Chicago (particularly the non-Catholic ones) had
complete segregation between blacks and whites, and Price (1966) described
the separate Negro cemeteries in southern Illinois. Francaviglia (1971) notes
that in some Oregon cemeteries blacks are given peripheral graves in 'bad'
neighbourhoods, and in some cases admission is completely denied. Segregation in cemeteries can also be based on religion. Pattison (1955) found that
26 out of the 70 cemeteries in Chicago were owned by churches, most of which
restricted burial to church members.

More commonly segregation is based on wealth, because the price of burial
lots generally varies a great deal, even within individual cemeteries. Kephart
observed how

> traditionally, class distinctions within the cemetery were based on size of
> lot and size of memorial or mausoleum. Historically, the rich man's grave
> was marked by a large memorial or mausoleum, the poor man's by a
> small head or footstone or perhaps by the absence of a stone.
>
> (Kephart 1950: 642)

Young (1960: 447) echoes the point and notes how 'cemetery lots are differentially priced and vary greatly in their elaboration. Grave markers, in
particular, can reflect the wealth and prestige of the buyer, and very likely a
whole family'. Mitford (1963: 128) discusses lot selection within cemeteries,
pointing out that 'there are "view lots" and "garden locations" for those who
aspire to be housed among the comfortable well-to-do, nice roomy "memorial
estates" for the really rich, and crowded plainer quarters for those accustomed
to tract housing'.

Hardwick, Claus and Rothwell (1971) note that prestige and status are
important factors in choosing a final resting place. They offer the example of
the military sections of some American cemeteries, where the most expensive
lots are those closest to the main flag pole. Other contributory factors include
personal preference, scenery, adjacent property improvements (such as religious
statues and landscaping), social class of the neighbours, and racial and religious
discrimination.

Analysis of plot sizes and locations within cemeteries can help in the delimitation of 'good' and 'bad' neighbourhoods within the cemeteries. There is
much potentially interesting geographical work to be done on such residential
segregation within cemeteries. As Francaviglia concludes,

> in the cemetery, architecture, 'town' planning, display of social status,
> and racial segregation, all mirror the living, not the dead. Cemeteries,

as the visual and spatial expression of death, may tell us a great deal about the living people who created them.

(Francaviglia 1971: 509)

Cemeteries in Louisiana: a case study

The cemeteries of Louisiana in the United States offer an interesting case study of cemetery form and practices that reflect both cultural diversity and topographic controls. One hallmark of the cemeteries in some parts of Louisiana is the above-ground burial vault, traditionally explained as a response to high water tables. The old St Louis Cemetery in New Orleans is a good example of an early cemetery that departed from the traditional churchyard plan, with its elaborate brick family vaults (used for several generations) which were usually plastered and whitewashed, and roofed in assorted period styles (Howett 1977).

Throughout Louisiana there are marked cultural influences on burial practices, described by Kniffen (1967). French and Catholic south Louisiana has large central cemeteries composed of above-surface white vaults on sanctified ground adjoining rural and small-town churches. These practices have largely extended to smaller congregations of Anglo-Saxon Protestants (particularly Episcopalians) within the general area. Throughout the rural north of Louisiana, however – as throughout the upland South – the dead are interred beneath the ground in graves with a general east-west orientation and plain markers (with no crosses). Urban cemeteries in both north and south Louisiana to a large extent reflect rural regional practices.

Detailed analyses of the cemeteries of Ascension Parish at the eastern end of French Louisiana, by Nakagawa (1988, 1989), demonstrate smaller-scale regional variations that again reflect both cultural and environmental influences. The population composition within Ascension Parish is not reflected in number of cemeteries of each denomination – only 1 of the 39 cemeteries is Catholic whilst over six out of ten people are, and most (31) cemeteries are Protestant (11 are for whites, 19 are for blacks, and 1 is mixed racially). The traditional burial form in this area is below-ground, in a grave 4–6 feet deep. In French Louisiana the traditional form is the above-ground burial tomb, some up to 5 storeys high. Detailed analysis of the distribution of the above-ground tombs suggests that cultural factors rather than a high water table provides the best explanation. Selective diffusion, particularly among Catholics, appears to have introduced the above-ground tombs into the area from New Orleans, initially into Ascension Catholic Cemetery. The French Catholic influence has been very strong, because even black Protestant cemeteries in the area have many Catholic crosses.

Further studies by Nakagawa (1990a, 1990c) have focused on Louisiana cemeteries as cultural artefacts. Detailed analysis of 236 cemeteries in the north and south of the state suggest that cemetery landscapes are expressions of culture, reflected through the group identity which individuals freely adopt and display. Tangible expressions of this identity include cemetery architecture, such as the distribution of grave sheds, shell decoration, toy decoration, flower-pot decoration, and photographs of the deceased. The results suggest that cemetery landscape is a revealing but hitherto largely unexplored source of information about culture and group identity.

Cemeteries and land-use planning

Cemeteries are a distinct type of land use that affects surrounding areas. Not everyone wishes to overlook a cemetery from their home, despite the fact that most burial grounds are quiet and dignified places that offer peace and tranquillity to both the living and the deceased. This is true particularly with publicly owned paupers' cemeteries, in which people are buried without flowers, clothes, graves or names, and – ultimately – without dignity (Kephart 1950).

Studies have shown that proximity to cemeteries can strongly influence residential attractiveness, which is reflected in land values. Hardwick, Claus and Rothwell (1971) estimate that (at 1970 prices), a cemetery may gross over $3 million an acre (based on around 3,000 lots per acre at $1,000 per lot, perpetual care included). These prices are higher than the cost of land at the prime intersection in the Central Business District.

Land-take for new cemeteries is a common land-use problem, particularly in areas with rapid population growth. Some of these problems are described by Pattison (1955) in a case study of the evolution of Chicago's 70 cemeteries. After the town was formally established in 1835, population increased tenfold in two years. The first burial grounds were located centrally, along the rivers and in wind-blown sands. But soon interment in the city area was forbidden on health grounds (after a cholera epidemic) and two new burial tracts were laid out north and south of the town. By 1850 the population had increased tenfold again (to 30,000), and the city was rapidly expanding outwards. A new cemetery was established two miles north of the city centre and the contents of the older municipal cemeteries were relocated there. Part of this new cemetery was bought by a Jewish burial society, and the Roman Catholic Church founded a separate cemetery. Each of the three 1850 cemeteries had closed and been converted by 1900 (the abandoned municipal cemetery, for example, became the core of Chicago's first public park). They were replaced by three privately owned non-sectarian cemeteries, and complemented by ten other privately operated, three new Roman Catholic and several Jewish cemeteries. The earlier cemeteries were evacuated and their contents removed to new sites away from the city centre. Between 1900 and 1950 the number of cemeteries grew from 27 to 70 and all but ten of the new ones were business (rather than church) enterprises. The total area of occupied burial land was increasing at a rate of 25 acres per year in the mid-1950s, and this figure is certain to be much higher today.

Jewish cemeteries also pose planning problems because in Judaism religious law strongly governs both the location of cemeteries and their internal spatial structure, as Newman's (1986) study of Tel Aviv clearly demonstrates. Jewish religious law (*halacha*) forbids the use of cremation for the disposal of corpses and requires that death be followed by internment in the ground (reflecting belief that the original man was created from the 'dust of the earth' and must return there). Jewish cemeteries are exclusively for use by Jews, and different sub-communities often have their own sections within them. Religious laws dictate that burial must take place as soon as possible following death, preferably on the same day, and forbid funerals on the Sabbath (Saturday) or on other Jewish holy days. Cemeteries must therefore be located close to major settlements, to avoid delays. But religious laws also require that cemeteries cannot be located adjacent to synagogues or close (within 23 yards) to houses.

Designation of a Jewish cemetery as consecrated ground is in perpetuity, and changes of land use are forbidden even if the cemetery is no longer functional (even on rediscovery after 2,000 years of non-use). Plots cannot be reused (unlike in Muslim burial grounds, where plots can be reused after about 30 years), so the land-take requirements continue through time, and recent searches for suitable land have concentrated on the urban periphery. Conflicts arise where proposals are made to designate new Jewish cemeteries, particularly in areas where the more powerful political and financial groups reside.

Cemeteries are sometimes viewed as externalities, though Newman (1986: 104) stresses that they are psychological rather than physical ones because 'death has a mystical aura about it, and residents prefer to be associated with the living and the vibrant rather than with the morbid and the stagnant'.

RELIGION AND THE GENERAL LANDSCAPE

The impact of religion is not confined to obvious landscape features such as places of worship and death, although these inevitably display the imprint of religious factors perhaps more clearly than many other types of landscape. In this section we will review some of the wider issues, including the impact of religion on settlement patterns and landscape planning, by looking at both historical and contemporary examples.

Study of the religious imprint on the cultural landscape is a fairly widely accepted area of inquiry within geography, as we saw in Chapter 1, (see pp. 2–3) so this theme has attracted the attention of many geographers. Indeed, Isaac (1960) regards it as central to the geography of religion. He illustrates this claim with the example of Japanese religious landscape design. In creating their gardens, the Japanese have traditionally used objects considered *kami* (miraculous and sacred in themselves) as pivots around which to mirror their religious conceptions and symbolism. Thus plants, rocks, mountains and ponds are used with both spiritual and aesthetic purposes, to great effect (see p. 207).

Some religious landscapes are highly distinctive and owe their origin and character to the religious imprint. The distinctive Hindu cultural landscape, for example, is described by de Blij and Muller (1986: 208) who emphasise how 'temples and shrines, holy animals by tens of millions, distinctively garbed holy men, and the sights and sounds of endless processions and rituals all contribute to an atmosphere without parallel; the faith is a visual as well as an emotional experience'.

Many religious imprints are much more subtle than the design of Zen gardens or Hindu temples. Even traditional town forms in the United States bear some hallmarks of religion because the original New England settlements (much copied by settlers as they migrated west) were designed to be large enough to support a church and its congregation yet small enough to allow rural folk to walk or ride to church conveniently (Lewis 1980).

Religion and settlement patterns

Settlement form and appearance in many areas have been strongly influenced by religious factors, both implicit and explicit. In this section we look at some examples that illustrate such influences in a variety of different contexts.

Indian villages

Traditional Indian villages provide one of the most graphic examples of the influence of religion on settlement form. As Mukerjee (1961: 396) points out, Indian 'villages and towns were often planned according to a religious symbolism, which governed the layout of wards and streets, the location of temples, monasteries and village halls, as well as of open spaces, tanks and gates'. The temple of Vishnu is traditionally erected at the centre of the town, with deities at its gates to protect the town from disease and anarchy. The temple is the spiritual source of energy for the town, but it also had important socio-cultural purposes because its assembly halls provides a venue for meetings of the royal court and town or village assembly. Nearby would be guest-houses where pilgrims would stay for a day or two, and many town centres also contain temple avenues and quadrangles where merchants sell their goods.

Malshe and Ghode (1989) describe the influence of religion on the structure, function and land use of Pandharpur, one of the most revered places in Maharashtra that attracts vast numbers of Hindu devotees. The site started with an ancient temple on a mound, high above likely river flood levels. This grew into a small rural settlement and from the fifth century onwards the town grew in religious importance. This historic religious evolution is well reflected in the zonation of land use within the town, with a religious core centred on the river banks. Here are concentrated a number of temples and sacred bathing places (*ghats*), along with shops selling religious goods (such as flowers and incense sticks).

Judaism and settlement form

Judaism generally has a much less obvious impact on settlement form than does Hinduism, but there are some striking examples. One of the most distinctive forms of settlement in Eastern Europe is the *shtetl* or Jewish small town or quarter. Bar-Gal (1985) describes a typical architectural pattern centred on the synagogue, a ritual bath, a religious school (*yeshiva*) and a market. The *shtetl* has narrow streets and densely clustered housing with an internal orientation based on interior courtyards. Large buildings contain multiple residential units, shielded from view by casual passers-by. Entry to the courtyard could be controlled, and each unit was organised and promoted by a religious leader around whom the settlers rallied. A traditional Jewish pattern of life continued within this sheltered environment, which perpetuated a close-knit social fabric.

The *shtetl* is an attempt to preserve religious identity and lifestyles by adopting a distinctive approach to physical planning. A radically different form of planning is relevant to major cities, particularly where Judaism is not necessarily the only (or the dominant) religion. Jerusalem is a particularly good example because, as Efrat and Noble stress, it is

> sacred to at least three major religions [and] has long been a source and a scene of contention among the adherents of these faiths and their political sponsors. During the past half century, each of the three religions, represented by a Christian, a Jewish and an Islamic polity, has attempted to determine the orientation of development in the city. Each effort has had only limited success.
>
> (Efrat and Noble 1988: 387)

Jerusalem contains three main areas. The Old City, defined by the massive encircling wall built in the early sixteenth century, comprises the Armenian, Christian, Jewish and Muslim quarters and the Temple Mount. East Jerusalem, beyond the City Walls, is mostly populated by Arabs. West Jerusalem, including Mount Zion, is mainly Jewish. The city has numerous well-defined residential neighbourhoods, even within the Jewish community (particularly where Hasidic Jews live). Each major phase of the city's recent history has left its mark on physical planning, including the British-designed city plan (after the First World War) which limited building height to 35 feet. Political and religious divisions from 1948 to 1967 inspired separate types of development in each political sector, and the ambitious 1959 master plan remained unfulfilled. An attempt was made in the (post-reunification) 1968 plan to restore Jewish prominence by reconstructing the Jewish quarter of the Old City that had been badly damaged during the 1948 fighting, building the new Hebrew University on Mount Scopus and establishing new Jewish neighbourhoods.

Monasteries and colonisation

The Christian Church has traditionally built monasteries as strategic bridge-heads from which to colonise and settle largely undeveloped lands, and these have often served as nuclei around which new settlements have evolved through time.

This was certainly the case in the Odenwald region east of the River Elbe in Germany where – as Nitz (1983) describes – the Benedictine Abbey of Lorsch played a key part in colonisation and opening up the area. Settlements were introduced as early as the late eighth century when the imperial monastery of Lorsch started to colonise a royal forest donated by Charlemagne. They were *Waldhufen*, row settlements with strip farms incorporating portions of the forest. Over a period of 350 years the abbey colonised about a hundred new settlements as forest was cleared and wilderness was converted into a cultural landscape. The abbey soon became the largest landowner in the area as aristocratic families made grants of fields, vineyards, farm-holdings and some-times even their entire property. A systematic network of central places and castle settlements was also established within the colonisation area. Through time the settlement forms spread by diffusion to other areas, often via personal contacts of territorial lords with Lorsch.

The evolution of settlements in many parts of England was also heavily influenced by the growth of monastic orders, particularly the Cistercians. Donkin (1967, 1969) describes how the Cistercians spread out from their original abbey at Citeaux, near Dijon in France (founded in 1098) and built monastic houses generally in remote areas in river valleys. By 1152 they had established 333 monastic houses throughout Europe. Settlements were estab-lished from Sicily to Sweden, from Portugal to Poland, and the monks and peasant colonists under their control cleared huge areas of forest and drained many marshes, especially in northern and central Europe. They were very effective pioneer farmers.

The monastic plantation in northern England was not numerically large (13 houses out of 75 in all of England and Wales, the first two – Rievaulx and Fountains – dating from 1131 and 1132 respectively), but it had a significant impact on landscape and settlement (Figure 7.4). Over a third of the initial sites were soon abandoned as unsuitable, most commonly because of harsh physical conditions and poor economic prospects (Donkin 1959). Cistercian land planning during the twelfth and thirteenth centuries, particularly in northern England, was based largely on granges. These were agricultural estates as satellites around the mother abbey, typically in the order of a hundred acres of arable and pasture for a great many sheep (Donkin 1964). It was not uncommon for existing populations to be evacuated in order to isolate the abbey and create a home grange. Most of the land colonised by the monks was secularised through time secularised and granges were converted to villages.

229

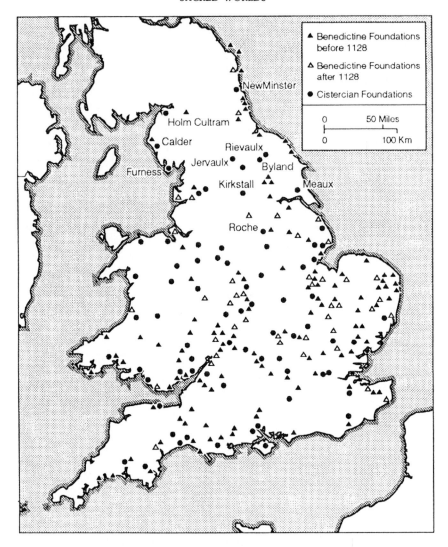

Figure 7.4 The distribution of medieval Benedictine and Cistercian monasteries in
England and Wales

Source: After Donkin (1969).

Case studies of Anabaptist landscapes

The Protestant Anabaptists emerged during the Reformation (see Chapter 4,
pp. 111–16) in Europe, initially as the Swiss Brethren in Switzerland.
Different groups evolved which shared the fundamental Anabaptist belief in
the literal interpretation of the Bible and in the practice of adult baptism but

placed different emphases on some beliefs and practices. In this section we look at three such groups – the Hutterites, Mennonites and Amish – who migrated to the United States to escape religious persecution in Europe, and took their lifestyles with them. Each group has created very characteristic farming landscapes that reflect its cultural heritage and the persistence of its strongly held religious beliefs. At the close of the twentieth century these religious groups and their distinctive lifestyles might appear anachronistic to many people, but they provide colourful testimonies of the association between religion and landscape.

The Hutterites

The Hutterites were an early separatist group within the Swiss Brethren (which started in 1525). Between 1529 and 1536 Jacob Hutter shaped his ideas and organised his break-away group in Moravia (in modern-day central Czechoslovakia). The Hutterites diffused to Slovakia, Transylvania and Russia, and after centuries of persecution a small group of them migrated to North America in the 1870s. Most of their descendants still live on communal farms in the Canadian provinces of Manitoba, Saskatchewan and Alberta and in the states of South Dakota and Alberta.

Simpson-Housley (1978) describes the religious lifestyle they took with them, which required close-knit communities and strongly shaped the design and layout of their buildings and farms. The group's lifestyle is rooted in the belief that obedience to community is the only way to God, and its ideology has been preserved by isolating itself from the outside world. The Hutterite communities have shaped a very distinctive settlement morphology in which spiritual activities are physically separated from secular ones. At the centre of the settlement are the kitchen complex, long houses and kindergarten – wooden buildings primarily associated with the colonies' daily and spiritual lives – which are often painted blue or white and blue. Other buildings – associated with secular, economic and trade functions – are situated at right angles or parallel to the long houses, and directly aligned with cardinal points. The secular buildings are painted different colours to the spiritual ones. Conspicuously absent from Hutterite settlements are worldly facilities such as commercial stores, taverns, movie houses or dance halls.

Hutterite farming practices are also quite distinctive, involving greater diversity of crops and pastoral products than nearby farms. This diversification is designed primarily to create more work opportunities within the community and lessen dependence on trade with the outside world. A sample of Hutterites interviewed by Simpson-Housley (1978) were much more positive in their attitudes towards farming than their neighbours, believing that their area possessed a climate favourable for farming, expecting a good wheat crop each year, regarding drought as less of a problem, and seeing no reason to insure their crops against failure.

The Mennonites

Jordan and Rowntree (1990) note that dispersed farmsteads are the typical North American rural settlement pattern, but cohesive religious groups have traditionally formed village settlements. The Hutterites conform to that model, and so too do the Mennonites from whom they evolved. The Mennonites are followers of the sixteenth-century Dutch Anabaptist leader Menno Simons, and descendants of several Mennonite groups that emigrated to North America survive in traditional communities.

Warkentin (1959) has described the Mennonite agricultural settlements of Southern Manitoba, established by a German-speaking group from Russia who settled there in 1875. They created similar farm villages in Canada to the ones they were familiar with in Russia. The North American rectangular survey system encouraged scattered farmsteads and unit-block holdings, but the Mennonites favoured fragmented land holdings and communal pasture. Many such villages later disappeared, but some survive. One characteristic feature is the long row of cottonwood trees lining the central street; 'apparently the cohesive bond of ethnicity encouraged these immigrants to live in clustered communities, where they could be in close daily contact with people of their own kind' (Jordan and Rowntree 1990: 322). Figure 7.5 shows the Mennonite street village of Neuhorst in Manitoba, which is typical of their clustered village settlements. Martin's (1988) study of Mennonite settlements established in Kansas between 1873 and 1884 highlights a distinctive cultural landscape defined as an arrangement of settlement, farm size and type, field patterns, housing types and other aspects of the built environment including schools and churches.

Mennonite farmers in Manitoba often adopt similar farming practices to their neighbours (Carlyle 1981), partly because of physical constraints (particularly climate and soils). Typical Mennonite farms are smaller and families larger than neighbouring ones, so land is used more intensively. Intensification in some areas has centred on livestock, especially pigs and poultry, whereas in other areas it has centred on sugar beets, sunflowers and vegetables.

The Amish

The Amish also evolved from Swiss Mennonite stock. Followers of Jakob Amman, who originally split from the Mennonites between 1693 and 1697, dispersed throughout the Rhineland and began emigrating to North America in the early 1700s. They settled first in parts of Pennsylvania, particularly in Lancaster and Somerset counties, then progressively moved west.

Different Amish traditions emerged, and the so-called Old Order Amish – still found in more than 60 scattered settlements in 18 states – are the most conservative. Like the Hutterites the Amish are an agrarian subculture. But the Hutterites have developed a socialistic way of life, living in communistic

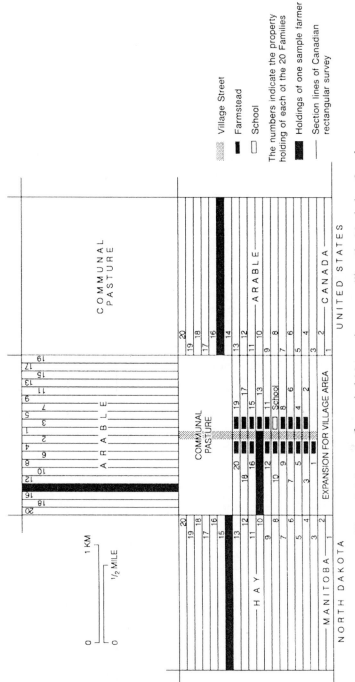

Figure 7.5 Layout of a typical Mennonite street village in Manitoba, Canada

Note: The plan shows the pattern of land use in the village of Neuhorst.
Source: After Warkentin (1959).

Village Street
Farmstead
School

The numbers indicate the property
holding of each of the 20 Families

Holdings of one sample farmer

Section lines of Canadian
rectangular survey

colonies, and they have accepted recent technology and state schools. The Old Order Amish still live on family-owned farms, avoid modern technology, and have their own schools (Stinner *et al.* 1989).

The Amish hold strong Bible-based views that radically shaped their landscape and gave it a distinctive Amish imprint. Theirs is a traditional farming landscape, simple and plain, used sustainably with little waste. As Kent and Neugebauer point out,

> the differences between the Amish and their neighbours in agricultural practices, transportation, dress, and other characteristics are evident to even the casual observer. The existence, persistence, and growth of these unique cultural landscapes in parts of the United States and Canada is partly responsible for the considerable attention the Amish receive from tourists, newspaper travel writers, and academics.
>
> (Kent and Neugebauer 1990: 426)

Distinctive Amish landscapes have been described in detail by Noble (1986) and Kent and Neugebauer (1990), whilst Landing (1969) and Crowley (1978) have documented the evolution and diffusion of Amish settlements within the United States.

Amongst the best-known are the Old Order Amish of Lancaster County, Pennsylvania. Glass notes how these so-called 'plain people' have

> clung tenaciously to a traditional form of agrarian life, believing that man cannot maintain a proper relationship with God unless he first establishes a proper relationship with the land. Although there is considerable doctrinal variety amongst the plain people, most agree that the urban 'English' (non-Deutsch) world is incompatible with a proper Christian life, and it should therefore be kept at arm's length. Traditional agricultural practices, long since abandoned by most American farmers, are assiduously retained.
>
> (Glass 1979a: 43)

The Old Order Amish have a very distinctive identity that is reflected

> most conspicuously through their adherence to plain clothing traditions, rejection of automobile ownership in favour of buggy transportation, their refusal to utilise commercial electricity, telephones, and church buildings (although there are some exceptions), wearing of beards by married men, horse powered farm machinery, and the maintenance of the Pennsylvania Dutch dialect as an in-group language.
>
> (Landing 1969: 238)

The Lancaster County Amish work small farms, using traditional implements. Glass (1979: 43) comments that the average Amish farm has such great variety of crops and animals that it 'sometimes looks like something out of Mother Goose, a landscape where the clock has been turned back two hundred years or more'. Not all Old Order Amish conform to this traditional model, however, because Mook's (1957) description of the Old Order Amish of the Kishacoquillas Valley in central Pennsylvania portrays family-type farming with a focus on dairying on much larger farms.

The Amish live in tight-knit communities that are cut off as much as possible from the rest of the world ('the English'), which is believed to be the source of sin. Separateness is maintained by a strict set of socio-religious rules called the *Ordnung*. These forbid behaviour that would encourage contact with outsiders (such as the use of automobiles for transportation and the connecting of Amish property to the modern world via telephones or electricity), and encourage behaviour that preserves group identity (such as dressing in a uniform simple style and using a German dialect in speech). To insure compliance with the *Ordnung*, the Amish practise the *Meidung* – a form of excommunication that requires that a transgressor be shunned or socially ostracised from fellowship with all other church members, including one's spouse (Glass 1979b). Marriage with non-Amish is forbidden, as is remarriage after divorce. Recent cultural changes, such as the use of automobiles, pose particular challenges to the traditional Amish way of life (Landing 1972).

Amish houses are very traditional and plain, especially when compared with other American houses. Noble (1986: 35) captures the atmosphere inside, where 'the absence of all pictures or portraits of present or past generations, of all wall decorations, including wallpaper, of rugs or carpets, of knickknacks or art objects, of fancy quilts, covers, or pillowcases heightens the impression of sobriety and austerity'.

Amish landscapes are quite distinct. One hallmark of this is the small compact communities based on church districts. Although most Amish groups meet in private houses not churches, they are still based on congregations. Church districts are small enough to allow members to travel to worship by horse and buggy or on foot, which might mean up to 200 individuals spread over several square miles (Kent and Neugebauer 1990). This encourages group members to live close together, and creates a close-knit community.

The physical appearance of Amish landscapes is generally very distinctive (Noble 1986, Kent and Neugebauer 1990). Amish barns are often bigger than those of other farmers, and a common type is the large, wooden, German-style barn with ventilation cut-outs high on the gable wall. Typical Amish houses are also quite large, partly because the typical Amish family is large but also because church services are held in the home. The stereotype is the large family farm house with solid colour (blue or green) cloth curtains drawn back to one side of the window frame during the day. An adjacent Grandfather House, smaller than the main family house but joined to it by a shared porch, caters for the needs of ageing parents. A typical Amish farm also has a kitchen garden visible from the road, whitewashed board fences, and a working windmill (because electrical power is restricted or forbidden).

Amish farms and fields tend to be smaller than their non-Amish neighbours because they rarely use tractors and mechanised harvesting equipment. Horse farming and hand labour limit the size of farming unit that is manageable, but they have also created an efficient low-input sustainable type of agriculture (Stinner *et al.* 1989). Most Amish farms are general rather than specialised, use horse-drawn ploughs, and use animal manure to build and maintain soil fertility. Crop rotations of 3 to 5 year cycles (e.g. hay–hay–corn–oats–winter wheat) are common, so Amish fields usually contain a variety of crops (Figure 7.6). Many Amish farmers also manage woodlots (usually hardwoods) for logs, maple syrup production and fuel. Some craftsmen make and sell wooden furniture or horse-drawn vehicles on the farm.

The small scale and great diversity of Amish farming produce a hetero-geneous landscape mosaic that is very distinctive. This, coupled with the numerous landscape features which are characteristically Amish, gives rise to a landscape which is both ancient and modern. Such a landscape preserves Amish traditions and religious heritage, whilst also highlighting the wisdom of self-sufficiency and sustainable resource management.

Mormon landscapes

Amongst the best documented 'religious landscapes' in the United States is the Mormon culture region (see Chapter 5, pp. 159–63) which retains many interesting landscape features. Indeed, Francaviglia (1970) argues that the Mormon landscape is a vital ingredient in the image many people have of the American West.

The Mormon region in Utah is viewed as sacred by Mormons, and even non-Mormons agree that its landscape is unique (Jackson 1978b). Characteristic elements of this landscape image include the Mormon village, irrigation agriculture, and distinctively Mormon architectural styles. Much of the traditional Mormon landscape was developed in the nineteenth century and nurtured in geographic isolation by a utilitarian pioneer society, but changes in agriculture and transportation – particularly since the 1920s – have introduced modern ingredients such as split-level housing, super highways and tall office buildings and apartments.

The most characteristic feature of the Mormon landscape in the Salt Lake Basin was the rural village. This unique village plan encompassed both residential areas and arable land, and it was based on

> a main street with side streets running perpendicular and parallel to the main street. The houses are usually of brick surrounded by well-kept lawns. Behind the house will be found a general farmyard with barns and other outbuildings, haystacks, and livestock pens. The church, social hall, and school are usually juxtaposed and in the centre of the community.

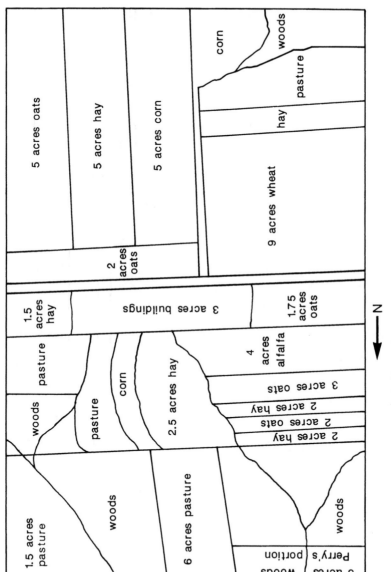

Figure 7.6 Pattern of land use on a typical Old Order Amish farm

Source: After Stinner *et al.* (1989).

Thus the community becomes a series of farmsteads grouped around a church, school and social hall.

(Seeman 1938: 302)

Mormon religious beliefs strongly influenced their views of planning, and this is reflected clearly in Joseph Smith's plan for a utopian society separate from the rest of the world. Lifshetz (1976) summarises the key ingredients of Smith's ideas. He saw the Mormons as a chosen people who must be segregated, and this required a place large enough to accommodate all who were worthy. Smith proposed that a central city (Zion, the New Jerusalem) should be established where God would direct the establishment of his kingdom on earth through his servants. In order to be truly worthy as a chosen people, he argued, poverty must be eliminated and the people lead a life of co-operation in all undertakings.

Joseph Smith developed his plan for the City of Zion to be built in stages, each part complete in itself but leading towards the final goal of a city large enough to be called a Kingdom. It was based on a city one mile square divided by a grid of streets, with building blocks at the centre reserved for public structures (Figure 7.7). Other blocks would be reserved for residences, their gardens, orchards and domestic animals. The plan resembled many East Coast towns based on a grid plan, but with much bigger dimensions; streets would be at least 130 feet wide, and building blocks would range from 10 to 15 acres.

The Mormons' beliefs also shaped their attitudes towards land and natural resources (Kay and Brown 1985) which in turn exerted a powerful influence over their farming practices. They had vision of a future earth where only the faithful would live, and saw it as their duty to reclaim this future home from the corrupt hands of non-believers. As Dyall (1989) points out, this propelled the Mormons towards independence and self-sustainability and was a powerful impetus for developing a form of subsistence agriculture based on sensitive stewardship of nature and protection of the environment. Joseph Smith taught that believers should work with God in transforming the earth and removing the curse from it, so that Mormons redeemed themselves by redeeming the earth. He also taught that Mormons should obtain a farm if they could, in order to become self-sustaining, and that large tracts (including large farms which could not be fully worked and looked after) should be divided up and used profitably. Thus the Mormon farming landscape, with small farms sustainably managed, was directly rooted in their religious beliefs, particularly the belief that God and people are in partnership in redeeming the earth.

Religion and planning constraints and ideals

Although most of the case studies outlined above describe what are essentially nineteenth-century landscapes that have survived, it is important to stress that the impact of religion on landscape is not simply a historical hangover and the

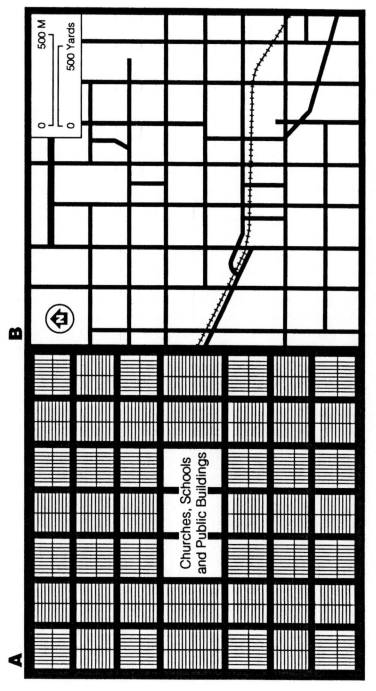

Figure 7.7 The general layout of Joseph Smith's Mormon City of Zion

Source: After Sopher (1967).

influence continues in various ways. Perhaps an obvious example is the planning of new churches and ecclesiastical buildings, which reflects the interplay between religious and secular objectives (Allon-Smith and Crouch 1979).

The impact of religion on landscape is most evident in cases where a new religious group enters an area that has already been settled and tries to modify the existing landscape to better suit its religious teachings or practices. Political, cultural and ideological clashes between newcomers and the existing population highlight conflicting values and attitudes. A graphic example of such forces at work is provided by the take-over of an established town – Antelope City in Oregon – by the Rajneesh community during the 1980s, which Buckwalter and Legler (1983) describe. The community was founded by followers of Rajneesh Chandra Mohan (India's most controversial guru, who changed his name to Bhagwan), who settled on the 70,000 acre 'Big Muddy' ranch near Antelope in 1981. He was immediately joined by 58 supporters who helped to prepare the farm for productive use and establish the Rajneesh Neo-Sannyas International Commune. Large numbers of mobile homes started to appear on the farm, and slowly the concept of a modern religious city (Rajneeshpuram) rising in the desert emerged and took root. Local residents grew increasingly concerned about rising numbers in this marginal environment, although by the summer of 1982 construction and development of the ranch and city facilities were well under way and Rajneesh followers were playing prominent roles in local politics and working hard to get their new city incorporated.

More pervasive but less obvious is the indirect impact which religion can exert on the development of landscape via public attitudes. Saltini (1978) suggests that some aspects of the rural landscape of Europe show a strong imprint of Roman Catholic teaching on human use of the natural world, which sees nature as a utilitarian rather than a sacred resource that is there to be tamed, fashioned and used. Saltini accounts for broad differences in the European landscape partly in religious terms. He distinguishes between areas mainly in the south where the Catholic influence has been strong and long-lasting and where landscapes have been heavily influenced by human activities, and areas mainly in the north where the pagan sense of the sanctity of natural places has survived and more natural landscapes persist.

In other situations the impact of religion on landscape is both direct and strong. Islamic law, for example, sets a clearly defined framework for urban land-use planning (Safak 1980). Islam accepts both public and private ownership rights, but the former get priority in cases of conflict. Specific requirements govern the form, pattern and building and environmental standards of cities, and other measures cover private property and family residence. Many traditional Islamic rules are very detailed, including exact measurements for markets, mosques, and residences. When a new Muslim town or city is established, the government court, central mosque and central square must be

240

located and built first. Then main roads are built outwards from the main square, and streets or lanes designed to lead from the main roads to the residential quarters. Islamic law makes special provision for clean water pipelines, which must be well paved and maintained, and for cleaning the city and its environment. Preservation of trees and green fields in and around the city is encouraged, as is public access to shorelines, riverbanks and dam margins. Individual buildings must be designed to conform with Islamic rules and order, both internally and externally.

Landscape architecture in Saudi Arabian cities clearly shows the strong influence of Islamic teachings (Filor 1988). Land-use patterns are mixed, with orchards and agricultural land penetrating into the very centre of the city. Streets and alleyways serve as social meeting areas as well as passageways. The urban form is inward looking – traditional form of building consists of blank exterior walls to the street and rooms opening on to internal courtyards – reflecting the need for privacy and protection from the scorching overhead sun.

Although Islamic law has traditionally defined the context within which physical planning takes place, recent rapid development in some Middle Eastern countries has relied heavily on the skills and designs of Western architects, engineers and builders. As Wells-Thorpe (1988) records, this has encouraged recent attempts to integrate traditional Islamic design ideas with modern technologies. Timeless Islamic architectural features include the sight and sound of water, the extensive use of screens that allow both privacy and air movement, an almost total elimination of window openings, an inward-looking appearance of most buildings whose street facade matters less than their internal appearance, a network of streets, and an acknowledgement of the supremacy within the townscape of the Mosque. This blend of old and new is well illustrated in the Airport complex at Jeddah (Saudi Arabia) where Teflon-coated fibreglass fabric canopies of the Hajj (pilgrimage) reception centre resemble traditional Arab tents and create a shaded village with natural air circulation.

The modern transformation of traditional Islamic cities is not without problems, however, because modernisation often goes hand-in-hand with demolition of Islamic townscapes and historic monuments. Nakabayashi (1989) documents the superimposition of modern European-style urban planning on the historic Islamic city of Aleppo in Syria, where new road construction has demolished many houses in the old town. He proposes that new cul-de-sac accessways be built, resembling the traditional blind alleys that are disappearing.

Religion and landscape names

One of the most enduring ways in which religion influences landscape is through place-names. In many countries it is not uncommon to find religious names for towns and topographic features. Jordan and Rowntree (1990: 222)

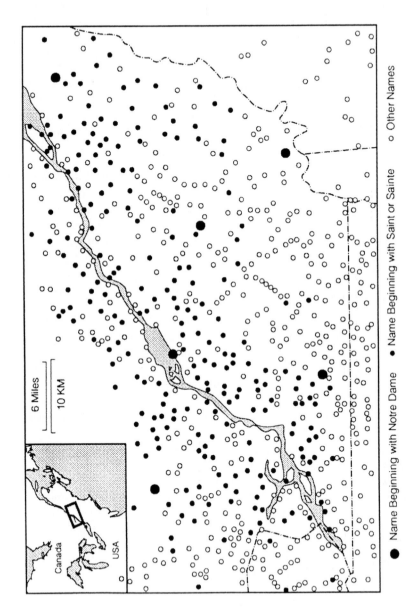

● Name Beginning with Notre Dame • Name Beginning with Saint or Sainte ○ Other Names

Figure 7.8 Distribution of religious place-names in the French Canadian province of Quebec

Source: After Jordan and Rowntree (1990).

note that Christian saints' names are commonly used for settlements in Roman Catholic and Greek Orthodox areas, particularly in overseas colonial places settled by Catholics, such as Latin America and French Canada. This helps to explain why so many towns and villages in the French Canadian province of Quebec have religious place-names (Figure 7.8). Religious place-names are much less common in Protestant areas, and Jordan (1973: 154) comments on the decrease in frequency of sacred toponyms in Europe from the Catholic south to the Protestant north.

The most detailed study of the geography of religious town names, by Brunn and Wheeler (1966), examined the pattern in the United States. They identified three types of place-names – those mentioned in the Bible, an individual's name mentioned in the Bible, and place-names having prefixes St, Ste, San or Santa. The three most commonly occurring Biblical names were Salem, Lebanon and Athens, although there was no evidence of any spatial clustering for these names. States with the greatest number of religious town names were mainly in the central and eastern Middle West, except for California and Texas which both contained numerous saint names (Table 7.2).

Two broad belts emerged in the map of religious town names – an east-central belt (stretching from eastern Pennsylvania westward into the Middle West, colonised by pioneers) and a south-western belt (stretching from Florida to California, reflecting Spanish and to a lesser extent French influences). These regions correspond to the early settled areas and transportation routes, where cultural groups like the Germans, English, Irish, Spanish and French left a strong imprint on the landscape. Cultural centres that were settled early, such as Yankee New England and the Deep South, show much less tendency towards religious influence on town names.

An interesting variant on the influence of religion on place-names is the use of religious truck names in Peru. Field studies by Kus (1979) suggest that between 35 and 40 per cent of all trucks in Peru have names painted on them, the percentage being higher in rural areas and lower in cities. Just over one in five (21.1 per cent) have religious names, with another 9.1 per cent being named for male saints and 2.4 per cent for female saints.

Table 7.2 Leading states with religious town names in the United States

State	Number	State	Number
California	66	Texas	32
Ohio	54	Indiana	30
Pennsylvania	46	New York	28
Illinois	42	Minnesota	27
Missouri	38		

Source: Brunn and Wheeler (1966: 200).

CONCLUSION

This chapter is longer than some of the earlier ones for two main reasons – there are a greater number and variety of published geographical studies to draw on, and the impacts of religion on landscape represent without doubt the most visible link between geography and religion. Geography's traditional concern with landscape provides a further rationale. The material and examples have been drawn from many different countries and most major religions, although the heavy reliance on case studies from the United States once again reflects the strength of such studies in that country.

The different types of landscape of worship we explored on pp. 199–213 provide what to many observers must be the most obvious physical signs of religion, dominated by centres of worship (particularly churches, temples). Yet, from a geographical point of view, the spatial patterns of such facilities are perhaps more inviting and illuminating, both in terms of the distribution and site characteristics of the worship centres and of the territorial sub-divisions associated with them (most notably in the Catholic and Anglican parish systems). Landscapes of death represent an area which relatively few geographers are keen to explore in great detail, despite the proven potential for such studies in reconstructing how attitudes to death, tastes in burial, and land requirements for dealing with the deceased have changed through time.

At the broader scale of detecting the impact of religion on the general landscape we must usually look much harder for clear evidence, and the best examples come from areas where religious persistence over long periods has left an indelible imprint on the (usually rural) landscape. The Mormon culture region and Anabaptist landscapes in parts of the United States provide the best documented case studies.

8

SACRED PLACES AND PILGRIMAGE

Religion's in the heart, not in the knees.
Douglas Jerrold, *The Devil's Ducat* (1830)

INTRODUCTION

One of the more prominent geographical dimensions of religious expression is the notion of sacred space. Most religions designate certain places as sacred or holy, and this designation often encourages believers to visit those places in pilgrimage and puts responsibilities on religious authorities to protect them for the benefit of future generations. Geographers have explored a wide range of questions within this area, such as why and on what basis space is defined as sacred, what implications this designation might have for the use and character of those areas, how believers respond to the idea of sacred space, and how is their response (especially through pilgrimage) reflected in geographical flows and patterns.

Much of the work on this theme builds upon the foundation established by Eliade (1959) in his influential book on *The sacred and the profane*. He explores how ordinary (profane) space is converted into holy (sacred) space, and suggests that this symbolic process reflects the spiritual characteristics associated with both the physical features and the deeper, abstract implications of delimiting a particular site as sacred. Designation of a site as sacred is generally a response to two types of events. Some events (which he calls *hierophanic*) involve a direct manifestation on earth of a deity, whereas in other (*theophanic*) events somebody receives a message from the deity and interprets it for others.

In this chapter we shall encounter examples of sacred space with reflect both of these types of event. We begin by looking at some contexts where religious belief in nature as the home of gods has elevated particular dimensions of the natural world to the status of sacred space.

245

RELIGIOUS ECOLOGY

Religious ecology is founded on the belief that the natural world is part of, not apart from, the deity or deities that created it. Many religions, particularly historic ones, have worshipped all or part of nature. Some have done so through the belief that god and nature are one and the same thing, others through the belief that nature is the dwelling place of the gods, and yet others through the belief that nature is a window to the gods that allows believers to see them at work and to have regular (and often intimate) contact with them.

Known and unknown worlds

It seems to be part of the human psyche to divide the world into a known realm and an unknown realm, because all cultures through history have done so. Religion in some ways serves as a bridge between the two realms, which Isaac (1967) proposes are separated by temporal and spatial gates. Rites of passage (particularly death) are gates in time, and sacred places (such as sacred rivers) are gates in space. Thus, he argues, specific features of the real world are also located in the mythical world and through sacred places we root the unknown world within the known world.

Christinger (1965) suggests that there are many ways of entering the unknown world, but that they are often beset with difficulties. As a result, religious cults evolved to enable mortal humans to pass through the entrance caves of sacred grottoes, ascend the sacred mountains or cross the sacred waters of death and rebirth. In these ways sacred space is much more than hallowed ground because it symbolises and sometimes embodies the gateway to the unknown. It has a practical as well as a conceptual relevance.

Veneration of nature

The most tangible expression of religious ecology is through the veneration of nature. Different dimensions of the natural world are designated as sacred in different religions, and they are held in deep respect in honour of their assumed holiness. Examples can be found in many cultures and throughout history. In Roman Britain, for example, the

> Romans and Britons both believed that various natural features of the landscape had divine associations, either as the homes of gods or as gods in themselves. In both peoples the same kinds of place attracted feelings of reverence, which the Romans expressed in terms of river gods, the nymphs or other deities of springs and fountains, the woodland god Silvanus, or the Genius of a particular place. Christians later denounced these practices . . .
>
> (Blagg 1986: 15)

Nature spirits play a central role in the Chinese and Korean Buddhist practice of geomancy, which is used to select spiritually appropriate sites for houses, villages, temples and graves. Geomancy is based on prophecy from the pattern made when a handful of earth is thrown down or dots are drawn at random and connected with lines. Yoon (1976) illustrates how geomancy has been used in Korea to regulate human ecology by influencing people to select auspicious environments and to build harmonious structures on them. *Feng Shui* is also rooted in belief in nature spirits, as Lai's (1974) study showed (see Chapter 7, pp. 217–18).

Particular environments are venerated in some religions. Perhaps the most obvious is the Hindu designation of the River Ganges as holy, which attracts vast numbers to bathe in it and makes it both symbolically and spiritually important to the Indian nation (Gopal 1988). Pre-Christian religions often regarded caves as holy places, and Jordan (1973: 151) speculates that this idea has survived in the grotto at Lourdes in France (where the Virgin Mary is said to have appeared in a vision) and in the grotto of St John on the Greek island of Patmos (where the biblical book of Revelation was composed). Holy status is even bestowed upon particular stones in some religions. By far the best example is the famous Black Stone at Mecca which Muslims believe was sent down from heaven by Allah, although geologists believe it to be a meteorite (Jordan and Rowntree 1990: 204).

Mountain peaks and other high places have been regarded as holy sites (often as homes of the gods) since ancient times. Good examples include Mount Olympus (the highest mountain in Greece), believed in Greek mythology to be the dwelling place of Zeus and other important gods, and Mount Fuji or Fujiyama (the highest mountain in Japan) which is sacred in Japanese Shintoism. Jordan (1973) notes how the veneration of high places has continued in Orthodox Christianity, with numerous monasteries built on hilltops throughout mainland Greece and the Aegean isles and chapels built on hilltops in parts of Austria and the former Yugoslav republic of Slovenia.

Forests were often seen as the haunts of woodland gods, and tribal religions of the ancient Germans, Slavs, Celts and Greeks venerated trees and forests. Some (such as Jordan 1973: 151) argue that the Christian use of evergreen trees in Christmas celebrations is a relic of the pre-Christian tree worship in heavily forested northern Europe. Yew trees were worshipped as sacred trees in pagan Britain, and one recent study (Wallace 1992) suggests that many Christian churchyards in England may originally have been built close to yew trees because of the aura of holiness and sanctity that the tree had for pre-Christian peoples.

A graphic example of how the veneration of nature can be translated into tangible changes in landscape is the evolution of locust cults in north-east China, described by Hsu (1969). This area has been repeatedly damaged through history by plagues of locusts, and through time eight religious cults emerged which worshipped locusts and locust-gods. The cults built 870

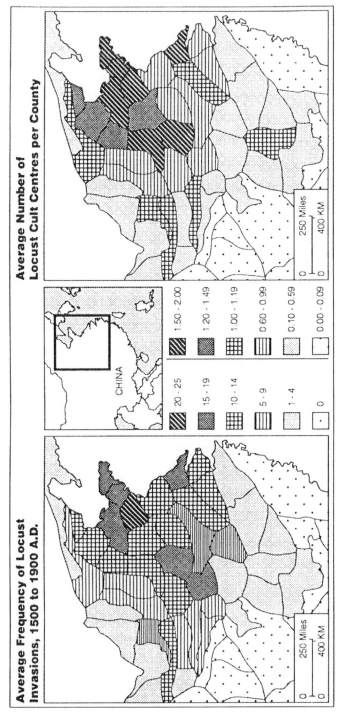

Figure 8.1 Locust cult centres and the pattern of locust infestations in China, 1500–1900

Source: After Hsu (1986).

temples where religious rituals designed to minimise locust plagues were enacted, and the distribution of cult temples is significantly correlated with frequency of locust infestations (Figure 8.1). The largest number of locust temples is in the North China Plain, the Lower Yangtze Valley and the eastern half of the north-western mountain belt.

Heavenly bodies – the sun, moon and stars – were widely worshipped throughout the pre-Christian world, and impressive ceremonial structures survive in some places. Recent excavations in central Mexico City (Park 1984) have unearthed remains of the Great Temple of the Aztecs at Tenochtitlan, which the Aztecs regarded as the centre of the universe. Regular mass human sacrifices were made from the top of the pyramidal temple to the Aztec war god Huitzilopochtli and rain god Tlaloc, and the site was abandoned in 1519 when the Spanish conquistadores arrived. Another example is Stonehenge in southern England, long used by Druids as a mystical and religious site, and believed to have been a sophisticated observatory. Preservation of this internationally recognised World Heritage site which continues to attract vast numbers of visitors, including New Age Travellers intent on celebrating the summer solstice, is important but problematic (Crouch and Colin 1992).

SACRED PLACES

The religious expression of sacred space varies greatly through space and time, and in this section we will explore how and why sacred space is designated and why it is important. Tyler (1990: 16) insists that by dividing the world into sacred and profane areas the great monotheistic religions actually created a religious geography, which has survived particularly where the attitudes and behaviour of believers are influenced by sacred places. Pilgrimage is one of the clearest manifestations of the significance of sacred space to believers, as we shall see later (pp. 258–84).

There is abundant evidence from many cultures that the notion of sacred space is deep-rooted and long-lived. Early pagan cultures had their own definition of sacred space that controlled where people went, what they did and how they did it. Archaeological evidence of ancient ditch systems in the Ohio Valley (Wright 1990), for example, suggests that sacred spaces were being defined in soil and water (with enclosures representing a material expression of the creation of the world as an historical event) some 2,300 years ago.

Properties of sacred space

Definition

Jackson and Henrie offer the useful definition of sacred space as

> that portion of the earth's surface which is recognised by individuals
> or groups as worthy of devotion, loyalty or esteem. Space is sharply

discriminated from the non-sacred or profane world around it. Sacred space does not exist naturally, but is assigned sanctity as man defines, limits and characterises it through his culture, experience and goals.

(Jackson and Henrie 1983: 94)

There is no clear answer to the central question of what defines the holiness or sanctity of a place. Yi Fu Tuan (1978: 84) argues that the true meaning of 'sacred' goes beyond stereotype images of temples and shrines, because 'at the level of experience, sacred phenomena are those that stand out from the commonplace and interrupt routine'. He puts an emphasis on qualities such as apartness, otherworldliness, orderliness and wholeness in defining what is sacred.

Isaac (1964: 28) cautions geographers to approach the definition of sacred places with sensitivity, because 'to broach the theme of holiness or the sanctity of place in geography always verges on the trite or the impertinent ... [and threatens] to intrude on a domain preempted by theology'. He emphasises the significance of awe and wonder in the experience of the holy, and points out (ibid.: 29) that 'on almost all levels of culture there are segregated, dedicated, fenced, hallowed spaces. The holy, or hallowed, means separated and dedicated'. Isaac also underlines the fact that sacred places are not transferable (they are valued because of their associated holiness) and they do not need to be re-established with each new generation (there is an inherited appreciation of the holiness of the site).

Sacred space to most religions means real places on the ground. To the exiled Jews, however, the notion of sacred space was not necessarily territorially defined. Maier (1975) explores the idea of Torah (the body of Jewish sacred writings and traditions) as movable territory, and suggests that it may have developed as a symbolic substitute for the loss of real territory. In explaining how Israel managed to exist as an identifiable political identity during thousands of years of exile, he points out that

when the people of Israel went into exile, the Shekhina (soul of God) accompanied them as token that they were not entirely abandoned by God. ... In accompanying the people of Israel into exile, the Shekhina-Torah is the promise of eventual return, when land and people, Torah and God are reunited in their proper dwelling place.

(Maier 1975: 21)

In effect, the Jews took their portable space with them into exile and only replaced it with a real territorial definition of space in recent centuries.

Classification

Not all sacred sites have equal status or perceived holiness, even amongst believers. Jackson and Henrie (1983) suggest a typology for categorising sacred space at three broad levels, based on a detailed empirical study of Mormon culture in the USA. So-called *mystico-religious sites* are perceived as most sacred because Mormons believe that God and man are in direct contact through them. Temples, shrines, cathedrals, sacred groves, mountains or trees may serve as the focus of such mystico-religious sacred space to Mormons, along with special places such as Bethlehem and the Holy Land. *Homelands*, the second level of sanctity, are sacred space because they represent the roots of each individual, family or people but they are sacred only to believers. Mormon homelands include Utah, the Rocky Mountains, and Jackson County (Missouri). At the lowest level of sacred space are the *historical sacred sites* that have been assigned sanctity as a result of an event occurring there. Mormon examples include Joseph Smith's birthplace and Kirtland (Ohio) where the first Mormon temple was built.

Sacred space is not always three-dimensional and areal. Stoddard (1987) argues that sacred points and lines, as well as areas, are significant components of the geography of sacred spaces. Believers recognise that sacred areas are endowed with divine meaning, which separates them qualitatively from secular or profane places. Migration patterns to such sacred areas (as in pilgrimage) are influenced more by spiritual than practical objectives, so that cost- or distance-minimisation are less important than the very fact of getting there or being there. Sacred points, which might possess distinctive site characteristics (e.g. hilltops or sites of historic events), are often associated with pilgrimages. Sacred paths are not necessarily synonymous with pilgrim routes, because many such routes are not in themselves regarded as sacred (unlike the sacred points along the way). Sacred processions and circumambulations, where the entire path exists as sacred space, are typical examples.

Selection

How are sacred sites selected? There is no clear answer because different religions select their sacred sites on different criteria, and the criteria used even within one religion can change through time. Most sacred sites persist, so the inventory at any one point in time is the outcome of many previous decisions by different people at different times in different places.

Some sacred sites are selected because they are associated with people who have some particular religious significance or credibility. The small kingdom of Bhutan, on the southern slopes of the Himalayas, was converted to Buddhism in the eighth century by Guru Rimpoche. Bailey describes how the temple of Kuje (literally 'body print') – the holiest spot in Bhutan – was built

> against a rock under which Guru Rimpoche sat in meditation for so long that the print of his body was impressed on the rock itself, while the pilgrim's staff which he stuck into the ground can now be seen grown as a large tree.

> (Bailey 1935: 96)

Many individual pilgrimage sites in Islam (see pp. 263–71) and Hinduism (see pp. 271–5) mark significant places in the lives of religious founders or leaders. Sites associated with the life of the Buddha – such as his birthplace at Lumbini in Nepal, Bodh-Gaya in India where he received enlightenment, and Sarnath (near Varanasi) where he first preached (Orland and Bellafiore 1990) – are both sacred and heavily visited.

Sacred sites are sometimes selected through an association with earlier myths and legends. Not all such sites are sacred in the sense that people pray or perform rituals there. Simpson (1986: 53) offers a catalogue of British sites which 'have religious significance, because they are places where traditional narratives assert that a supernatural event once occurred which endorsed Christian values, and which left a lasting mark upon the landscape'. Many landscape legends invoke the Devil acting as God's agent of retribution, and a lot of prehistoric monuments (stone circles and alignments, single standing stones, remains of chamber graves) have traditionally been explained this way. One of the most popular type of local legend in Europe is that which deals with buildings or towns submerged by the sea, a lake or swamp because of their wicked inhabitants (such as Llyn Syfadon – Llangorse Lake – in Wales).

Other sacred sites are recycled earlier religious sites. There are many examples (Grinsell 1986), including Christian chapels in Egypt converted from pre-Christian rock-tombs, ancient Egyptian temples converted to Christian use, and early Christian churches built within ancient temples in Egypt and Cyprus. Many megalithic monuments in Brittany (north-west France) were incorporated into early Christian churches, including parts of the standing stone alignments of Carnac. Similarly many early British churches were sited either on or adjoining prehistoric or other pagan monuments. The re-use of existing sacred places greatly assisted the early spread of Christianity amongst non-believers, but it also ensured that they survived as sacred space (albeit with a different religious orientation).

However, the differences between religious sites and sacred sites must not be overlooked. Places can be of historical significance in a religion without being imbued with the quality of sanctity, and places used for worship can be sacred space but are not necessarily so (a Quaker meeting house is not the 'house of God', for example).

Designation

Most sacred space and sacred places are designated in perpetuity even if – like the movable Torah – they sometimes uproot their territorial foundation. Most such places are also clearly defined, and their geographical location remains fixed through time.

There are examples, however, of sacred places that appear to have been loosely defined through time and might even have shifted position on occasions. This seems to have been the case with the so-called '75-sacred-place view' in the Omine Sacred Mountain Area in Japan, which Oda (1989) describes. The Omine Mountains have traditionally represented sacred mountain areas throughout Japan, and at one time it is believed that 75 sacred places (*nabiki*) were designated. Analysis of sacred place names reveals that the precise location of all 75 sites is not definite, nor is it clear that all traditional ritual sites are listed.

Occasionally the opportunity arises to study how sacred space is designated at the time of designation, and this can throw light on the processes involved. The establishment of a Hare Krishna temple in West Virginia between 1973 and 1981 is described by Prorok (1986), who emphasises the central role of symbolic acts in defining sacred space. The temple, on a 99-acre hilltop site surrounded by forest and pasture land south of Wheeling, was built in honour of Swami Prabhupada, founder of the International Society for Krishna Consciousness (ISKCON) (Figure 8.2). It started as a home for Prabhupada and a small spiritual community, and through time evolved into an elaborate Hindu temple. Prorok notes how

> ritual activity and acts of founding and settling the community have attracted Krishna to their site, thereby making it his home. By making it both God's home and that of his representative, Prabhupada, devotees believe they attract more converts. ... By worshipping, meditating and experiencing other religious rigors here, the site's sacredness is enhanced, which attracts disciples and, in turn, makes the site more sacred.
>
> (Prorok 1986: 137–8)

Protection and preservation

Designation of particular places as sacred can be a mixed blessing, because whilst this special status normally makes such places top priority for preservation and protection, it also encourages large numbers of visitors who can damage the very thing they want to see and experience.

Buddhist sites in India are visited by many Indians and growing numbers from South-east Asia, and face mounting pressure from Government-guided tourism development. Orland and Bellafiore (1990) directed a development study of Sarnath, 7 km from Varanasi, where the Buddha gave his first sermon. They graphically describe the constraints which visitors to this important sacred site face – including the lack of interpretative material and signposting, the unsightly clutter of souvenir shops and carts selling hand-crafts and food, the badly worn lawns, the noisy and bustling streets and large congregations of beggars – and highlight the growing conflicts between religious and recreational users of the site (one of the few open spaces in the Varanasi region). Varanasi must be typical of many of the world's heavily visited sacred places, and its plight is a salutary reminder of the need for careful management of both site and people.

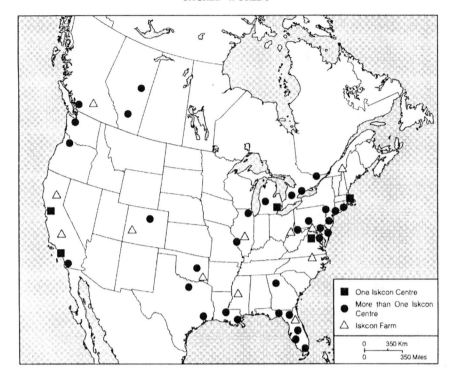

Figure 8.2 The distribution of ISKCON centres in North America in 1981
Source: After Prorok (1982).

The opposite side of the coin is illustrated by sacred places that have been carefully managed and have made significant contributions to the preservation of local landscapes. Most of the islands in Japan's Setonaikai (Inland Sea) have been extensively developed in the post-war period, but some have been preserved as sacred islands (Kondo 1991). Few visitors have entered the uninhabited sacred islands that remain richly wooded (unlike the other islands where most trees have been felled). One sacred island, Ikishima Island in the bay of Sakoshi, was designated a national natural monument in 1924 because of its rare laurel forest grove. The only visitors land on the island during an annual autumn religious festival. Preservation of the island's sacred status is further encouraged by local legends that foretell insanity for anyone who cuts a tree on Ikishima and guarantee local fishermen's catches so long as the area remains sacred.

Temple forests in India provide another example of sacred places that also benefit wildlife and landscape. Worship of trees and plants has been part of religious practice in India since about 600 CE, although it is has recently declined. Hindu theology requires people to worship deities in order to appease them (the deity Vanadevi is believed to be responsible for forests), and the very act of planting of specific plants or trees is considered to be an act of worship in itself. Hindus consider particular trees as well as forests to be sacred, such as *Ficus religiosa* (the Bodhi tree) which is planted in most Indian villages. The Buddha was sitting under a Bodhi tree when he attained Nirvana, and it is believed to be associated with forgiveness of wrongdoing.

Chandrakanth *et al.* (1990) describe a variety of different temple forest types. These include star forests (created to facilitate the worship of stars personified in the form of trees rather than idols), nine planet temple forests (more common than star forests, smaller in size and established by individuals rather than communities), and zodiac forests (a square forest in which all twelve zodiac signs are exclusively represented by trees in a particular configuration). They propose the establishment of new temple forests to assist soil and water conservation in the headwaters of large watersheds and to benefit agricultural lands downstream. As well as fulfilling their traditional religious role, existing temple forests could also provide herbal medicines and act as reserves of trees to help create climax forest through succession (Chandrakanth 1989).

Hindu holy sites

A number of studies have focused on Hindu holy sites, and together they shed some light on broader themes such as the geographical distribution of such sites at a variety of scales.

Stoddard (1968: 148) notes how 'the distribution of holy sites presents an interesting geographic problem ... [because] no tangible object *per se* creates the quality of sanctity at an earth location because the existence of a holy site depends upon human sanctification through thought and deed'. Despite this human dependence, there is not necessarily any close correlation between the distributions of holy sites and people because the location of sacred sites (Figure 8.3) largely reflects historic and topographic factors. Stoddart examined 21 major Hindu holy sites of national importance in India and tested whether 'the major Hindu holy sites are distributed optimally relative to the Hindu population, where optimality is defined in terms of minimising aggregate travel distance ... when each individual of an unevenly distributed population travels to the nearest site' (ibid.: 150), using 1961 census data. Not surprisingly, he found that 'the arrangement of actual holy places is not the one that yields the minimum aggregate travel' (ibid.: 152) and concluded that major Hindu holy sites are not distributed optimally relative to the contemporary Hindu population.

One topographic factor of particular importance in Hinduism is proximity to water. Many sacred sites are concentrated along the seven sacred rivers of the Hindus – the Ganga (Ganges), the Yamuna, the Saraswati, the Narmada, the Indus (Sindhu), the Cauvery and the Godavari. The Ganges is India's holiest of holy rivers and there are many sacred shrines on its banks. Datta (1962)

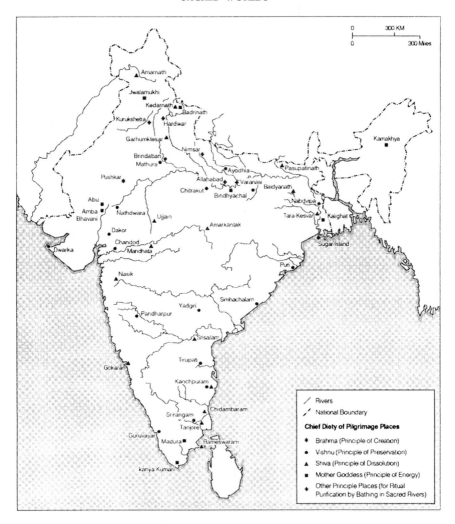

Figure 8.3 Hindu pilgrimage places in India
Source: After Broek and Webb (1973).

tested the spatial association between the distribution of Brahmins (orthodox Hindus who firmly believe the Hindu religious texts) and sacred rivers, along the Ganges, using 1931 census data. He found a marked concentration of Brahmins close to the Ganges in almost every district, supporting the hypothesis that places close to sacred rivers are regarded as more sacred than places away from them.

The distribution of sacred places in Tamil Nad, in south-east India, is also heavily influenced by proximity to water. Chettiar's (1941) analysis of temples there found a range of factors important in site selection, including birthplaces of and visits by saints, royal patronage, pious acts by newly arriving migrant tribes, and thankfulness for divine mercy (such as recovery from disease). Topographic factors were also important, with most temples concentrated along river valleys and others built along the seashore.

Varanasi – previously Benares – in the north (Figure 8.3) is the sacred centre of Hinduism in India. It attracts vast numbers of pilgrims, and has a clearly defined religious geography that reflects the locations of sacred sites and the routeways of sacred walks. Singh (1987b) describes the city's pilgrimage mandala – a circular pilgrimage route that symbolises the universe – which reflects both body symbolism (related to the divinity) and water symbolism (mostly in reference to the Ganga river). This creates what he calls a *faithscape* that encompasses sacred space, sacred time and sacred numbers (all three are central to Hindu spirituality) and embodies both symbolic and tangible elements in 'an attempt to interlink man and divinity with an aim to realise man's identity in the cosmos' (ibid.: 522).

Although Varanasi is a major Hindu centre a quarter of its population is Muslim, and Kumar (1987) has studied how the city is constructed in the Muslim mind. The city contains many graves and mausoleums built to the *shahids* (Islamic martyrs), most of which continue to serve as active centres of popular worship and entertainment. This Muslim sacred geography of the city, notes Kumar (1987: 266), 'allows for the construction of a certain kind of history [though] whether it is in fact "true" may be open to debate'.

Sacred directions and orientations

The impacts of religion on the orientation of ceremonial centres and the use of sacred directions have been explored in some interesting geographical studies.

Some ancient cultures developed remarkable abilities to navigate by the stars, to pinpoint precise locations by astronomical survey, and to orientate ceremonial structures in preferred directions. The Maya American Indians are a good example because they built vast, carefully planned and well-engineered ceremonial centres (up to 900 CE). Fuson (1969: 511) argues that their engineering abilities simply reflect the Classical Mayan emphasis on order and precision, and their centres represent 'a cultural expression of the integration of architecture, sculpture, painting, hieroglyphics, mathematics, astronomy and religion'. Mapping of the centres shows that structures within a ceremonial complex are hardly ever aligned with each other. Building orientation appears at first sight to be random, but field surveys show that many Maya structures are aligned to astronomical positions or roughly towards magnetic north (reflecting the fact that the ruling priest class were masters of astronomy and mathematics).

Religious orientation is also evident in mythical thinking in ancient Egypt. Klimeit (1975) shows how the ancient Egyptians saw their land as a place where their gods resided but also where the creation of the world began. Egypt was viewed as the centre of the world – the 'kingdom in the middle' (ibid.: 271) – and map making in Egypt was developed for both secular and religious purposes (to guide the soul through the realms of the afterworld). Sacred structures were sometimes oriented in relationships with what was assumed to be the original primal hill (the source and centre of the world).

Gordon (1971) describes a number of contexts in which religious factors favoured certain sacred directions and orientations. Ancient religions based on sun worship had particular reverence for east. Old Testament passages show that the ancient Jews also favoured the direction of Jerusalem (the City of God) and regarded north as unfavourable. The Prophet Mohammed originally followed Jewish tradition and prayed towards Jerusalem, until he received a revelation from God instructing him to turn his back upon it and face Mecca. Since then the sense of Holy Direction (towards Mecca) has pervaded most dimensions of the Muslim's everyday life. Throughout the world of Islam the faithful turn towards Mecca to pray, and they are forbidden to spit or relieve nature facing in that sacred direction.

Sacred directions are also reflected in the orientation of churches, mosques and synagogues (Gordon 1971). In the West, Jewish synagogues are mostly aligned from west to east with worshippers facing the Ark towards Jerusalem (in the east they are aligned in the opposite direction towards Jerusalem). Since the eighth century Christian churches have been oriented with the altar (viewed as paradise) facing east. Orthodox Christian churches also have their altar at the eastern end. In Muslim mosques, a special niche (the *mihrab*) is built into a wall so that the prayers of those facing it will be addressed towards Mecca.

PILGRIMAGE

The notion of sacred space is clearly very important in both theory and practice. It demarcates certain places and spaces as having some particular religious association, and by definition sets them apart from the rest of geographical space. The designation might reflect historic or topographic factors, and it might be instrumental in encouraging large numbers of religious believers to visit the site or area. The distribution of sacred space, as we have seen above, reveals important clues about religious meanings and symbolism and about why certain places are regarded as special. But the dynamics of sacred space are perhaps even more interesting to geographers, who have shown great interest in how and why pilgrims travel to sacred sites, and how their pilgrimages affect environment and society, particularly in and around their destinations.

The nature of pilgrimage

The *Collins English Dictionary* (1979) defines *pilgrimage* as 'a journey to a shrine or other sacred place', and a *pilgrim* as 'a person who undertakes a journey to a sacred place as an act of religious devotion'. Such journeys often involve large numbers of people, who travel long distances by a variety of means, often for specific religious festivals. Pilgrimages are typical of both ethnic and universalising religions, and they are found in the major historical religions – Christianity, Islam, Judaism, Hinduism, Buddhism, Confucianism, Taoism and Shintoism (Turner 1973).

Pilgrimage sites vary a great deal in importance, from small shrines that attract the faithful from the immediate area to world-famous places visited by believers from many countries (Jordan 1973). Among the best examples of the latter – which we return to later in this chapter – are the Arabian city of Mecca that is a centre of pilgrimage within Islam, and the French town of Lourdes that is visited by millions of Roman Catholics. Pilgrimage sites vary a great deal in character. Some have been the setting for religious miracles, others are believed to house gods or contain sacred physical features (such as rivers), and yet others represent religious source areas of administrative centres (Jordan and Rowntree 1990).

Eade and Sallnow (1991: 6) point out that 'a pilgrimage locale is typically a site associated either with the manifestation of the divine to human beings or with the human propensity to approach the divine'. Motives for taking part in pilgrimages vary from one religion to another, but the main benefit most pilgrims enjoy is spiritual satisfaction and comfort. This might mean many different things, including the purification of souls or the promise of some desired objective (such as wealth, longevity or happiness). Turner (1973) draws the important distinction between pilgrimage that is obligatory (as in modern Islamic pilgrimage to Mecca) and pilgrimage that is a voluntary act involving a vow or promise (such as early Christian sacred travel to Palestine or Rome). Obligatory pilgrimage inevitably involves larger numbers, guarantees the survival of the pilgrimage route and destinations, and has its own inbuilt dynamics.

A review of the literature on pilgrimage, by Shair (1979), highlights the difference between pilgrimage journeys and other types of journey. The journey is a religious act in its own right, it is made for a specific (spiritual) purpose, and it represents a pathway to particular sacred places. Unlike other types of journey, the pilgrimage usually introduces social, economic and physical difficulties or sacrifices for the pilgrims, who usually accept them with resignation as part of the special nature of pilgrimage. Shair (1979) also suggests that pilgrimage represents a particular religious rite of passage, which involves separation (leaving home), transition (travel to the sacred place) and incorporation (arrival).

To talk of pilgrimage as if it was a uniform phenomenon is misleading because, as Rinschede and Sievers (1987) and others point out, pilgrimage is very heterogeneous. It occurs at a variety of scales (including small, regionally confined places of pilgrimage and large-scale pilgrimage centres) and involves different size groups from the individual to the mass pilgrimage. Moreover, some pilgrimages are deliberately difficult and require great commitment to complete, whereas others require little sacrifice and can be made by almost anyone.

One of the themes underlined by Sopher (1967) in his review of pilgrimage is that these mass circulations of people often have very significant economic as well as religious impacts. He commented, for example, on how the Muslim pilgrimage to Mecca 'promotes secondary flows of trade, cultural exchange, social mixing, and political integration, as well as certain less desirable flows, such as the spread of epidemic disease' (ibid.: 52)

Pilgrimage can also have significant impacts on some areas by promoting particular forms of tourism. Rinschede and Sievers (1987) argue that pilgrimage can have a major effect on local economies by encouraging the development of infrastructure such as shrines, shops selling devotional articles, and facilities for overnight accommodation (including dormitories and camp sites). In some places pilgrimage is the dominant form of tourism, although it is often very seasonal and short-lasting (Rinschede 1990). Wiebe (1980) has assessed the potential role of religious tourism in Afghanistan, and underlines the inherent conflict between Islamic and Hindu religious use of sacred sites and non-religious use by visitors. Sievers (1987) takes the opposite line and argues that pilgrimage tourism in Sri Lanka, involving Buddhist, Muslim, Hindu and Christian destinations, holds great potential for the local and national economies.

Journey verses destination

The word 'pilgrim' comes from the Latin *peregrinus*, which literally means foreign, travelling or migratory. Tuan (1984: 5) sees religious pilgrimage as a ritual by which we break up 'the drowsiness of routine' that dictates the pattern of our daily life. He contrasts being 'in place' and 'out of place', suggesting that we spend most of our lives *in* place (surrounded by the security of familiar relationships, habits and routines), but we have a periodic need as individuals and as society to transcend place (and then be *out* of place). These rituals that break up our routines expand our horizons – if only fleetingly – to embrace the cosmos, not just our own 'seemingly immutable social place'. By detaching ourselves from place during pilgrimage, he argues, we see place for what it really is – 'a temporary abode, not an enduring city' (ibid.: 9).

This notion of pilgrimage as an escape from the normal is echoed by other writers. Turner (1973), for example, emphasises the role of pilgrimage in changing the pilgrims. They begin in a Familiar Place (at home), journey to a Far Place (the pilgrimage shrines, which are usually distant and peripheral to the rest of their lives), then return – ideally changed – to the Familiar Place.

Both Tuan and Turner suggest that in pilgrimage it is the journey itself that really matters, perhaps just as much as arrival at the destination. Sopher (1987) agrees, and he sees detachment from place as a means of release in preparation for the final release on death as a critical goal of pilgrimage. The Exodus of the ancient Jews and their journeys in search of the Promised Land (as documented in the Old Testament) were essentially a form of pilgrimage (Davies 1979), and here again the journey itself is as important as the destination.

The role of the journey relative to that of the destination is reflected also in the phenomenon of multi-religion pilgrimages. It is generally assumed that sacred places of each religion are patronised only by believers of that faith, but studies by Bhardwaj (1987a) have shown that 'there is a large number of shrines in India that attract pilgrims of diverse religion'. For example, many Sikhs make regular pilgrimages to the Hindu shrines at Mansa Devi (near Chandigarh), Naina Devi and Chintapurni (in Himanchal Pradesh), many Hindus from Punjab used to visit the Sikh holy shrine (Golden Temple) at Amritsar, there are many holy sites in Uttar Pradesh that are Islamic in origin but regularly visited by pilgrims of other faiths, and the Dargh Sharif at Nagore (Tamil Nad) has long attracted Muslims, Hindus and Christians. Sievers (1987) describes some sacred sites in Sri Lanka that are described as holy by Buddhists, Hindus, Muslims and Christians. They include Sri Pada (Adam's Peak) where a strange hollow in the rock at the summit is 'simultaneously honoured as being the footprint of Buddha, Adam and Thomas, the apostle of India' (ibid.: 434) and Kataragama on the holy Menik Ganga that has temples containing Hindu, Buddhist and Muslim shrines.

One explanation of such multi-religion pilgrimage is the attraction of some shrines because of inherent qualities such as association with healing. Bhardwaj points to the fact that

> the goal of pilgrimage is frequently rather specific and very much related to this existential reality [healing], not to an abstract or unknown here-after. In fact, the goal is to get well. But even if that is not possible, to at least feel better in the process.
>
> (Bhardwaj 1987a: 463)

Dynamics of pilgrimage

One of the most important geographical dimensions of pilgrimage is the centripetal flow or circulation of large numbers of people to a religious centre (or centres). As Rinschede and Sievers (1987: 214) point out, 'the core of the geographical pilgrimage phenomenon is . . . of socio-geographical nature. The pilgrims primarily are a spatially effective social group who, as a rule, belong to one of the big religious communities'. The geographical importance of the phenomenon lies in the interrelation between the pilgrims (the people, their origin, social groupings and motivations) on the one hand and the pilgrim destination (the shrines, their setting, location and development) on the other.

Detailed geographical studies have focused on a number of different types of pilgrimage, as we shall see in later sections. One small-scale study that reflects some important aspects of the dynamics of pilgrimage is Gurgel's (1976) analysis of the travel patterns of Canadian visitors to the Mormon culture hearth. The area, within the rolling hills of western New York state, is centred on Sacred Grove and Hill Cumorah (where Joseph Smith had many of the divine visions that led him to establish the Mormon Church). It received an estimated 200,000 visitors in 1973. Visitor surveys revealed that most (84 per cent) visitors to Joseph Smith's home were Mormons, half travelled in family groups, and most came from within a 300 km radius and returned home the same day. Statistical analysis showed that numbers of visitors were influenced by the distance-decay effect, by size and intensity of the visitors' religious community, and by socio-economic factors (such as occupation and income). The Hill Cumorah pageant, an annual passion play based on the appearance of Jesus to his followers in the New World (a belief unique to Mormons), attracted three-quarters of all visitors to the Mormon hearth in 1973. Many of them were non-Mormons and female visitors from further away.

Postmodernism and pilgrimage

Some recent writers have suggested postmodern interpretations of religious pilgrimage. Such interpretations are based on the idea that pilgrimage is an arena in which competing discourses take place and from which pilgrims recover religious meanings.

Eade and Sallnow contrast traditional and postmodern conceptions of pilgrimage to sacred places. In the traditional view, after Eliade and others,

> the power of a miraculous shrine is seen to derive solely from its inherent capacity to exert a devotional magnetism over pilgrims from far and wide, and to exude of itself potent meanings and significances for its worshippers ... its power is internally generated and its meanings are largely predetermined.
>
> (Eade and Sallnow 1991: 9)

In the postmodern view,

> a pilgrimage shrine, while apparently emanating an intrinsic religious significance of its own, at the same time provides a ritual space for the expression of a diversity of perceptions and meanings which the pilgrims themselves bring to the shrine and impose upon it. As such, a cult can contain within itself a plethora of religious discourses.
>
> (Eade and Sallnow 1991: 10)

They offer the example of Roman Catholic pilgrimage to Lourdes, some of which go in pursuit of divine favour, some as a tangible sign that God and the Virgin Mary have their individual interests at heart, while others go as an act of sacrifice (a small-scale personalised replay of Christ's own sacrifice on the cross). Another illustration is the pilgrimages to Jerusalem and the Holy Land by Greek Orthodox, Roman Catholics and Christian Zionists.

> Each group brings to Jerusalem their own entrenched understandings of the sacred; nothing unites them save their sequential – and sometimes simultaneous – presence at the same holy sites. For the Greek Orthodox pilgrims, indeed, the precise identification of the site itself is largely irrelevant; it is the icons on display which are the principal focus of attention. For the Roman Catholics, the site is important in that it is illustrative of a particular biblical text relating to the life of Jesus, but it is important only in a historical sense, as confirming the truth of past events. Only for the Christian Zionist does the Holy Land itself carry any present and future significance, and here they find a curious kinship with indigenous Jews.
>
> (Eade and Sallnow 1991: 14)

Bowman supports the postmodern view, arguing that

> pilgrimages are journeys to the sacred, but the sacred is not something which stands beyond the domain of the cultural; it is imagined, defined, and articulated within cultural practice. ... It is at the sites whence the pilgrims set out on their searches for the centre that pilgrims learn what they desire to find. At the centres where they go in expectation of fulfilling that desire pilgrims experience little other than that which they already expect to encounter.
>
> (Bowman 1991: 120–1)

Islamic pilgrimage to Mecca

The annual pilgrimage of Muslims to Mecca – the so-called *hajj* – is a remarkable movement of people in the Middle East in terms of both size and durability. As King (1972: 62) emphasises, 'the pilgrimage is one of the world's greatest gatherings of different races and languages. It has endured the 13 centuries of Islam virtually without interruption'. Rowley (1989) describes the *hajj* as the axis of Islam's tendency to centralise and unify sacred space, around which all other religious rites revolve.

Its influence extends to all the countries of Islam, and for one month every year the city of Mecca in Saudi Arabia (with a resident population of around 150,000) has more visitors (over a million) than any other city in the world. The *hajj* is a major source of income for Saudi Arabia (the third largest earner after oil exports and spending by oil companies). Indeed, before oil was discovered in Saudi Arabia in 1938, spending by pilgrims was the country's largest source of foreign exchange earnings.

Role of pilgrimage within Islam

To Muslims the pilgrimage to Mecca (Figure 8.4) is not simply an act of religious obedience, it is a duty. It is the fifth pillar (foundation of faith) of Islam – along with declaration of faith, prayer, charity and fasting – although it is the only one that is not obligatory. Islam requires that every adult Muslim perform the pilgrimage to Mecca and to nearby Arafat and Mina – where they receive the grace of Allah – at least once in a lifetime. But the obligation is deferred for four groups of people – for those who cannot afford to make the pilgrimage, for those who are constrained by physical disability, hazardous conditions, or political barriers, for slaves and those of unsound minds, and for women without a husband or male relative to accompany them (Brooke 1987).

None the less most Muslims do make the pilgrimage at least once, and for many of them the trip is the culmination of a lifetime's saving. For many Muslims (*hajjis*) the pilgrimage is a time of great hardship and personal suffering, and until recently many pilgrims died along the way (from

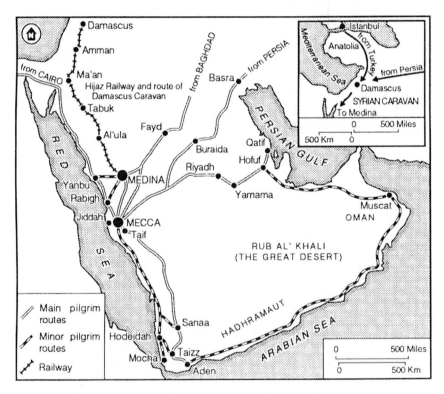

Figure 8.4 Isamic pilgrimage routes to Mecca in Saudi Arabia
Note: The main overland routes up to the beginning of the twentieth century are shown.
Source: After King (1972).

exhaustion, hunger, thirst, disease). Death during the pilgrimage is regarded as particularly honourable and is believed to guarantee entry into the afterlife.

In a typical year about 0.1 per cent of the world's Muslims may be in Mecca (King 1972). Official figures reveal that 1.85 million pilgrims took part in 1984, and the 1986 total was 2.25 million according to unofficial estimates (Brooke 1987).

Both Mecca and Medina are forbidden to non-Muslims, and boundary stones on all routes leading into the cities mark the point (30 km out) beyond which non-believers must not pass.

Large numbers of animals are slaughtered annually during the *hajj*. Brooke (1987) estimates that about a million animals (mainly sheep, goats, camels and cattle) are transported to Mina (near Mecca) and slaughtered there according to strict rituals. Disposal of the vast number of carcasses, within seven days, has to be carefully planned and managed to avoid sanitary problems in the hot, dry environment.

The *hajj* commences on the eighth day of the twelfth month (*Dhu'l-Hijja*) of the Muslim lunar year and ends on the thirteenth day of *Dhu'l-Hijja*. Prescribed rites are performed which follow the order of the farewell pilgrimage in prayers and physical movement to the various sites as performed by the Prophet Mohammed in 632 CE (Rowley and El-Hamdan 1978). The rites and rituals are performed in a tightly defined sequence (Table 8.1).

Table 8.1 Rites and rituals performed during the *hajj* in Mecca

1   Days 1–3; pilgrims arrive at Mecca in a state of ritual purity (*ihram*), and perform the arrival *tawaf* (walk round the Kaaba holy shrine seven times) and the sa'i (literally 'the running' – seven one-way trips, jogging and walking between two small hills about 400 metres apart near the Haram Mosque)

2   Day 3; pilgrims leave Mecca to arrive in Mina (about 5 km north) before noon; they offer five prayers and spend the night there

3   Day 4; pilgrims leave Mina to arrive in Arafat (about 15 km further north) before noon to perform *wuquf* (literally 'the standing' – a standing vigil at Arafat from noon to sunset); after sunset they leave for 'the rushing' to Muzdalifah, where they spend several hours in prayer and collect 49 pebbles to throw at the Jamarat pillars in Mina

5   Day 5; the pilgrims arrive in Mina early morning; they stone the largest of the Jamarat pillars, and may then sacrifice an animal; then they end *ihram* (the state of ritual purity) by *tahallul* (the ritual haircut); in the afternoon they return to Mecca for another *tawaf* (walk round the Kaaba seven times) and then return to Mina to spend at least two nights there; each day at Mina they stone the three pillars of Jamarat

5   Not earlier than day 7; pilgrims return to Mecca and perform the departure *tawaf* (walk round the Kaaba seven times); most pilgrims then leave for Medina or to return home

Source: Summarised from Brooke (1987).

The *hajj* pilgrimage is multi-dimensional, involving the visit to and walk around the Kaaba (the holy shrine in Mecca, containing the black stone), visits to various other holy sites in and around Mecca, the walk between the two hills of al-Safa and al-Marwah, and finally the return to Mecca for a last visit to the Kabaa.

Most pilgrims stay in Mecca for about a month, although the actual ceremonies take only a few days. Pilgrims who have travelled far to reach Mecca often stay a year or longer, also visiting Medina – Islam's second holy city, 300 km north of Mecca, where the prophet Mohammed died and is buried (King 1972).

History and evolution

The history of pilgrimage in the Mecca area is described by Rutter (1929), King (1972) and Shair and Karan (1979).

Mecca was already a prosperous commercial centre when Mohammed was born there in 570 CE. It had long been an important city, with an annual gathering of Arab nomads of the region who engaged in ritual celebration and exchanged news and goods. Arab tribes travelled to worship stone idols situated in or near the Kaaba area of Mecca (Wolf 1951). Mecca was the main station for the Arab caravans coming from Yemen and leading north to Damascus and other cities of the Levant. Major commercial caravan routes from Egypt, Syria, Yemen and Persia converged at Mecca.

Pilgrim circulations increased greatly after Mohammed established the Islamic State in Medina and as more and more people accepted the new religion. From the sixth to the tenth centuries commercial trade dropped significantly, and pilgrimage became more dominant. The volume of Muslim pilgrims increased further throughout most of the Ottoman period from the sixteenth century onwards.

Because the sacred city of Mecca is closed to non-Muslims, until recently relatively few Westerners had witnessed the *hajj*. Lady Evelyn Cobbold was the first English woman to make the journey, which she describes in graphic detail:

> it was with a feeling of awe and reverence that I joined the vast throngs gathered together from the far-flung lands of Islam. . . . I stood in the Haram (great mosque), whose long arcades stretched away into the dusky distance, while the Holy of Holies, the mighty cube of the Kaaba, rose in simple majesty from the centre of the huge quadrangle. Broad paths led to it, and on each side of them lay exhausted pilgrims stretched asleep on the gravel. . . . The night before the Great Pilgrimage, Mecca was a seething mass of hadjis and camels. Sleep was impossible. . . . When we passed through the stone pillars marking the end of the forbidden territory I felt I had not only performed a sacred duty, but I had also seen and lived the greatest pageant of history.
>
> (Cobbold 1935: 107–16)

Sources, numbers and pathways

The dynamics of the *hajj* pilgrimage are described in a number of geographical papers (including Anon 1937, Isaac 1973, and Rowley and El-Hamdan 1977). King (1972) offers the most detailed description.

Most pilgrims travelled overland until the nineteenth century when the Suez Canal was cut and East Indies steamship lines brought vast numbers of Far East pilgrims by sea. Damascus and Cairo were the two most important collecting points for the pilgrims *en route* to Mecca (Figure 8.4). Despite transport improvements during the nineteenth century (particularly in shipping), numbers of pilgrims declined mainly because of Bedouin attacks on pilgrim caravans, and political instability caused by Turkish war efforts.

A revival of the *hajj* during the early twentieth century was assisted by the construction of a special Hejaz railway connecting Damascus with Medina, starting in 1900 (McLoughlin 1958). The railway was built by German engineers and operated by a *waqf* (self-perpetuating non-profit religious endowment) and paid for largely by donations from Muslims around the world. Its impact was to be short-lived, because the line was blown up in 1917 by Colonel T.E. Lawrence and his Bedouin Arab supporters.

Many pilgrims travelled overland to Mecca, using two main caravan routes, from Syria and Egypt. A popular pilgrim caravan travelled across Central Africa from the west coast (Rutter 1929, Birks 1977) eastwards to Nigeria. Many African pilgrims spent up to three years on their journey, trading, working or begging along the way, travelling mostly on foot with their families. It was not uncommon for children to be born along the way, and for many pilgrims to die before they reached their holy goal.

The growth in significance of the *hajj* has affected transport in a number of ways (King 1972). New pilgrimage routes were established linking Mecca with Iraq, Iran and Oman, and the overall pattern of transport within Saudi Arabia became highly focused on Mecca. Pilgrim traffic is heavily concentrated at one time in the year, and it is unidirectional in nature (towards Mecca before the pilgrimage, away from Mecca afterwards). The movement of vast numbers of pilgrims towards Mecca has also encouraged the expansion of settlements and oases along pilgrim routes.

The pattern of source areas of pilgrims (Figure 8.5) suggests a fairly strong distance-decay effect, with most travelling relatively short distances to Saudi Arabia. Middle Eastern Arab countries such as Yemen and Turkey provided many pilgrims in 1968, and an outer ring providing smaller numbers includes Libya, Egypt and Sudan in North Africa and Iran, Pakistan and India to the east. More distant source areas generally provided smaller numbers of pilgrims, although there were some notable exceptions (such as Indonesia and Malaysia).

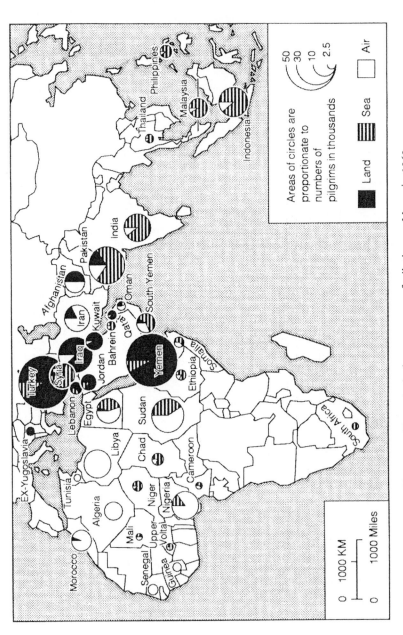

Figure 8.5 The principal source areas of pilgrims to Mecca in 1968

Source: After King (1972).

Changing character of the pilgrimage

Numbers attending the *hajj* have fluctuated through time, largely in harmony with waves of economic and political change around the world. Data summarised by King (1972) show that the estimated 152,000 pilgrims in 1929 had fallen to 20,000 in 1933 because of world depression, and then recovered to 67,000 in 1936 and 100,00 in 1937. The Second World War saw a fall in the number of pilgrims (there were an estimated 9,000 non-Arab pilgrims in 1939). Since 1945 numbers have risen progressively (Figure 8.6), with minor down-turns associated with Arab wars (such as the 1967 Arab-Israeli War, when many Muslims are reported to have given their *hajj* savings to the Arab cause).

Shair and Karan (1979) report a number of changes in the pattern of pilgrims since 1945, including their volume, sources of origin and mode of transport. Increasing volumes particularly since the early 1970s largely reflect greater ease of travel and higher per capita incomes in the oil-rich Middle Eastern Arab countries. Rowley (1989) reports an estimated 2.25 million pilgrims in 1986 (1.64 million of them from outside Saudi Arabia). The main source area of pilgrims continues to be the Arab countries (which provided over half in 1974), and non-Arab Asian and African countries. Numbers from particular donor countries (such as Nigeria, Yemen, Iraq, Pakistan, Sudan, Iran, Algeria, Egypt and Indonesia) vary through time in response to the changing domestic economic and political climate. Travel by air has opened up the *hajj* to travellers from much further afield (1975 saw 55 per cent of pilgrims travel by air, 13 per cent by sea and 32 per cent overland) – particularly from Iran, Bangladesh, Thailand, Singapore and Afghanistan. Most use Jiddah airport, which has been enlarged and developed by the Saudi Arabian government. New good roads have also encouraged more overland pilgrims in recent years (the number tripled between 1958 and 1969).

King (1972) emphasises that one of the most important improvements in the *hajj* has been medical improvements, which have reduced the traditionally high mortality rates amongst pilgrims associated with unsanitary and crowded conditions in the holy cities, and exposure to many contagious diseases (particularly smallpox, cholera and malaria).

Explanatory models

Two interesting geographical attempts have been made to develop statistical models of pilgrim movements associated with the *hajj*. Rowley and El-Hamdan (1978) sought to explain the contemporary pattern of *hajjis* and predict future numbers for 1983 and 1993, using empirical data. They tested the hypothesis that the total number of pilgrims coming from a country is a function of the size of the country's Muslim population, its per capita income, its distance from Mecca, the cost of travel between it and Mecca, and the time it takes to travel between it and Mecca. The results, based on data from 36 countries,

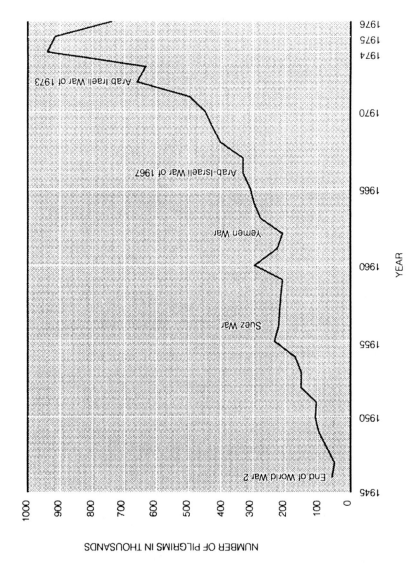

Figure 8.6 Growth in the number of pilgrims to Mecca, 1946–76

Source: After Shair (1979).

highlight two important factors (relative time and cost) which influence pilgrims' choice of mode of travel (by air or sea). The longer the duration of the journey from a country the higher is the percentage of its pilgrims who travel by air. The higher the cost of air travel relative to sea travel from a particular country, the lower is the percentage of its pilgrims who choose the air mode. Their logarithmic multiple regression model predicted that in 1993 more pilgrims would arrive by air than the total number (including Saudis) that undertook the *hajj* in 1973.

Shair and Karan (1979, see also Shair 1979) similarly developed a multiple regression model to explain the spatial pattern of pilgrimage to Mecca. This used the ratio of pilgrims per thousand Muslim population (the mean for 1965–75) as the dependent variable, and distance to Mecca, annual per capita national income, and percentage of Muslim population in the nation as independent variables. Per capita national income emerged as the most important factor determining the number of pilgrims (explaining 61 per cent of the variations for Arab countries, 74 per cent for non-Arab Asians, 72 per cent for non-Arab Africans, and 100 per cent for the North African countries).

Residuals from the regression could then be mapped (Figure 8.7). Positive residuals – with higher than predicted numbers of pilgrims – were explained in terms of high per capita national income (Bahrain, Qatar and Libya), ease of travel (Jordan, Yemen and the Democratic Yemen) and the presence of rich merchants who place a high value on pilgrimage (South Africa). Negative residuals – with lower than predicted numbers of pilgrims – were found in former Yugoslavia, Kuwait, the United Arab Emirates, Greece, Iran, Egypt, Palestine, Turkey, Pakistan and Tunisia. A variety of explanations were offered, including a lower value placed on pilgrimage, inequalities in wealth distribution (few very rich, many poor), government restrictions, and political instability.

Hindu and Buddhist pilgrimage

Pilgrimage also figures prominently in Buddhism and Hinduism, and in this section we examine some of the geographical studies that have focused on the patterns and dynamics of these population flows.

Buddhist pilgrimage

Amongst the most detailed and accessible geographical studies of Buddhist pilgrimage are those by Tanaka (1977, 1981). He examined the traditional circular pilgrimage on Shikoku Island, Japan, which involves a distance of 1,540 km (970 miles) and attracts pilgrims of both sexes, all ages and all classes of society.

The focus of this pilgrimage is a series of 88 Buddhist temple compounds where pilgrims carry out set religious rituals. Historical records show that the

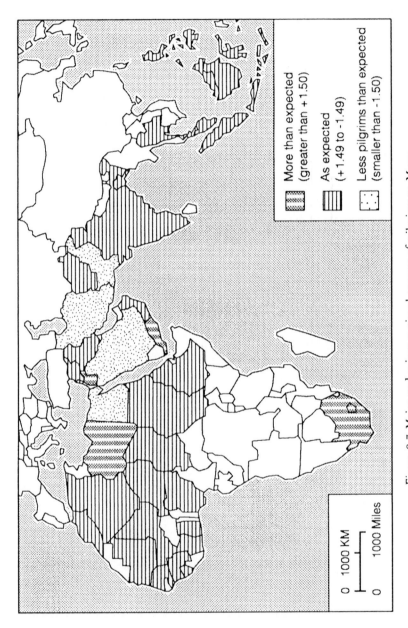

Figure 8.7 Major and minor national sources of pilgrims to Mecca

Note: The map shows the pattern of residuals from a multiple regression, as described by Shair (1979).

positions of the sacred places are not necessarily absolutely fixed, because fourteen of the sacred places have shifted locations since the mid–sixteenth century. Tanaka (1977: 113) argues that 'the geographic setting of the sacred place is characterised not solely by the site as it exists unaltered by man, but rather primarily by the assemblage of landscape markers that have been invested with special meaning'. He identifies 35 types of physical feature of the 88 Buddhist compounds (including halls, priests' residences, gates, statues, sacred water and pilgrim road signs) and 13 discrete sets of pilgrim activities (which define the 'ritual units').

The evidence suggests that the sanctity that defines the sacred places has some mobility and it 'is expressed in the landscape through the physical setting and the associated pilgrim ritual, both of which seem to conform to relatively fixed formal patterns' (Tanaka 1977: 131). Spatial adjustments in the pilgrimage circuit have evolved as a response to the evolution of the pilgrimage itself (Tanaka 1981) – at least 14 sacred sites have changed location since the middle of the sixteenth century, the point of embarkation has not been constant through time (temples were only numbered after the end of the seventeenth century), most but not all pilgrims visit the sites in a clockwise direction, and travel times have been reduced from two months (on foot) to two weeks by the use of horse-drawn wagons, local railways and charter buses. Tanaka (1981: 250) concludes that 'the pilgrimage sites as cultural symbols emerge, decline, and sometimes diffuse. Spatial patterns of pilgrimage evolve over time'.

Compatible evidence of the evolution of Buddhist pilgrim routes is offered in Onodera's (1990) study of travellers' records left in the Kanto district between 1706 and 1939. New routes were introduced at the start of the eighteenth century, partly as a result of increased pilgrimage to Konpira shrine but also from a change of travellers' consciousness towards the pilgrimage.

Hindu pilgrimage

Pilgrimage occupies a central position within Hinduism and it is a prominent feature of the religious landscape of India (Bhardwaj 1973). Hindu pilgrimage is undertaken for a variety of reasons, including seeking a cure for ill-health, washing away sins, and fulfilling a vow or promise to the deity for favours granted.

Bhardwaj (1985) outlines the diversity of the Hindu pilgrimage system that contains many different types of sacred place (*tirthas*), particularly places associated with temples (*mandir tirtha*) and those associated with sacred water (*jala tirtha*). Local deities and saints are worshipped at some sites, while other sacred sites are dedicated to male gods of lesser stature than the trinity (Shiva, Vishnu, Brahma) or to goddesses. Important sacred centres are often associated with the life and activities of the major gods (particularly Vishnu).

An estimated 100,000 pilgrims visit the Hindu shrine at Badrinath – a small hamlet in the Central Himalayas – each year between May and November (Watson 1961). The shrine is dedicated to Vishnu, and the pilgrims regard every rock, river and mountain they encounter along the way as sacred. Unlike many Hindu pilgrimage centres, where fairs and cheerful company are part of the pattern, the Badrinath pilgrimage is simple and personal. Pilgrims (many of them elderly) travel on their own, or in small groups or families.

Hindu pilgrimage to Muktinath in the Nepal Himalayas near Tibet has been described in detail by Messerschmidt (1989a, 1989b). Eliade (1959) argued that choice of holy space in most religions is not random, but is found and identified by the help of mysterious signs. In Muktinath these include the high mountain location and headwater site, and the presence of certain natural elements such as fossils and fires. The main access to Muktinath for Hindus is from the south along the steep, deeply incised valley of the Kali Gandaki river that contains numerous lesser religious sites and shrines, often in strategic places such as the junctions of rivers. The smaller sacred places are stopping-off points *en route* to the main temple complex at Muktinath, centred on the Mandir (temple) of Vishnu. Messerschmidt notes how

> each stop along the way is, in itself, a link of greater or lesser importance between earth and heaven, man and god, the profane and the sacred, the mundane world left behind and the sacred world at the apex of the pilgrimage, on the mountainside at Muktinath. Each step provides one of several sacred thresholds, in series, through which the pilgrim must pass to attain the final goal, the devotional climax.
>
> (Messerschmidt 1989a: 90)

The central part of Muktinath houses the temple complex. The main temple (Vishnu Mandir) is a three-tiered pagoda-style structure that houses an image of Vishnu. The temple complex also includes associated pilgrim rest houses and is associated with the presence of spring water, natural gas fires, fossils, and a sacred grove of poplar trees.

> The natural features . . . are representative of the deities or of their magical activities, and are looked upon by pilgrims with awe and wonder as nature made supernatural. In addition, the combination of an arduous pilgrim trek, the passage across the thresholds of sacred sites en route, and the ultimate belief that salvation is obtained through 'site vision' of Lord Vishnu at the site, makes Muktinath a powerfully evocative place.
>
> (Messerschmidt 1989a: 104)

Interviews with a sample of Muktinath pilgrims revealed that some aspects of the pilgrimage ritual serve to affirm social status, so that 'the Hindu pilgrimage may be viewed . . . as an enhanced or encapsulated image of Brahminical society in which structure-affirming behaviour is expected, is the norm' (Messerschmidt 1989b: 117).

Micro-scale studies of attendance patterns at Meenakshi Temple at Madurai in India, by Noble *et al.* (1987), reveal interesting detail of the dynamics of pilgrim behaviour. The temple is one of the most important centres of Hindu worship, attracting pilgrims from all over India. Visitor behaviour was monitored during a festival week in June–July 1977, when an average of around 40,000 people visited the temple each day. Peak attendances were recorded before 9.30 a.m. (coinciding with the morning journey to work, market and school) and between 7 and 8 p.m. (as people return home from various activities). Some days, which Hindus regard as auspicious on astrological grounds, attracted more visitors to pray or take part in religious ceremonies at the temple.

Like the Islamic pilgrimage to Mecca (see pp. 263–71), Hindu pilgrimage circulation has changed in recent decades as transport improvements make such journeys easier. Sopher (1968) emphasises the impact of such increased mobility on pilgrim circulation in Gujarat, where cars are replacing walking and some sections of society are following pilgrim routes for secular, tourist reasons rather than out of religious duty. But there is also an opposing trend, towards 'Sanskritization', among middle-level rural populations who have increasing opportunities to travel. This has encouraged an increase in the popularity of the traditional all-India religious tour, involving the circulation of religious specialists. This curious blend of secularisation and sacrilisation illustrates how dynamic pilgrimage can be in practice.

Roman Catholic pilgrimage in Europe

There are many religious sites within Europe that attract vast numbers of pilgrims, most of them associated with Roman Catholicism. Nolan and Nolan's (1989) book *Religious pilgrimage in modern Western Europe* is a useful source of information. This synthesises much of the material published by Nolan (1983, 1986, 1987a, 1987b) which provides the core of this section.

Table 8.2 Number of Catholic shrines in different European countries

Country	Number	Country	Number
Ireland	113	Portugal	337
France	1,035	Switzerland	283
Austria	925	Benelux	199
West Germany	831	England, Scotland & Wales	152
Italy	661	Scandinavia	10
Spain	584		
	Total number of shrines = 5,130		

Source: Nolan (1983).

Inventory and characteristics

There are more than 5,000 pilgrimage sites within Europe that are visited by an estimated 70 to 100 million people per year. France has the greatest concentration of Catholic shrines, but there are also large numbers in Austria, the former West Germany, Italy and Spain (Table 8.2). The shrines vary from major international pilgrimage centres, which attract millions of visitors each year from around the world, to local sites such as small chapels, roadside crosses and holy wells visited by the faithful from surrounding villages.

The pilgrimage sites are not evenly distributed (Figure 8.8), even within Catholic areas, and the distribution is not closely correlated with population density. Nolan (1987a) provides a detailed inventory of 6,150 Christian pilgrimage shrines in Western Europe that updates the figures in Table 8.2. Over 1,000 active places of pilgrimage are reported in both France and Spain, 938 in former West Germany and 121 in Ireland. The inventory shows that pilgrimage seems particularly strong in Catholic parts of Europe's Germanic lands, and that Austria and Switzerland have the most shrines per Roman Catholic citizen.

Jackowski (1987) describes the geography of pilgrimage in Poland which has around 500 pilgrimage sanctuaries, the vast majority of which (430) are dedicated to the Virgin Mary. Over half (65 per cent) of the sanctuaries have local significance, 30 per cent are of regional importance, and the rest have national and international importance. One of the most important is the eighteenth-century Marian shrine at Jasna Gora (Czestochowa), visited by an estimated 3.5 million pilgrims each year. Some 700,000 of the Jasna Gora pilgrims are from outside Poland, representing an estimated 45 countries. The other internationally important pilgrimage site in Poland is Niepokalanow that is visited by around 10,000 pilgrims a year from 23 countries.

Many of the Christian shrines dotted throughout Europe date back nearly two millennia. Nolan's (1987a) inventory shows that the first such shrines developed at the tombs of apostles and martyrs possibly as early as the end of the first century CE, and that about 4 per cent of the shrines were holy places in pre-Christian pagan times. Evidence of dates of shrine formation and decline 'suggested patterns of ebb and increase in shrine-formation activities that provide a good fit with historical periods of major socio-cultural change' (ibid.: 231), with major phases of shrine formation during the High Medieval period (1100–1399) and in the post-Reformation era (1530–1779) (Table 8.3).

Comparison with the much smaller number of Christian shrines in India (Nolan 1987b: 372) suggests that 'in the aggregate, Christian pilgrimage in India is both old and very recent'. A similar proportion of the shrines was established before 1100, few shrines were established over the following seven centuries, and most active Christian shrines in India are less than 200 years old (Table 8.3).

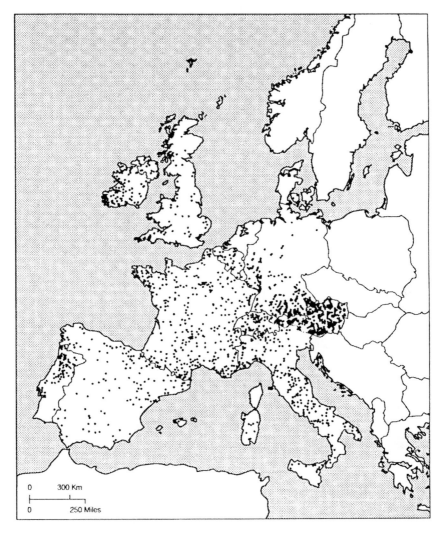

Figure 8.8 The distribution of active Roman Catholic shrines in Western Europe
with sacred site features

Source: After Nolan (1986).

Most Christian pilgrimage focuses on a specific historical person and by far
the most common subject of devotion in modern Catholic pilgrimage is the
Virgin Mary (Mother of Jesus), to which two-thirds of Europe's shrines are
dedicated. Christ is venerated at only one in twelve sites, and just over a quarter
of the sites is dedicated to saints (Table 8.4). Mary is the primary subject of
devotion at half the Christian shrines in India, but the tendency to focus on

Table 8.3 Periods of cult establishment at dated shrines in Western Europe and India

| | Per cent of dated shrines | |
| | *Western Europe* | *India* |
Period	*(n = 4,049)*	*(n = 71)*
Early (pre-1099)	16	15
High Medieval (1100–1399)	22	3
Renaissance (1400–1529)	15	0
Post-Reformation (1530–1779)	32	18
Modern (1780–1899)	14	63

Source: Nolan (1987b).

Table 8.4 Primary subjects of devotion at religious shrines in Western Europe and India

| | Per cent of all shrines | |
| | *Western Europe* | *India* |
Subject	*(n = 6,051)*	*(n = 145)*
Mary	66	48
Christ	8	7
Saints	27	45

Source: Nolan (1987b).

saints (particularly male saints) is much stronger in India than in Western Europe.

Nolan (1987a: 233) reports that many of the Catholic shrines contain cult objects, defined as 'potentially movable objects that are either the remains of deceased humans or are made by human hands'. Common cult objects include the physical remains of deceased holy people, and objects touched by or associated with such people. In the European tradition, the actual image is not so important as what it represents – 'the image is symbolic of a particular manifestation of the power of the divine individual, usually as expressed in a particular place' (ibid.: 234).

Just under half of Europe's active shrines are associated with sacred features of site, particularly height, water, trees and groves, caves and stones (Table 8.5). Such environmental features were important aspects of pilgrimage shrine location in pre-Christian Europe, and suggest that many Christian shrines may have either reused earlier holy sites or adopted those with similar environments. Sacred site features are important at a fifth of the Indian Christian shrines, with high place location the most common feature and presence of water more common in India than in Western Europe.

278

Table 8.5 Specific features at shrines with sacred environmental aspects of site

| Site feature | Per cent of all shrines | |
	Western Europe (n = 2,022)	India (n = 29)
Height	48	52
Water	35	48
Trees	21	14
Stones	10	17
Caves	10	10

Source: Nolan (1987b).

Nolan (1986) traces the origins of pilgrimage traditions based on the veneration of nature back to pre-Christian pagan nature worship. Such traditions are usually founded on the assumed involvement of animals and plants in revealing sacred places, and on the assumed relationship between natural features and the sanctity of religious pilgrimage sites. Plants are venerated at 463 Christian shrines in Europe, with the major focus on individual tree associations (331 shrines) and forests and groves (98 shrines). Domesticated animals are venerated at 152 shrines, with nearly half (74) associated with cattle and most of the rest (59) associated with horses, mules and donkeys. Wild animals are venerated at 56 sites, accounted for mostly by birds (29) and deer (15).

Pilgrimage in Ireland

Although Ireland has less Catholic shrines than many other countries in Western Europe (Table 8.2), pilgrimage is an important feature of its religious landscape. Nolan (1983) argues that Irish pilgrimage is unique within Europe because of its strong roots in antiquity (many shrines occupy pagan and early Celtic Christian sites). Most shrines in Ireland were established more than 1,000 years ago and only about 13 per cent of active Irish shrines post-date the Reformation. Whilst there are a number of important shrines dedicated to the Virgin Mary, most Irish pilgrimage focuses on the saints (mostly males). The Marian shrines are often superimposed on earlier shrines dedicated to male saints.

Other unique aspects of Irish pilgrimage – more like Hindu pilgrimage in India than Christian pilgrimage in the rest of Europe – include its strong orientation towards masculine symbols of divinity and its strong focus on natural site features. Many more shrines in Ireland (92 per cent) than in Western Europe (42 per cent) are associated with sacred site features, particularly sacred stones (53 per cent versus 9 per cent) and holy wells, springs and other water features (86 per cent versus 34 per cent). Human remains, relics or images are focal points for pilgrim veneration at 63 per cent of the shrines throughout Western Europe, but are only so in 9 per cent of Irish shrines.

Whilst most Irish sacred sites are ancient, some recent examples attract vast numbers of pilgrims. The best example is Ireland's largest shrine, located at Knock in County Mayo. Nolan describes how

> here millions of pilgrims, including Pope John Paul II, came to pay their respects during the 1979 centenary of the Virgin Mary's appearance to 15 awed villagers near the gable end of a modest parish church. Owing in part to the efforts of the [then] current shrine administrator, [the former] Monsignor James Horan, Knock has become one of the great mainstream pilgrimage shrines of modern Europe. This 'Lourdes of Ireland' is fully representative of late twentieth-century European pilgrimage shrines with its vast expanse of parking lots, modernistic basilica, special services and accommodations for the sick, and its emphasis on interpretation of the meaning of pilgrimage as understood by contemporary theologians.
>
> (Nolan 1983: 426)

The tiny village of Knock (population 400) has an international airport! The airport was conceived as a gateway to the nearby Marian shrine in a vision by the late Mgr Horan, funded by international donations, and opened in 1986. It also serves as a gateway to Reek Croagh Patrick – Ireland's holiest mountain – a bus journey away near Westport. Folklore describes it as a pagan site consecrated by St Patrick who fasted 40 days and nights on the summit, wrestling with demons and throwing snakes into the sea. Burke (1992) describes how an estimated one million pilgrims climb to the summit each year, some 40,000 on the last Sunday in July. The ascent is seen as an act of penance for wrongdoing, and many pilgrims scramble barefoot over rough stones up a steep, zigzag path during their three-hour climb to the top; once there the penitents crawl seven times round the small stone church, then descend again.

Penitential pilgrimage – acts of personal discomfort or sacrifice, showing regret for one's wrongdoing – is a feature of a number of Irish pilgrimages. The most extreme penitential pilgrimage in Ireland, possibly in the whole of Western Europe (Nolan 1983), is the St Patrick's Purgatory pilgrimage at Station Island in Lough Derg, County Donegal. It involves a three-day ordeal of physical and mental stress and can only be undertaken by pilgrims over the age of 13. Pilgrims take part in a 24-hour vigil, fast on just one daily meal of toast and black tea, walk around barefoot on rough stones and pavements, and the perform a series of ritualistic exercises. The island is closed to all other visitors during the summer pilgrimage season (between 1 June and 15 August).

McGrath (1989) interviewed a sample of 250 Lough Derg pilgrims in 1985 and established an interesting user-profile. Many pilgrims had been before (often up to four times), reflecting a strong traditional belief that three visits to Lough Derg will ensure a happy death and a place in heaven. Most (93 per cent) pilgrims heard about the pilgrimage from family and friends, and two-thirds arrived there by car. Participation rates were highest in nearby and readily accessible areas. Few (15 per cent) pilgrims went alone, most went with friends (47 per cent), family (27 per cent) or organised groups (10 per cent). More than half (65 per cent) of the pilgrims were female, many were middle-class and well-educated, and pilgrims in the 15–25 and 35–45 years old classes were over-represented relative to the overall population. Reasons for under-taking the pilgrimage varied greatly. Many pilgrims mentioned spiritual reasons (including penance, asking God for special favours, and thanksgiving for good health, recovery from illness, and good exam results), but others went out of tradition or to enjoy the pilgrimage atmosphere.

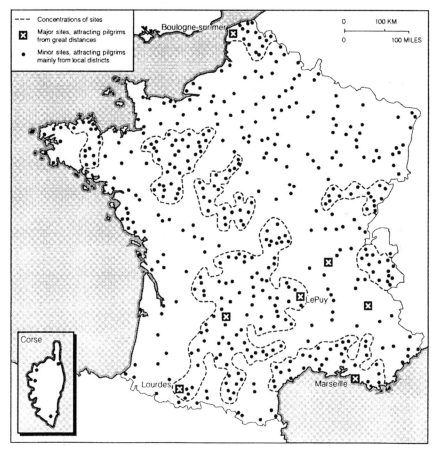

Figure 8.9 The distribution of pilgrimage sites of the Virgin Mary in France
Source: After Deffontaines (1948).

Lourdes

Without doubt the best-known pilgrimage centre in Europe is Lourdes in south-west France, at the foot of the Pyrenees close to the Spanish border (Figure 8.9). It is a major tourist centre, with the second largest number of hotels in France after Paris. Yet the pattern of religious tourism is unusual because most visitors go there between April and October and most visits are

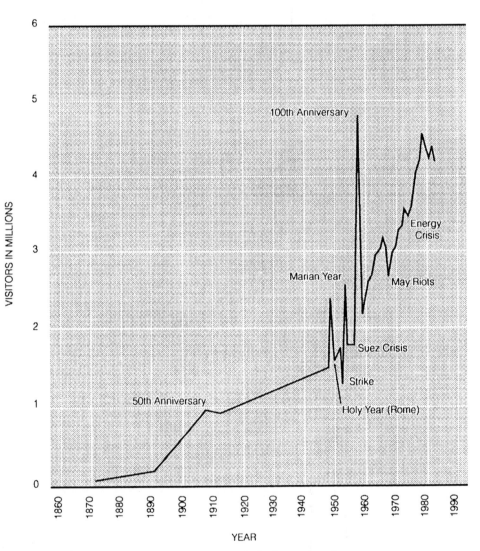

Figure 8.10 Growth in the number of pilgrims to Lourdes, 1872–1983
Source: After Rinschede (1986).

very short. Lourdes has been the focus of a number of geographical studies, including those by Lassere (1930) and Rinschede (1986, 1989).

This small town, with a population of around 18,000, attracts up to 5 million pilgrims each year (Jordan and Rowntree 1990). Many pilgrims seek miraculous cures at the famous grotto where the Virgin Mary is said to have appeared before 14-year-old Bernadette Soubirous in a series of 18 visions between February and July 1858 (Carroll 1985). Lourdes was a small rural town in the first half of the nineteenth century, but it developed rapidly as a pilgrimage centre after 1858. By 1872 more than 60,000 people had visited the site, and by 1908 the total had passed a million. In 1958 – the centenary of the visions – a record number of 4.8 million people visited the shrine (Figure 8.10). During the early 1980s an average of about 4 million pilgrims visited the shrine at Lourdes, compared with about a million visitors to Europe's other major Mary shrine at Fatima in Portugal (Vessels 1980). Between 1858 and the early 1980s, it is estimated that more than 200 million pilgrims had been to Lourdes.

Rinschede (1986, 1989) outlines the pattern of pilgrimage to Lourdes. In its early years Lourdes attracted mainly local believers, but now it attracts pilgrims from around the world. Six out of ten organised visitors in 1979 were foreign (compared with less than one in ten in 1895), and in 1978 visitors came from 111 different countries. Eight countries in West Europe accounted for almost all (97 per cent) of pilgrims in 1978 (Figure 8.11); France providing just over a third (37 per cent) of the total. On a per capita basis (the number of pilgrims per 1,000 Catholics) Ireland, Belgium and Britain provide more pilgrims to Lourdes than does France.

Most pilgrims (71 per cent in 1978) travel to the shrine on their own. Others go with small groups organised by private travel agencies, youth organisations and various religious institutions (11 per cent), or in large groups – often more than 1,000 people – organised by national agencies (about 18 per cent). Over two-thirds (69 per cent) of the pilgrims are female, and two out of three members of large organised groups are over 45 years of age. Labourers and rural people are strongly represented, while self-employed and highly educated people are under-represented. A small proportion (2 per cent in 1978) of the pilgrims is physically disabled, although this group grew significantly in size during the 1970s and comes mainly (65 per cent) from beyond France.

The pilgrim traffic to Lourdes is strongly seasonal. Most pilgrims visit between April and October, when weather conditions are most suitable for open-air activities at the shrine. This seasonality is reflected in the pattern of air and road traffic, and in the volume of postcards and letters handled by the local post office. Pilgrims travel to Lourdes by rail, bus, private car and (since 1948) by plane. In recent decades around two-thirds of all pilgrims arrived by train. The development of modern and faster means of transport, better mass transportation, and organised pilgrimages in recent years have all contributed to a marked increase in the numbers of pilgrims (Figure 8.10).

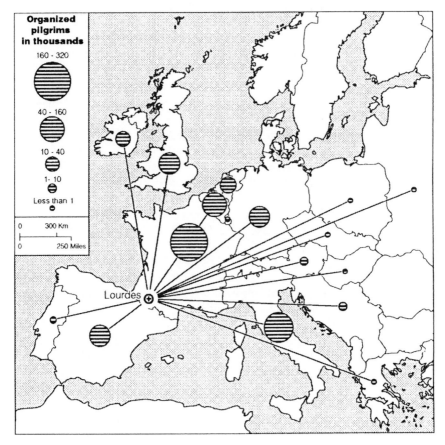

Figure 8.11 European origins of the group-organised pilgrims to Lourdes in 1978
Source: After Rinschede (1986).

The religious centre of Lourdes is the 'Domain of the Grotto'. This contains the grotto of the Marian Apparition, a spring and baths, a three-storey basilica built over the baths, and the subterranean Basilica of Pius X. Other prominent features of the religious landscape include the esplanade (open forecourt), hospitals and various administrative buildings. The religious area is surrounded by hotels, guest houses, and shops selling devotional articles.

By the mid-1980s Lourdes could provide 90,000 places for pilgrims to spend the night (a third of them in hotels and guest-houses, almost half in military and youth camps, and the rest in private quarters, flats, 'religious' houses, hospitals and camp sites). The growth in pilgrim traffic has been accompanied by changes in the townscape, including the widening of streets, and the renovating and demolition of old buildings. New hotels have been built, along with new public buildings (including a railroad station, town hospital, school and parish church, market halls and law courts).

CONCLUSION

The definition and demarcation of sacred places within what many people see as a secular world gives quite distinct territorial expression to religious belief and behaviour. Pilgrimage represents the main physical manifestation of the abiding pull of such sacred places, sometimes involving vast numbers of people travelling by various means from around the world. Few secular places can regularly attract as many visitors as Mecca and Lourdes, and the economic significance of pilgrims to such places must not be underestimated. But whilst most pilgrim centres are nowhere near so important and well-known as Mecca and Lourdes, they still have great meaning and significance to the faithful.

We have seen in this chapter that pilgrimage is motivated by different factors in different places. Some pilgrim trips are made out of duty, whereas others are made in the hope of receiving special blessings or healings. Yet others are made to increase personal holiness, or just simply to escape temporarily from the pressures of modern society.

Whilst motives vary, there are underlying patterns and processes which tie different pilgrimage traditions together, and deserve much closer geographical study and analysis. Such study could have at least three important benefits. The most obvious is to describe visible manifestations of pilgrimage, where the emphasis might be on population dynamics, flows and pathways. Closely related to this is the examination of the broader socio-economic implications of such population movements, for example on transport networks and local infrastructure. By no means the least important benefit is to increase our understanding of the experiential dimensions of pilgrimage for pilgrims. What does pilgrimage mean to them, how does it affect them, how relevant is it to the rest of their lives?

EPILOGUE

This book represents an exploration of some of the more important threads within the subject of geography religion, and I am as conscious of what it does not include as I am satisfied with what it does. It makes sense to offer a brief retrospective on why the book was written at all, and why it was written in this particular way!

I set out to provide a broad review of the work which geographers have done to date in this embryonic field, rather than to create an agenda for the future, and so the choice of what to include has been dictated largely by what published work is available. After an exhaustive search through the literature, and adopting a fairly liberal definition of 'religion' so as to reduce the risk of overlooking potentially interesting work, it was clear that the bulk of available studies are essentially empirical rather than theoretical. Inevitably, therefore, the structure of the book had to reflect this provenance of the material. Much of the published work was in the form of case studies, and in an attempt to paint as broad a canvas as possible I have erred on the side of inclusiveness and built as many as I could into the text. Readers will doubtless think that some case studies have been overplayed and some underplayed, and they will also be aware of others that I have not included. My apologies to any who feel that I have done particular themes a disservice in my treatment, and I encourage them to write to me and help me to get a better balance in future editions.

The main limitation of my chosen approach – reviewing what has been done rather than trying to sketch out a landscape from scratch – is the lack of an underlying theoretical framework. Each major theme covered in the book has a theoretical context, and I have tried to point this out (implicitly or explicitly) wherever relevant. To have adopted a particular theoretical framework and structured the body of the book around it might well have constrained the breadth of material which could be included, and would certainly have reduced the utility of the book as an introductory text.

In final analysis, I hope that what emerges from the reviews of major themes in successive chapters is an impression that the field of geography and religion is a legitimate one for geographers to work within, which now has a strong foundation to build upon, explores interesting and relevant issues, and has great potential.

If this book encourages others to look afresh at the field and to start breaking new ground by focusing geographical research on some of the key themes identified here, then this labour of love will have been well rewarded.

BIBLIOGRAPHY

Aay, H. (1972) Confronting the ecological crisis; the Kingdom of God in geographical perspective. *Vanguard*, Nov; 7–13, 27

Aay, H. (1976) Geography; calling and curriculum – I. *Christian Educators Journal*, Nov; 16–19

Aay, H. (1977) Geography; calling and curriculum – II. *Christian Educators Journal*, Jan; 11–14

Abler, R., J.S. Adams and P. Gould (1972) *Spatial Organization*. London, Prentice-Hall

Agius, F. (1990) Health and social inequalities in Malta. *Social Science & Medicine* 31; 313–18

Ahern, G. and G. Davie (1987) *Inner City God*. London, Hodder & Stoughton

Akoto, E. (1990) Christianity and inequalities in infant mortality rates in sub-Saharan Africa (in French). *Population (Paris)* 45; 971–92

Alba, R.D. (1981) The twilight of ethnicity among American Catholics of European ancestry. *Annals of the American Academy of Political and Social Science* 454; 86–97

al-Faruqi, I. and D. Sopher (eds) (1974) *Historical Atlas of the Religions of the World*. New York, Macmillan

Al-Haj (1988) The Arab internal refugees in Israel; the emergence of a minority within the minority. *Immigrants and Minorities* 7; 149–65

Allon-Smith, R. and D. Crouch (1979) Planning and the church; concepts and processes. *Chelmer Working Papers in Environmental Planning 2*, 30 pp.

Amir, S. (1990) Evaluation of environmental impacts of large scale physical developments in central Galilee: role of experts and policy-makers. *Environmental Management* 14; 823–32

Andre, I.M. and C. Patricio (1988) Catholicism in Portugal: ecclesiastic organisation and religious practices: a regional analysis. *Finisterra* 23; 225–49

Anon (1937) Pilgrims progress to Mecca. *National Geographical Magazine* 72 (5); 627–42

Anon (1990) *Faith in the countryside: report of the Archbishops' Commission on Rural Areas*. Worthing, Churchman Publishing

Archer, J. (1975) Puritan town planning in New Haven. *Journal of the Society of Architectural Historians* 34; 140–9

Ariyaratne, A.T. (1980) The role of Buddhist monks in development. *World Development* 8 (7–8); 587–9

Arreola, D.D. (1988) Mexican American housescapes. *Geographical Review* 78; 299–315

Attfield, R. (1983) Christian attitudes to nature. *Journal of the History of Ideas* 44; 369–86

288

Bailey, F.M. (1935) Bhutan: a land of exquisite politeness. *The Geographical Magazine* 1; 85–97

Bainbridge, W.S. (1982) Shaker demographics 1840–1900; an example of the use of US census enumeration schedules. *Journal for the Scientific Study of Religion* 21; 352–65

Bainbridge, W.S. and R. Stark (1980) Sectarian tension. *Review of Religious Research* 22; 105–24

Baker, J.N.L. (1928) Nathanael Carpenter and English geography in the seventeenth century. *Geographical Journal* 71; 261–71

Baly, D. (1957) *The Geography of the Bible*. London

Baly, D. (1979) Jerusalem, city of our solemnites; politics of the Holy City. *Geographical Perspectives* 42; 8–15

Bar-Gal, Y. (1985) The Shtetl; the Jewish small town in Eastern Europe. *Journal of Cultural Geography* 5; 17–30

Barrett, D.B. (ed.) (1982) *World Christian Encyclopedia; a Comparative Study of Churches and Religions in the Modern World, AD 1900–2000*. Nairobi, Oxford University Press

Barrows, H.H. (1923) Geography as human ecology. *Annals of the Association of American Geographers* 13; 1–14

Beaumont, P., G.H. Blake and J. M. Wagstaff (1976) *The Middle East: a Geographical Study*. London, Wiley

Begag, A. (1990) The 'Beurs', children of North-African immigrants in France; the issue of integration. *Journal of Ethnic Studies* 18; 1–14

Bellah, R.N. (1964) Religious evolution. *American Sociological Review* 29; 38–74

Bellah, R.N. (1967) Civil religion in America. *Daedalus* 96; 1–21

Ben-Arieh, Y. (1975) The growth of Jerusalem in the nineteenth century. *Annals of the Association of American Geographers* 65; 252–69

Ben-Arieh, Y. (1976) Patterns of Christian activity and dispersion in nineteenth century Jerusalem. *Journal of Historical Geography* 2; 49–69

Ben-Arieh, Y. (1978) Jerusalem as a religious city. *York University Department of Geography Discussion Paper Series 21*, 33 pp.

Ben-Arieh, Y. (1989) Nineteenth-century historical geographies of the Holy Land. *Journal of Historical Geography* 15; 69–79

Berdichevsky, N. (1980) New Orlean's churches; an index of changing urban social geography. *Ecumene* 12; 44–54

Berry, R.J. (1991) Christianity and the environment: escapist mysticism or responsible stewardship. *Science and Christian Belief* 3; 3–18

Bhardwaj, S. (1973) *Hindu Places of Pilgrimage in India; a Study in Cultural Geography*. Berkeley, University of California Press

Bhardwaj, S.M. (1985) Religion and circulation: Hindu pilgrimage; pp. 241–61 in R.M. Prothero and M. Chapman (eds) *Circulation in Third World Countries*. London, Routledge & Kegan Paul

Bhardwaj, S.M. (1987a) Single religion shrines, multireligion pilgrimages. *National Geographical Journal of India* 33; 457–68

Bhardwaj, S.M. (1987b) Geography and the 'hereafter'; pp. 321–33 in V.S. Datye *et al.* (eds) *Explorations in the Tropics*. Pune University of Poona

Bhatt, B.L. (1977) The religious geography of south Asia; some reflections. *National Geographical Journal of India* 23; 26–39

Bhattacharya, A.N. (1961) Geography and Indian religion. *National Geographer (India)* 4; 12–17

Bibby, R.W. and M.L. Brinkerhoff (1974) Sources of religious involvement. *Review of Religious Research* 15; 71–9

Bigelow, B. (1986) The Disciples of Christ in Antebellum Indiana: geographical indicator of the border south. *Journal of Cultural Geography* 7; 49–58

Birchall, J. (1983) The senior Christians who stay in Turkey. *Geographical Magazine* 55; 76–9

Birks, J.S. (1977) Overland pilgrimages from West Africa to Mecca; anachronism or fashion? *Geography* 62; 215–17

Biswas, L. (1984) Evolution of Hindu temples in Calcutta. *Journal of Cultural Geography* 4; 73–85

Bjorklund, E.M. (1964) Ideology and culture exemplified in southwestern Michigan. *Annals of the Association of American Geographers* 54; 227–41

Blachford, K. (1979) Morals and values in geographic education: toward a metaphysic of the environment. *Geographical Education* 3; 423–57

Blagg, T. (1986) Roman religious sites in the British landscape. *Landscape History* 8; 15–26

Blake, J. (1984) Catholicism and fertility: on attitudes of young Americans. *Population & Development Review* 10; 329–40

Boal, F.W. (1969) Territoriality on the Shankill–Falls Divide, Belfast. *Irish Geography* 6; 30–50

Boal, F.W. (1972) The urban residential sub-community; a conflict interpretation. *Area* 4; 164–8

Boal, F.W. (1976) Ethnic residential segregation; pp. 41–79 in D.T. Herbert and R.J. Johnston (eds) *Social Areas in the City. Volume 1: Spatial Processes and Form*. London, Wiley

Boal, F.W. and R.H. Buchanan (1969) Conflict in Northern Ireland. *The Geographical Magazine* (February); 331–6

Bohle, H.G. (1987) Spatial planning and ritual politics: the evolution of temple towns and urban systems in medieval South India; pp. 280–9 in V.S. Datye *et al.* (eds) *Explorations in the Tropics*. Pune, University of Poona

Bonine, M.E. (1987) Islam and commerce; Waqf and the Bazaar of Yadz, Iran. *Erdkunde* 41; 182–96

Boone, J.L., J.E. Myers and C.L. Redman (1990) Archaeological and historical approaches to complex societies; the Islamic states of Medieval Morocco. *American Anthropologist* 92; 630–66

Bourke, J. (1986) Piety or poverty: Catholic fertility in Australia and New Zealand 1911, 1926, 1936. *New Zealand Population Review* 12: 18–31

Bowen, W.A. (1976) American ethnic regions, 1880. *Proceedings of the Association of American Geographers* 8; 44–6

Bowman, G. (1991) Christian ideology and the image of a holy land; the place of Jerusalem pilgrimage in the various Christianities; pp. 98–121 in J. Eade and M. Sallnow (eds) *Contesting the Sacred*. London, Routledge

Boyd, J.M. (1984) The role of religion in conservation. *Environmentalist. Supplement* 7; 40–4

Boyer, P. and S. Nissenbaum (1974) *Salem Possessed; Social Origins of Witchcraft*. Cambridge, Mass., Harvard University Press

Bradley, I. (1987) Religious revival. *New Society* (6 November); 16–17

Bradshaw, M. (1990) The Christian and geographical explanation; pp. 376–82 in L. Francis and A. Thatcher (eds) *Christian Perspectives for Education*. London, Fowler Wright Books

Bragdon, K.J. (1989) The material culture of the Christian Indians of New England 1650–1775; pp. 126–31 in M.C.Beudry (ed.) *Documentary Archaeology in the New World*. London, CUP

Brah, A. (1987) Women of South Asian origin in Britain; issues and concerns. *South Asia Research* 7; 39–54

Breton, R. (1988) Religion and demographic change in India. *Population (Paris)* 43; 1089–122 (in French)

Brett-Crowther, M.R. (1985) Human ecology and development. *International Journal of Environmental Studies* 24: 187–204

Brierley, P. (1991) *Christian England*. London, Marc Europe

Brinkerhoff, M.B. and R.W. Bibby (1985) Circulation of the Saints in South America; a comparative study. *Journal for the Scientific Study of Religion* 24; 39–55

Broek, J.O.M. and J.W. Webb (1973) *A Geography of Mankind*. (Chapter 6 Religions; origins and dispersals, pp. 133–63) New York, McGraw Hill

Brook, S. (1979) Ethnic, racial and religious structure of the world population. *Population and Development Review* 5(3); 505–34

Brooke, C. (1987) Sacred slaughter; the sacrificing of animals at the Hajj and Id al-Adha. *Journal of Cultural Geography* 7; 67–88

Bruneau, T.C. (1980) The Catholic Church and development in Latin America; the role of the Basic Christian Communities. *World Development* 8 (7–8); 535–44

Brunhes, J. (1920) *Human Geography; an Attempt at a Positive Classification, Principles and Examples*. Chicago, Rand McNally

Brunn, S.D. and J.O. Wheeler (1966) Notes on the geography of religious town names in the United States. *Names; Journal of the American Name Society* 14; 197–202

Brush, J.E. (1949) The distribution of religious communities in India. *Annals of the Association of American Geographers* 39; 81–98

Buckwalter, D.W. and Legler, J.I. (1983) Antelope and Rajneeshpuram, Oregon – clash of cultures: a case study. *Urbanism Past and Present* 16: 1–13

Bullock, A. and O. Stallybrass (eds) (1981) *The Fontana Dictionary of Modern Thought*. London, Fontana/Collins

Burke, R. (1992) Pilgrims' uphill progress. *Weekend Guardian* (4–5 January); 29

Butlin, R. (1988) George Adam Smith and the historical geography of the Holy Land: contents, contexts and connections. *Journal of Historical Geography* 14; 381–404

Buttimer, A. (1976) Grasping the dynamism of lifeworld. *Annals of the Association of American Geographers* 66; 277–92

Buttner, M. (1974) Religion and geography; impulses for a new dialogue between *Religionswissenschaftlern* and geography. *Numen* 21; 163–96

Buttner, M. (1979) The significance of the Reformation for the re-orientation of Geography in Lutheran Germany. *History of Science* 17; 151–69

Buttner, M. (1980) Survey article on the history and philosophy of the geography of religion in Germany. *Religion*; 86–119

Buttner, M. (1987) Kasche and Kant on the physicotheological approach to the geography of religion. *National Geographical Journal of India* 33; 218–28

Cantrell, R., J. Krile and G. Donahue (1982) The community involvement of yoked parishes. *Rural Sociology* 47; 81–90

Carlyle, W.J. (1981) Mennonite agriculture in Manitoba. *Canadian Ethnic Studies* 13; 72–9

Carroll, M.P. (1985) The Virgin Mary at LaSalette and Lourdes; Whom did the children see? *Journal for the Scientific Study of Religion* 24; 56–74

Casper, D.E. (1983) *Religious Groups in Urban America: a Bibliography*. Vance Bibliographies, Public Administration Series: Bibliography P-1225

Cesarani, D. (1989) An embattled minority; the Jews in Britain during the First World War. *Immigrants & Minorities* 8; 61–81

Chandrakanth, M.G. (1989) Temple forests – their role in forestry development in India; pp. 319–25 in J. Krecek *et al.* (eds) *Headwater Control.* Prague, WASWC/IUFRO/CSVTS

Chandrakanth, M.G. *et al.* (1990) Temple forests in India's forest development. *Agroforestry Systems* 11; 199–211

Chapman, G. (1990) Religious vs regional determinism: India, Pakistan and Bangladesh as inheritors of empire. Chapter 6 in M. Chisholm and D.M. Smith (eds) *Shared Space, Divided Space.* London, Unwin Hyman

Chettiar, C.M.R. (1941) Geographical distribution of religious places in Tamil Nad. *Indian Geographical Journal* 16; 42–50

Christinger, R. (1965) Notions preliminaires d'une geographie mythique. *Le Globe* 105; 119–59 (in French)

Clark, A.H. (1960) Old World origins and religious adherence in Nova Scotia. *Geographical Review* 50; 317–44

Clar, G. (1984) *Innovation diffusion: contemporary geographical approaches.* Norwich, Geobooks

Clark, M. (1991) Developments in human geography: niches for a Christian contribution. *Area* 23; 339–44

Clarke, C.J. (1985) Religion and regional culture; the changing pattern of religious affiliation in the Cajun region of southwest Louisiana. *Journal for the Scientific Study of Religion* 24; 384–95

Clarke, J.I. (1985) Islamic populations; limited demographic transition. *Geography* 70; 118–28

Cloke, P., C. Philo and D. Sadler (1991) *Approaching Human Geography; an Introduction to Contemporary Theoretical Debates.* London, Paul Chapman Publishing

Cobbold, E. (1935) Pilgrim to Mecca. *Geographical Magazine* 1 (2); 107–16

Coburn, C.K. (1988) Ethnicity, religion and gender; the women of Block, Kansas, 1868–1940. *Great Plains Quarterly* 8; 222–32

Coleman, B.I. (1983) Southern England in the Census of Religious Worship, 1851. *Southern History* 5: 154–88

Compton, P.A. (1978) *Northern Ireland; a Census Atlas.* Dublin, Gill & Macmillan

Compton, P.A (1985) An evaluation of the changing religious composition of the population of Northern Ireland. *Economic & Social Review* 16: 201–44

Compton, P.A. and J. Coward (1989) *Fertility and Family Planning in Northern Ireland.* Aldershot, Avebury Press

Compton, P.A. and J.P. Power (1986) Estimates of the religious composition of Northern Ireland local government districts in 1981 and change in the geographical pattern of religious composition between 1971 and 1981. *Economic & Social Review* 17: 87–105

Conde, S.G. (1980) Moslem contributions to geography. *Philippine Geographical Journal* 24(2); 91–3

Cooke, P. (1990a) Modern urban theory in question. *Transactions, Institute of British Geographers* 15; 331–43

Cooke, P. (1990b) *Back to the Future; Modernity, Postmodernity and Locality.* London, Unwin Hyman

Coon, A.G. (1990) Development plans in the West Bank. *GeoJournal* 21; 363–73

Coones, P. (1986) *Euroclydon: a Tempestuous Wind.* Research Paper, University of Oxford School of Geography 36, 20 pp.

Cooney, G. (1990) The place of megalithic tomb cemeteries in Ireland. *Antiquity* 64; 741–53

Cooper, A. (1990) Geographies and religious commitment in a small coastal parish. *Geography of Religions and Belief Systems* 12; 3–5

Cooper, A. (1991a) Religio-geographical research and public policy. *Geography of Religions and Belief Systems* 13; 3–6

Cooper, A. (1991b) The geography of religion at the Association of American Geographers. *IBG Newsletter* (June); 10–11

Cooper, A. (1992) New directions in the geography of religion. *Area* 24; 123–9

Cornell G.L. (1985) The influence of Native Americans on modern conservationists. *Environmental Review* 9; 104–17

Cosgrove, D. (1990) Environmental thought and action: pre-modern and post-modern. *Transactions of the Institute of British Geographers* 15; 344–58

Costa, F.J. and A.G. Noble (1986) Planning Arabic towns. *Geographical Review* 76; 160–72

Creevey, L. (1980) Religious attitudes and development in Dakar, Senegal. *World Development* 8(7–8); 503–12

Croad, M. (1992) Easter promise. *Weekend Guardian* (27–28 June); 8–9, 26

Crouch, D. and A. Colin (1992) Rocks, rights and rituals. *Geographical Magazine* (June); 14–19

Crowley, W.K. (1978) Old Order Amish settlements; diffusion and growth. *Annals of the Association of American Geographers* 68; 249–64

Daaniel, O. (1990) The historical role of the Muslim community in Albania. *Central Asian Survey* 9; 1–28

Dahlan, A.S. (1990) Housing demolition and refugee resettlement schemes in the Gaza Strip. *GeoJournal* 21; 385–95

Darden, J.T. (1972) Factors in the location of Pittsburgh's cemeteries. *The Virginia Geographer* 7; 3–8

Datta, J.M. (1962) Influence of religious beliefs on the geographical distribution of Brahmins in Bengal. *Man in India* 42; 89–103

Davie, G. (1990) An ordinary god; the paradox of religion in contemporary Britain. *British Journal of Sociology* 41; 395–421

Davies, G.I. (1979) The way of the wilderness; a geographical study of the wilderness itineraries in the Old Testament. *Society for Old Testament Study Monograph Series* 5, Cambridge University Press, 138 pp.

Davis, K. (1980) Christian praxis for change; a Caribbean experiment. *World Development* 8(7–8); 591–601

de Almeida, J.F. (1986) Rural religious observance in Portugal. *Sociologia Ruralis* 26: 70–83 (in Portuguese)

Dear, M. (1988) The postmodern challenge: reconstructing human geography. *Transactions of the Institute of British Geographers* 13; 262–74

de Blij, H.J. and P.O. Muller (1986) *Human Geography; Culture, Society and Space.* New York, Wiley

Deffontaines, P. (1948) *Geographie et religions.* Paris, Gallimard

Deffontaines, P. (1953a) The religious factor in human geography; its force and its limits. *Diogenes* 2; 22–8

Deffontaines, P. (1953b) The place of believing. *Landscape* 3; 22–8

Degh, L. (1980) Folk religion as ideology for ethnic survival; the Hungarians of Kipling, Saskatchewan; pp. 129–46 in F.C. Luebke (ed.) *Ethnicity on the Great Plains.* Lincoln, University of Nebraska Press

de la Blanche, P.V. (1926) *Principles of Human Geography.* London, Constable

Delano Smith, C. (1987) Maps in bibles in the 16th century. *Map Collector* 39; 2–14

Delaruelle, E. (1943) Contribution a l'etude de la geographie religieuse du Sud-Ouest. *Revue Geographique des Pyrenees et du Sud-Ouest* 14; 48–78

Delavaud, A.C. (1981) The geographical role of the contemporary religious missions in the Orinoco delta and the Gran Savana of Venezuela. *Revista Geografica* 94: 53–65

Demerath, N.J. and W.C. Roof (1976) Religion – recent strands in research. *Annual Review of Sociology* 2; 19–33

Dempsey, M.W. (ed.) (1982) *Everyman's Factfinder*. Melbourne, J.M. Dent

Dethlefsen, E. and J. Deetz (1965–66) Death's heads, cherubs and willow trees: experimental archaeology in colonial cemeteries. *American Antiquity* 31; 502–10

De Vaus, D. (1982) The impact of geographical mobility on adolescent religious orientation; an Australian study. *Review of Religious Research* 23; 391–403

Dhesi, A.S. and H. Singh (1989) Education, labour market distortions and relative earnings of different religion-caste categories in India (A case study of Delhi). *Canadian Journal of Development Studies* 10; 75–89

Dhruvarajan, V. (1990) Religious ideology, Hindu women and development in India. *Journal of Social Issues* 46; 57–69

Dietrich, D.J. (1980) Christianity and conservation; an alternative to environmental exploitation. *Man-Environment Systems* 10; 3–10

Dillon, M. (1990) Perceptions of the causes of The Troubles in Northern Ireland. *Economic & Social Review* 21: 299–310

Din, K.H. (1989) Islam and tourism: patterns, issues and options. *Annals of Tourism Research* 16: 542–63

Doeppers, D.F. (1976) The evolution of the geography of religious adherence in the Philippines before 1898. *Journal of Historical Geography* 2; 95–110

Donaldson, P.J. (1988) American Catholicism and the international family planning movement. *Population Studies* 42; 367–73

Donkin, R.A. (1959) The site changes of Medieval Cistercian monasteries. *Geography* 44; 252–8

Donkin, R.A. (1964) The Cistercian grange in England in the twelfth and thirteenth centuries. *Studia Monastica* 6; 95–144

Donkin, R.A. (1967) The growth and distribution of the Cistercian Order in Medieval Europe. *Studia Monastica* 9; 275–86

Donkin, R.A. (1969) The Cistercian order and the settlement of Northern England. *Geographical Review* 59; 403–16

Doughty, R.W. (1981) Environmental theology; trends and prospects in Christian thought. *Progress in Human Geography* 5: 234–48

Dove, J. (1992) A life of their own. *Geographical Magazine* (August); 34–6

Dubey, D.P. (1985) The sacred geography of Prayaga (Allahabad); identification of holy spots. *National Geographical Journal of India* 31; 319–40

Dufourcq, E. (1988) A demographic approach to the spread of women's religious orders of French origin outside Europe. *Population (Paris)* 43; 45–76 (in French)

Duncan, J.S. (1980) The superorganic in American cultural geography. *Annals, Association of American Geographers* 70; 181–92

Durkheim, E. (1915) *The Elementary Forms of the Religious Life*. London, George Allen & Unwin

Dutt, A.K. and S. Davgun (1977) Diffusion of Sikhism and recent migration patterns of Sikhs in India. *GeoJournal* 1(5); 81–9

Dutt, A.K. and S. Davgun (1979) Religious pattern of India with a factorial regionalisation. *GeoJournal* 3; 201–14

Dyall, D.H. (1989) Mormon pursuit of the agrarian ideal. *Agricultural History* 63: 19–35

Eade, J. and M. Sallnow (eds) (1991) *Contesting the Sacred*. London, Routledge

Efrat, E. and A.G. Noble (1988) Planning Jerusalem. *Geographical Review* 78; 387–404

Eirinberg, K. (1992) Culture under fire. *Geographical Magazine* (December); 24-8

Eliade, M. (1959) *The Sacred and the Profane; the Nature of Religion*. New York, Harcourt, Brace and World

Elkind, D. (1970) The origins of religion in the child. *Review of Religious Research* 12; 35-42

Elsdon, R. (1989) A still-bent world; some reflections on current environmental problems. *Science and Christian Belief* 1; 99-122

Entrikin, J.N. (1976) Contemporary humanism in geography. *Annals of the Association of American Geographers* 66; 615-32

Epstein, R.H. (1992) Divergent paths. *Geographical Magazine* (April); 34-8

Evans, R. (1989a) Islam! From invasion to indenture. *Geographical Magazine* 61 (December); 8-11

Evans, R. (1989b) Fundamentalists flood the world. *Geographical Magazine* 61 (March); 10-14

Evans, R. (1990) Whose promised land? *Geographical Magazine* 62; 38-41

Evens, T.M.S. (1982) On the social anthropology of religion. *The Journal of Religion* 62; 366-91

Eyre, L.A. (1985) Biblical symbolism and the role of fantasy geography among the Rastafarians of Jamaica. *Journal of Geography* 84; 144-8

Fahy, G. (1974) Geography in the early Irish Monastic schools; a brief review of Airbheartach MacCosse's Geographical Poems. *Geographical Viewpoint* 3; 31-3

Falah, G. (1989) Israeli 'Judaization' policy in Galilee and its impact on local Arab urbanization. *Political Geography Quarterly* 8; 229-53

Falah, G. (1990) Arabs versus Jews in Galilee: competition for regional resources. *GeoJournal* 21; 325-36

Fecharaki, P. (1977) The Bahais of Nadjaf-Abad. *Revue Geographique de l'Est* 17; 89-91

Fenn, R.F. (1969) The secularisation of values. *Journal for the Scientific Study of Religion* 8; 112-24

Fickeler, P. (1962) Fundamental questions in the geography of religions; pp. 94-117 in P.L. Wagner and M.W. Mikesell (eds) *Readings in Cultural Geography*. Chicago, University of Chicago Press

Filor, S.W. (1988) Landscape architecture in Saudi Arabia. *Landscape Research* 13; 23-8

Finke, R. and R. Stark (1986) Turning pews into people; estimating 19th century church membership. *Journal for the Scientific Study of Religion* 25; 180-92

Fischer, E. (1957) Religions; their distribution and role in political geography; pp. 405-39 in H.W. Weigert (ed.) *Principles of Political Geography*. New York, Appleton

Fleure, H.S. (1951) The geographical distribution of the major religions. *Bulletin de la Société Royale de Géographie d'Egypte* 24; 1-18

Foster, R.H. (1981) Recycling rural churches in southern and central Minnesota. *Bulletin of the Association of North Dakota Geographers* 31; 1-10

Foster, R.H. (1983) Changing use of rural churches: examples from Minnesota and Manitoba. *Yearbook – Association of Pacific Coast Geographers* 45; 55-70

Francaviglia, R.V. (1970) The Mormon landscape; definition of an image in the American west. *Proceedings, Association of American Geographers* 2; 59-61

Francavilgia, R.V. (1971) The cemetery as an evolving cultural landscape. *Annals, Association of American Geographers* 61; 501

Frend, W.H.C. (1978) The Christian period in Mediterranean Africa, c. AD 200 to 700; pp. 410-89 in J.D. Fage (ed.) *The Cambridge History of Africa*, volume 2. Cambridge, Cambridge University Press

Fuson, R.H. (1969) The orientation of Mayan ceremonial centres. *Annals of the Association of American Geographers* 59; 494–511

Gale, S. (1977) Ideological man in a nonideological society. *Annals of the Association of American Geographers* 67; 267–72

Gamwell, F.I. (1982) Religion and the public purpose. *Journal of Religion* 62; 272–88

Gastil, R.D. (1975) *Cultural regions of the United States*. Washington, DC, University of Washington Press

Gaustad, E.S. (1976) *Historical Atlas of Religion in America*. New York, Harper & Row

Gay, J. (1971) *Geography of Religion in England*. London, Duckworth

Gerlach, L.P. and V.H. Hine (1968) Five factors crucial to the growth and spread of a modern religious movement. *Journal for the Scientific Study of Religion* 7; 23–40

Gezairy, H. *et al.* (1989) Eastern Mediterranean. *World Health* (July), 31 pp.

Gibbs, N. (1991) America's holy war. *Time* (9 December); 52–4

Gilbert, A. (1980) *The Making of Post-Christian Britain; a History of the Secularisation of Modern Society*. Harlow, Longman

Gilbert, C.P. (1991) Religion, neighbourhood environments and partisan behaviour: a contextual analysis. *Political Geography Quarterly* 10; 110–31

Gilbert, E.W. (1962) Geographie is better than divinity. *Geographical Journal* 128; 494–7

Gilsenen, M. (1990) *Recognizing Islam; Religion and Society in the Modern Middle East*. London, Tauris

Glacken, C.J. (1967) *Traces on the Rhodian Shore*. Berkeley, University of California

Glasner, P.E. (1977) *The Sociology of Secularisation; a Critique of a Concept*. London, Routledge

Glass, J.W. (1979a) Be ye separate, saith the Lord, pp. 51–63 in R.A. Cybriwsky (ed.) *The Philadelphia Region. Selected Essays and Field Trip Itineraries*. Washington, DC, Association of American Geographers

Glass, J.W. (1979b) Old Order Amish in Lancaster county; a self-guided tour, pp. 64–73 in R.A. Cybriwsky (ed.) *The Philadelphia Region. Selected Essays and Field Trip Itineraries*. Washington, DC, Association of American Geographers

Glock, C.Y. and R. Stark (1965) *Religion and Society in Tension*. Chicago, Rand McNally

Golde, G. (1982) Voting patterns, social context and religious affiliation in southwest Germany. *Comparative Studies in Society and History* 24; 25–56

Goldscheider, C. and W.D. Mosher (1988) Religious affiliation and contraceptive use; changing American patterns, 1955–82. *Studies in Family Planning* 19; 48–57

Gopal, B. (1988) Holy Mother Ganges. *Geographical Magazine* (May); 38–43

Gordon, B.L. (1971) Sacred directions, orientation and the top of the map. *History of Religions* 10; 211–27

Gosal, G.S. (1965) Religious composition of Punjab's population changes, 1951–61. *The Economic Weekly* 17; 119–24

Goulet, D. (1980) Development experts; the one-eyed giants. *World Development* 8(7–8); 481–9

Graber, L. (1976) *Wilderness as Sacred Space*. Washington, DC, Association of American Geographers Monograph Series

Gregorius, Bishop (1982) Christianity, the Coptic religion and ethnic minorities in Egypt. *GeoJournal* 6; 57–62

Gregory, D. (1981) Human agency and human geography. *Transactions of the Institute of British Geographers* 6; 1–18

Gregory, D. (1988) More at stake than good and evil. *Geographical Magazine* 60(9); 24–8

Gregory, D. and J. Urry (eds) (1985) *Social Relations and Spatial Structure*. London, Macmillan

Grinsell, L. (1986) The Christianisation of prehistoric and other pagan sites. *Landscape History* 8; 27–37

Grossman, D. (1991) Arab and Jewish settlement processes in west Samaria. *Pennsylvania Geographer* 29; 29–37

Gurgel, K.D. (1976) Travel patterns of Canadian visitors to the Mormon culture hearth. *Canadian Geographer* 20; 405–18

Gutmann, M.P. (1990) Denomination and fertility decline: the Catholics and Protestants of Gillespie County, Texas. *Continuity and Change* 5; 391–416

Hall, R. (1983) The teaching of moral values in geography. *Journal of Geography in Higher Education* 7; 3–13

Halsell, G. (1990) Islam in a Communist state. *Aramco World* 41; 34–42

Halvorson, P.L. and W.M. Newman (1978) *Atlas of Religious Change in America, 1952–1971*. Washington, DC, Glenmary Research Centre

Hamidullah, M. (1938) The city state of Mecca. *Islamic Culture* 12(3); 254–76

Hammond, N. (1972) The planning of a Maya ceremonial center. *Scientific American* 226; 82–91

Hannemann, M. (1975) *The Diffusion of the Reformation in Southwestern Germany, 1518–1534*. University of Chicago Department of Geography Research Paper 167.

Hardwick, W.G., R.J. Claus and D.C. Rothwell (1971) Cemeteries and urban land value. *Professional Geographer* 23; 19–21

Hargrove, E.C. (ed.) (1986) *Religion and Environmental Ethics*. Atlanta, University of Georgia Press

Harke, H. (1990) 'Warrior graves'? The background to the Anglo-Saxon burial rite. *Past and Present* 126; 22–43

Harrison, R.T. and D.N. Livingstone (1980) Philosophy and problems in human geography; a presuppositional approach. *Area* 12; 25–31

Harvey, D. (1979) Monument and myth. *Annals of the Association of American Geographers* 69; 362–81

Harvey, D. (1989) *The Condition of Postmodernity*. Oxford, Blackwell

Hassal, G. *et al.* (1990) Muslims in the USSR. *Aramco World Magazine* (Jan-Feb) 48 pp.

Hayward, D.F. (1982) The storm-tossed voyages of Saint Paul. *Geographical Magazine* 54; 664–5

Heatwole, C.A. (1978) The Bible Belt; a problem of regional definition. *Journal of Geography* 77; 50–5

Heatwole, C.A. (1985) The unchurched in the Southeast, 1980. *Southeastern Geographer* 25; 1–15

Heatwole, C.A. (1986) A geography of the African Methodist Episcopal Zion Church. *Southeastern Geographer* 26; 1–11

Heckenberger, M.J., J.B. Petersen and L.A. Basa (1990) Early woodland period ritual use of personal adornment at the Boucher site. *Annals of the Carnegie Museum* 59; 173–217

Hempton, D. (1984) *Methodism and Politics in British Society*. London, Hutchinson

Henkel, H. (1989) *Christian Missions in Africa: a Social Geographical Study of the Impact of their Activities in Zambia*. Zambia Geographical Association Occasional Study 16, 236 pp.

Hennessey, M. (1988) The priory and hospital of New Gate; the evolution and decline of a medieval monastic estate, pp. 41–54 in W.J. Smyth and K. Whelan (eds) *Common ground; essays on the historical geography of Ireland*. Cork, Cork University Press

Hershkowitz, S. (1987) Residential segregation by religion: a conceptual framework. *Tijdschrift voor Economische en Sociale Geografie* 78: 44–52

Hewitt, C. (1981) Catholic grievances, Catholic nationalism and violence in Northern Ireland during the civil rights period: a reconsideration. *British Journal of Sociology* 32: 362–80

Hey, D.G. (1973) The pattern of nonconformity in South Yorkshire, 1660–1851. *Northern History* 8; 86–118

Hilty, D.M., R.L. Morgan and J.E. Burns (1984) King and Hunt revisited: dimensions of religious involvement. *Journal for the Scientific Study of Religion* 23; 252–66

Hinnells, J.R. (ed.) (1984a) *The Penguin Dictionary of Religions*. Harmondsworth, Penguin

Hinnells, J.R. (ed.) (1984b) *A Handbook of Living Religions*. Harmondsworth, Penguin

Homan, R. and G. Rowley (1979) The location of institutions during the process of urban growth; a case study of churches and chapels in nineteenth-century Sheffield. *East Midland Geographer* 7(4); 137–52

Hookyas, R. (1957) *Religion and the Rise of Modern Science*. Edinburgh, Scottish Academic Press

Horton, M.C. (1987) Early Muslim trading settlements on the East African coast; new evidence from Shanga. *Antiquaries Journal* 67; 290–323

Hostetler, J.A. (1980) The Old Order Amish on the Great Plains; a study in cultural vulnerability; pp. 92–108 in F.C. Luebke (ed.) *Ethnicity on the Great Plains*. Lincoln, University of Nebraska Press

Houston, J.M. (1978) The concepts of 'Place' and 'Land' in the Judeo-Christian Tradition, p. 224–37 in D. Ley and M. Samuels (eds) *Humanistic Geography; Prospects and Problems*. Chicago, Maaroufa Press

Houtart, F. (1980) Attitudes towards development among Catholics in Sri Lanka. *World Development* 8(7–8); 603–12

Howett, C. (1977) Living landscapes for the dead. *Landscape* 21; 9–17

Hsu, Shin-Yi (1969) The cultural ecology of the Locust Cult in traditional China. *Annals of the Association of American Geographers* 59; 730–52

Hultkrantz, A. (1966) An ecological approach to religion. *Ethnos* 31; 131–50

Huntingdon, E. (1945) *Mainsprings of Civilization*. New York, John Wiley & Son

Huntingdon, E. (1951) *Principles of Human Geography*. London

Ibrahim, F.N. (1982) Social and economic geographical analysis of the Egyptian Copts. *GeoJournal* 6: 63–7

Isaac, E. (1959a) Influence of religion on the spread of citrus. *Science* 129; 179–86

Isaac, E. (1959b) The citron in the Mediterranean; a study in religious influences. *Economic Geography* 35; 71–8

Isaac, E. (1960) Religion, landscape and space. *Landscape* 9; 14–18

Isaac, E. (1962) The act and the covenant; the impact of religion on the landscape. *Landscape* 11; 12–17

Isaac, E. (1964) God's acre – property in land; a sacred origin? *Landscape* 14; 28–32

Isaac, E. (1965) Religious geography and the geography of religions. *Man and the Earth*, University of Colorado Studies, Series in Earth Sciences 3. Boulder, Colorado. University of Colorado Press, 1–14

Isaac, E. (1967) Mythical geography. *Geographical Review* 57; 123–5

Isaac, E. (1973) The pilgrimage to Mecca. *Geographical Review* 63; 406–9

Jackowski, A. (1987) Geography of pilgrimage in Poland, *National Geographical Journal of India*, 33; 422–9

Jackson, J.B. (1952) What we want. *Landscape* 1; 2–5

Jackson, J.B. (1967–8) From monument to place. *Landscape* 17; 22–6

Jackson, R.H. (1978a) Mormon perception and settlement. *Annals of the Association of American Geographers* 68; 317–34

Jackson, R.H. (1978b) Religion and landscape in the Mormon cultural region; pp. 100–27 in K.W. Butzer (ed.) *Dimensions of Human Geography*. University of Chicago Department of Geography Research Paper 186.

Jackson, R.H. and R. Henrie (1983) Perception of sacred space. *Journal of Cultural Geography* 3; 94–107

Jacobson, J.L. (1990) *The Global Politics of Abortion*. Worldwatch Paper 97, 69 pp.

Jeans, D.N. and E. Kofman (1972) Religious adherence and population mobility in nineteenth century New South Wales. *Australian Geographical Studies* 10; 193–202

Jeffery, A. (1929) Christians at Mecca. *Moslem World* 22(3); 109–16

Johnson, D.W., P.R. Picard and B. Quinn (eds) (1974) *Churches and Church Membership in the United States, 1971*. Washington, DC, Glenmary Research Centre

Johnson, H.B. (1967) The location of Christian missions in Africa. *Geographical Review* 57; 168–202

Johnson, J.H. (1962) The political distinctiveness of Northern Ireland. *Geographical Review* 52; 78–91

Johnson, S.D. and J.B. Tamney (1984) Support for the Moral Majority: a test of a model. *Journal for the Scientific Study of Religions* 23; 183–96

Johnston, R.J. (1983) *Geography and geographers*. London, Arnold

Johnston, R.J. (1986) *On Human Geography*. Oxford, Basil Blackwell

Jolles, C.J. (1989) Salvation on St Lawrence Island; Protestant conversion among the Sivuqaghhmiit. *Arctic Anthropology* 26; 12–27

Jones, E. (1960) Problems of partition and segregation in Northern Ireland. *Journal of Conflict Resolution* 4; 96–105

Jones, E. and J. Eyles (1977) *An Introduction to Social Geography*. London, Oxford University Press

Jones, G.R.J (1985) Churches and secular settlements in ancient Gwynedd. *Cambria* 12; 33–53

Jones, P.N. (1976) Baptist chapels as an index of cultural transition in the South Wales coalfield before 1914. *Journal of Historical Geography* 2; 347–60

Jordan, T.G. (1970) Population origin groups in rural Texas. *Annals of the Association of American Geographers* 60; 404–5

Jordan, T.G. (1973) *The European culture area*. New York, Harper & Row

Jordan, T.G. (1976) Forest folk, prairie folk; rural religious cultures in North Texas. *Southwestern Historical Quarterly* 80; 135–62

Jordan, T.G. (1980) A religious geography of the Hill Country Germans of Texas; pp. 109–28 in F.C. Luebke (ed.) *Ethnicity on the Great Plains*. Lincoln, University of Nebraska Press

Jordan, T.G. and L. Rowntree (1990) *The Human Mosaic; a Thematic Introduction to Cultural Geography*. New York, Harper and Row

Kan, S.H. and Y. Kim (1981) Religious affiliation and migration intentions in non-metropolitan Utah. *Rural Sociology* 46: 669–87

Karadawi, A. (1991) The smuggling of the Ethiopian Falasha to Israel through Sudan. *African Affairs* 90; 23–49

Kasche, G.H. (1795) *Ideas about Religious Geography*. Lubeck (in German)

Katz, Y. and S. Neuman (1990) Agricultural land transactions in Palestine, 1900–1914; a quantitative analysis. *Explorations in Economic History* 27; 29–45

Kay, J. (1988) Concepts of nature in the Hebrew Bible. *Environmental Ethics* 10; 309–27

Kay, J. (1989) Human dominion over nature in the Hebrew Bible. *Annals of the Association of American Geographers* 79: 214–32

Kay, J. and C.J. Brown (1985) Mormon beliefs about land and natural resources, 1847–1877. *Journal of Historical Geography* 11: 253–67

Kay, J.H. (1982) Last of the Shakers. *Historic Preservation* 34; 14–21

Keely, C.B. (1989) The Catholic Church and the integration of immigrants. *Migration World* 17: 30–3

Kennedy, L. (1978) The Roman Catholic Church and economic growth in nineteenth century Ireland. *Economic and Social Review* 10(1); 45–60

Kent, R.B. and R.J. Neugebauer (1990) Identification of ethnic settlement regions: Amish-Mennonites in Ohio. *Rural Sociology* 55; 425–41

Kephart, W.M. (1950) Status after death. *American Sociological Review* 15; 640–50

Kessner, T. (1977) *The Golden Door; Italian and Jewish Mobility in New York City 1880–1915*. New York, Oxford University Press

King, R. (1972) The pilgrimage to Mecca; some geographical and historical aspects. *Erdkunde* 26; 61–73

Kipnis, B.A. and I. Schnell (1978) Changes in the distribution of Arabs in mixed Jewish-Arab cities in Israel. *Economic Geography* 54(2); 168–80

Kirk, W. (1975) The role of India in the diffusion of early cultures. *Geographical Journal* 141; 19–34

Kish, G. (ed.) (1978) *A Source Book in Geography*. Boston, Harvard University Press

Kitching, C. (1986) Re-roofing old St Paul's Cathedral, 1561–66. *London Journal* 12; 123–33

Klimeit, H.J. (1975) Spatial orientation in mythical thinking as exemplified in Ancient Egypt: consideration toward a geography of religion. *History of Religions* 14; 266–81

Kling, Z. (1989) Sarawak Malay culture. *Sarawak Museum Journal, Special Issue* 40 (Part 2); 1–43 (in Malaysian)

Kliot, N. (1989) Accommodation and adjustment to ethnic demands; the Mediterranean framework. *Journal of Ethnic Studies* 17; 45–70

Kniffen, F.B. (1967) Necrogeography in the United States. *Geographical Review* 57; 426–7

Knippenberg, H., C.M. Stoppelenburg and H.H. Van der Wusten (1989) The zone of orthodox-protestants in the Netherlands, 1920–1985/86. *Geografisch Tijdschrift* 23; 12–22

Knudsen, J.P. (1986) Culture, power and periphery – the Christian lay movement in Norway. *Norsk Geografisk Tidsskrift* 40; 1–14

Kondo, M. (1991) The formation of sacred places as a factor of the environmental preservation: the case of Setonaikai (Inland Sea) in Japan. *Marine Pollution Bulletin* 23; 649–52

Kong, L. (1990) Geography and religion; trends and prospects. *Progress in Human Geography* 14; 355–71

Kumar, N. (1987) The Mazars of Banaras; a new perspective on the city's sacred geography. *National Geographical Journal of India* 33; 263–7

Kus, J.S. (1979) Peruvian religious truck names. *Names; Journal of the American Name Society* 27(3); 179–87

Laatsch, W.G. and C.F. Calkins (1986) The Belgian roadside chapels of Wisconsin's door peninsula. *Journal of Cultural Geography* 7; 117–28

Labaki, G.T. (1989) Struggle of the Maronite. *Migration World* 17: 24–7

LaFreniere, G.L. (1985) World views and environmental ethics. *Environmental Review* 9; 307–22

Lai, C.D. (1974) A Feng Shui model as a location index. *Annals of the Association of American Geographers* 64; 506–13

Lamme, A.J. (1971) From Boston in one hundred years; Christian Science, 1970. *Professional Geographer* 23; 329–32

Landing, J. (1969) Geographic models of the Old Order Amish settlements. *Professional Geographer* 21; 238–43

Landing, J. (1972) The Amish, the automobile and social interaction. *Journal of Geography* 71; 52–7

Landing, J. (1982) A case study in the geography of religion; the Jehovah's Witnesses in Spain, 1921–1946. *Bulletin, Association of North Dakota Geographers* 32; 42–7

Lass, H. (1987) Songhai. *Africa Insight* 17; 216–17

Lassere, P. (1930) Lourdes; étude géographique. *Revue Géographique des Pyrenees et du Sud-Ouest* 1; 5–41

Le Bras, G. (1945) La géographie religieuse. *Annales d'Histoire Sociale* 8; 87–112

Lemon, J.T. (1966) The agricultural practices of national groups in eighteenth-century South-eastern Pennsylvania. *Geographical Review* 56; 467–96

Leong, G.C. and G.C. Morgan (1973) *Human and Economic Geography*. London, Oxford University Press

Levine, G.J. (1986) On the geography of religion. *Transactions of the Institute of British Geographers* 11: 428–40

Lewis, P. (1980) When America was English. *Geographical Magazine* 52(5); 342–8

Lewthwaite, G. (1972) Geography; pp. 175–84 in R.W. Smith (ed.) *Christ and the Modern Mind*. Leicester, Inter-Varsity Press.

Ley, D. (1974) The City and Good and Evil; reflections on Christian and Marxist interpretations. *Antipode* 6; 66–74

Ley, D. (1980) *Geography Without Man; a Humanistic Critique*. University of Oxford, School of Geography, Research Paper 24

Ley, D. (1981) Cultural/humanistic geography. *Progress in Human Geography* 5; 249–57

Ley, D. (1982) Rediscovering man's place. *Transactions of the Institute of British Geographers* 7; 248–53

Ley, D. and M. Samuels (eds) (1978) *Humanistic Geography; Prospects and Problems*. Croom Helm, London

Lifshetz, R. (1976) Inspired planning; Mormon and Fourierist communities in the nineteenth century. *Landscape* 20; 29–35

Ling, T. (1980) Buddhist values and development problems; a case study of Sri Lanka. *World Development* 8(7–8); 577–86

Little, D. (1981) Land use and the common good: religious backgrounds. *Environmental Education and Information* 1; 209–23

Livingstone, D.N. (1983) Environmental theology; prospect in retrospect. *Progress in Human Geography* 7; 133–40

Livingstone, D.N. (1984a) The science of history and the history of geography: interactions and implications. *History of Science* 22; 271–302

Livingstone, D.L. (1984b) Natural theology and neo-Lamarckism: the changing context of nineteenth-century geography in the United States and Great Britain. *Annals of the Association of American Geographers* 74; 9–28

Livingstone, D.N. (1988) Science, magic and religion: a contextual reassessment of geography in the sixteenth and seventeenth centuries. *History of Science* 26; 271–302

Livingstone, D.N. (1990) Geography, tradition and the Scientific Revolution: an interpretative essay. *Transactions, Institute of British Geographers* 15; 359–73

Louder, D.R. (1975) A simulation approach to the diffusion of the Mormon Church. *Proceedings of the Association of American Geographers* 7; 126–30

MacDonald, K. (1990) Mechanisms of sexual egalitarianism in Western Europe. *Ethology & Sociobiology* 11; 195–237

McGrath, F. (1989) Characteristics of pilgrims to Lough Derg. *Irish Geography* 22; 44–7

McKay, J. (1985) Religious diversity and ethnic cohesion: a three generational analysis of Syrian–Lebanese Christians in Sydney. *International Migration Review* 19; 318–34

McKee, A. (1991) A Christian commentary on LDC debt. *International Journal of Social Economics* 18; 25–36

McLoughlin, B. (1958) The Hejaz railroad. *Geographical Journal* 124; 282–3

Maclellan, D. (1983) The ecumenical start of New France in Arcadia. *Canadian Geographic* 103: 66–73

Maier, E. (1975) Torah as movable territory. *Annals of the Association of American Geographers* 65; 18–23

Malshe, P.T. and S.K. Ghode (1989) Pandharpur: a study in geography of religion. *Transactions of the Institute of Indian Geographers* 11; 46–62

Malvido, E. (1990) Migration patterns of the novices of the Order of San Francisco in Mexico City, 1649–1749; pp. 182–92 in D.J. Robinson (ed.) *Migration in Colonial Spanish America*. New York, CUP

Mamoria, C.B. (1956) Religious composition of population in India. *The Modern Review* (September); 189–98

Manyo, J.T. (1983) Italian-American yard shrines. *Journal of Cultural Geography* 4; 119–25

Martin, B.S. (1988) The impact of Mennonite settlement on the cultural landscape of Kansas. MA thesis, Department of Geography, Kansas State University

Martin, D. (1973) The secularisation question. *Theology* (Feb); 81–7

Martin, R.B. (1989) Faith without focus; neighbourhood transition and religious change in inner-city Vancouver. MA thesis, University of British Columbia, Department of Geography.

Mathisen, J.A. (1989) Twenty years after Bellah; whatever happened to American Civil Religion. *Sociological Analysis* 50; 129–46

Meine, F.J. (ed.) (1957) *The Consolidated–Webster Encyclopedic Dictionary*. Chicago, Consolidated Book Publishers

Meinig, D.W. (1965) The Mormon culture region; strategies and patterns in the geography of the American West, 1847–1964. *Annals of the Association of American Geographers* 55; 191–220

Melton, J.G. (ed.) (1989) *The Encyclopedia of American Religions*. Detroit, Michigan, Gale Research Inc

Messerschmidt, D.A. (1989a) The Hindu pilgrimage to Muktinath, Nepal. Part 1. Natural and supernatural attributes of the sacred field. *Mountain Research and Development* 9; 89–104

Messerschmidt, D.A. (1989b) The Hindu pilgrimage to Muktinath, Nepal. Part 2. Vaishnava devotees and status reaffirmation. *Mountain Research and Development* 9; 105–118

Mewett, P.G. (1985) Sabbath observance and the social construction of religious belief in a Scottish Calvinist community. *Studies in Third World Societies* 26: 183–200

Meyer, J.W. (1975) Ethnicity, theology and immigrant church expansion. *Geographical Review* 65; 180–97

Milbank, J. (1990) *Theology and Social Theory; Beyond Secular Reason*. Oxford, Blackwell

Miller, R. (1989) Legislating for fair employment: the Fair Employment (Northern Ireland) Bill, 1988. *Journal of Social Policy* 18: 253–64

Mills, W.J. (1982) Metaphorical vision: changes in Western attitudes to the environment. *Annals of the Association of American Geographers* 72: 237–53

Milspaw, Y. (1980) Plain walls and little angles; pioneer churches in central Pennsylvania. *Pioneer America* 12(2); 76–96

Mitford, J. (1963) *The American Way of Death.* New York, Simon & Schuster

Moberg, D.O. (1982) The salience of religion in everyday life; selected evidence from survey research in Sweden and America. *Sociological Analysis* 43; 205–17

Mook, M.A. (1957) The Amish and their land. *Landscape* 6; 24–9

Morrill, R. (1984) The responsibility of geography. *Annals of the Association of American Geographers* 74; 1–8

Morrill, R.L. and J.M. Dormitzer (1979) *The Spatial Order.* Massachusetts, Duxbury Press

Morris, C.T. and I. Adelman (1980) The religious factor in economic development. *World Development* 8(7–8); 491–501

Mosher, W.D. and C. Goldscheider (1984) Contraceptive patterns of religious and racial groups in the United States, 1955–76. *Studies in Family Planning* 15: 101–11

Moss, R.P. (1973) *Earth in our Hands.* Leicester, Inter-Varsity Press

Moss, R.P. (1975) Responsibility in the use of nature – I. *Christian Graduate* 28; 69–80

Moss, R.P. (1976) Responsibility in the use of nature – II. *Christian Graduate* 29; 5–14

Moss, R.P. (1978) Environmental problems and the Christian ethic; pp. 63–86 in C.F.H. Henry (ed.) *Horizons of Science.* New York, Harper & Row

Moss, R.P. (1985) The ethical underpinnings of man's management of nature. *Faith and Thought* 111; 23–56

Mukerjee, R. (1961) Ways of dwelling in the communities of India; pp. 390–401 in G.A. Theodorson (ed.) *Studies in Human Ecology.* New York, Harper & Row

Muller, C. (1990) Religious observance in Lower Normandy. *Etudes Normandes* 4; 41–63

Murphy, A.B. (1989) Territorial policies in multiethnic states. *Geographical Review* 79; 410–21

Murphy, Francis X. (1981) Catholic perspectives on population issues. *Population Bulletin* 35; 43 pp.

Murvar, V. (1975) Toward a sociological theory of religious movements. *Journal for the Scientific Study of Religions* 14; 229–56

Myres, J.N.L. (1978) The origin of the Jersey parishes; some suggestions. *Bulletin Société Jersiaise* 22 (2); 163–75

Nakabayashi, I. (1989) Urban structure of the Islamic city and its modern trans-formation; a case study of Aleppo, Syria. *Geographical Reports – Tokyo Metropolitan University* 24; 1–14

Nakagawa, T. (1988) Cemetery forms of Ascension parish, Louisiana; a necrogeography. *Tsukaba Studies in Human Geography* 12; 113–37 (in Japanese)

Nakagawa, T. (1989) Spatial variation of Ascension Parish cemeteries, Louisiana. *Annual Report, University of Tsukuba, Institute of Geoscience* 15; 4–9

Nakagawa, T. (1990a) Louisiana cemeteries as cultural artifacts. *Geographical Review of Japan, Series B* 63; 139–55

Nakagawa, T. (1990b) Cemetery landscape evolution of a Japanese rural community. *Annual Report – University of Tsukuba, Institute of Geoscience* 16; 8–12

Nakagawa, T. (1990c) Grave structures and decorations of Louisiana cemeteries. *Tsukuba Studies in Human Geography* 14; 145–68 (in Japanese)

Nash, M. (1980) Islam in Iran; turmoil, transformation or transcendence? *World Development* 8(7–8); 555–61

Naughton, T. and B. Shanmugam (1990) Interest-free banking; a case study of Malaysia. *Quarterly Review – National Westminster Bank* (February); 16–32

Newman, D. (1985) Integration and ethnic spatial concentration: the changing distribution of the Anglo-Jewish community. *Transactions of the Institute of British Geographers* 10; 360–76

Newman, D. (1986) Culture, conflict and cemeteries: Lebenstraum for the dead. *Journal of Cultural Geography* 7; 99–115

Newman, W.M. and P.L. Halvorson (1979) American Jews; patterns of geographic distribution and change, 1952–1971. *Journal for the Scientific Study of Religion* 18; 183–93

Newman, W.M. and P.L. Halvorson (1984) Religion and regional culture; patterns of concentration and change among American religious denominations, 1952–1980. *Journal for the Scientific Study of Religion* 23; 304–15

Nicolas, G. (1989) Humanism and world views in geography. *Cahiers de Géographie du Quebec* 33: 379–85

Nitz, H.J (1983) The Church as colonist: the Benedictine Abbey of Lorsch and planned *Waldhufen* colonisation in the Odenwald. *Journal of Historical Geography* 9: 105–26

Noack, E.F. (1979) Caravan route to the Muztagh Karakoram. *Geographical Magazine* 52; 132–39

Noble, A.G. (1986) Landscape of piety/landscape of profit: the Amish-Mennonite and derived landscapes of northeastern Ohio. *East Lakes Geographer* 21: 34–48

Noble, A.G., A.K. Dutt and P. Vishnukumari (1987) Daily and diurnal fluctuation in the attendance patterns of the Meenakshi Temple, Madurai, India; pp. 290–4 in V.S. Datye *et al.* (eds) *Explorations in the Tropics.* University of Poona

Nolan, M.L. (1983) Irish pilgrimage: the different tradition. *Annals, Association of American Geographers* 73: 421–38

Nolan, M.L. (1986) Pilgrimage traditions and the nature mystique in western European culture. *Journal of Cultural Geography* 7; 5–20

Nolan, M.L. (1987a) A profile of Christian pilgrimage shrines in Western Europe. *National Geographical Journal of India* 33; 229–38

Nolan, M.L. (1987b) Christian pilgrimage shrines in Western Europe and India; a preliminary comparison. *National Geographical Journal of India* 33; 370–8

Nolan, M.L. and S. Nolan (1989) *Religious Pilgrimage in Modern Western Europe.* Chapel Hill, University of North Carolina Press

Norris, R.E. (1982) San Antonio after the Alamo. *Geographical Magazine* (April); 193–6

Nyrop, R.F. (ed.) (1979) *Israel; a Country Study.* Washington, DC, The American University

Oda, M. (1989) The formation and its meaning of the 75-sacred-place view in the Omine Sacred Mountain Area. *Jimbun Chiri/Human Geography, Kyoto* 41; 512–28 (in Japanese)

O'Flanagan, P. (1988) Urban minorities and majorities; Catholics and Protestants in Munster towns c. 1659–1850; pp. 124–48 in W.J. Smyth and K. Whelan (eds) *Common Ground; Essays on the Historical Geography of Ireland.* Cork, Cork University Press

Olliver, A. (1989) Christian Geographers' Fellowship conference report. *Area* 21; 106–8

Onodera, A. (1990) Patterns and changes in pilgrim routes to Ise shrine; a study based on travellers' records left in the Kanto district. *Tsukuba Studies in Human Geography* 14; 231–55 (in Japanese)

Oomen, T.K. (1990) State and religion in multi-religious nation-states: the case of South Asia. *South Asia Journal* 4; 17–33

Orland, B. and V.J. Bellafiore (1990) Development directions for a sacred site in India. *Landscape and Urban Planning* 19; 181–96

Osborne, R.D. and R.J. Cormack (1986) Unemployment and religion in Northern Ireland. *Economic & Social Review* 17: 215–55

Pacione, M. (1990) The ecclesiastical community of interest as a response to urban poverty and deprivation. *Transactions of the Institute of British Geographers* 15; 193–204

Pacione, M. (1991) The Church Urban Fund; a religio-geographical perspective. *Area* 23; 101–10

Palmer, H. and T. Palmer (1982) The religious ethic and the spirit of immigration: the Dutch in Alberta. *Prairie Forum* 7: 237–65

Park, C.C. (1984) Seat of the Aztecs. *Geographical Magazine* (August); 396–7

Park, C.C. (1992) *Caring for Creation: a Christian Way Forward*. London, Marshall Pickering

Parsons, T. (1966) Religion in a modern pluralistic society. *Review of Religious Research* 7; 125–46

Pattison, W.D. (1955) The cemeteries of Chicago; a phase of land utilization. *Annals of the Association of American Geographers* 45; 245–57

Perrin, R.D. (1989) American religion in the post-Aquarian age; values and demographic factors in church growth and decline. *Journal for the Scientific Study of Religion* 28; 75–89

Pieper, J. (1979) The monastic settlements of the Yellow Church in Ladakh, central places in a nomadic habitat. *GeoJournal* 1; 41–54

Piggott, C.A. (1980) A geography of religion in Scotland. *Scottish Geographical Magazine* 96(3); 130–40

Pillsbury, R. (1971) The religious geography of Pennsylvania; a factor analytic approach. *Proceedings of the Association of American Geographers* 3; 130–4

Pollins, H. (1989) Immigrants and minorities – the outsiders in business. *Immigrants and Minorities* 8; 252–70

Poole, M. and F.W. Boal (1973) Religious residential segregation in Belfast in mid-1969; a multi-level analysis; pp. 1–40 in B.D. Clarke and M.B. Gleeve (eds) *Social Patterns in Cities*. Institute of British Geographers Special Publication No. 5. London, IBG

Portugali, J. (1991) Jewish settlement in the occupied territories: Israel's settlement structure and the Palestinians. *Political Geography Quarterly* 10; 26–53

Pratt, V. (1970) *Religion and Secularization*. London, Macmillan

Price, L.W. (1966) Some results and implications of a cemetery study. *Professional Geographer* 18; 201–7

Pringle, D. (1987) The planning of some pilgrimage churches in Crusader Palestine. *World Archaeology* 18; 341–62

Pringle, D. (1990) Separation and integration: the case of Ireland; ch. 8 in M. Chisholm and D.M. Smith (eds) *Shared Space, Divided Space*. London, Unwin Hyman.

Prorok, C.V. (1986) The Hare Krishna's transformation of space in West Virginia. *Journal of Cultural Geography* 7; 129–40

Prorok, C.V. (1987) The Canadian Presbyterian Mission in Trinidad: John Morton's work among East Indians. *The National Geographical Journal of India* 33; 253–62

Prorok, C.V. (1988) Hindu temples in Trinidad; a cultural geography of religious structures and ethnic identity. Ph.D. thesis, Department of Geography and Anthropology, Louisiana State University

Prorok, C.V. (1991) Evolution of the Hindu temple in Trinidad. *Caribbean Geographer* 3; 73–93

Prothero, R.M. (1971) Nomads, pilgrims and commuters. *Geographical Magazine* 43; 271–8

Proudfoot, L.J. (1983) The extension of parish churches in medieval Warwickshire. *Journal of Historical Geography* 9; 231–46

Pullinger, D.J. (ed.) (1989) *With Scorching Heat and Drought? A Report on the Greenhouse Effect*. Church of Scotland Society, Religion and Technology Project. St Andrew Press, Edinburgh

Qureshi, S. (1980) Islam and development; the Zia regime in Pakistan. *World Development* 8(7–8); 563–75

Raitz, K.B. (1978) Ethnic maps of North America. *Geographical Review* 68; 335–50

Ravenhill, W.R. (1985) The making and mapping of the parish; the Cornish experience. *Cambria* 12; 55–72

Rele, J.R. and T. Kanithar (1977) Fertility differentials by religion in greater Bombay; role of explanatory variables, pp. 371–84 in L.T. Ruzicka (ed.) *The Economic and Social Supports for High Fertility*. Canberra, Australian National University

Rinschede, G. (1986) The pilgrimage town of Lourdes. *Journal of Cultural Geography* 7; 21–33

Rinschede, G. (1989) The pilgrimage town of Lourdes. *National Geographical Journal of India* 35; 379–421

Rinschede, G. (1990) Religious tourism. *Geographische Rundscau* 42; 14–20 (in German)

Rinschede, G. and A. Sievers (1987) The pilgrimage phenomenon in socio-geographical research. *National Geographical Journal of India* 33; 213–17

Robbins, T. and D. Anthony (1979) Sociology of contemporary religious movements. *Annual Review of Sociology* 5; 75–89

Rogers, A. (1992) Key themes and debates; pp. 233–52 in A. Rogers *et al.* (eds) *The Student's Companion to Geography*. Oxford, Blackwell

Romann, M. (1990) Territory and demography; the case of the Jewish–Arab national struggle. *Middle Eastern Studies* 26; 371–82

Roof, W.D. and J.L. Roof (1984) Review of the polls; images on God among Americans. *Journal for the Scientific Study of Religion* 23; 201–5

Rooney, J.F., W. Zelinsky and D.R. Louder (eds) (1982) *This Remarkable Continent; an Atlas of United States and Canadian Society and Cultures*. College Station, Tex., Texas A&M University Press

Rose, C. (1992) Altared images. *The Guardian* (29 July)

Rose, G.S. (1986) Quakers, North Carolinians and blacks in Indiana's settlement patterns. *Journal of Cultural Geography* 7; 35–48

Rowley, G. (1984) Divisions in a holy city. *Geographical Magazine* 56: 196–202

Rowley, G. (1989) The centrality of Islam; space, form and process. *GeoJournal* 18; 351–9

Rowley, G. (1990) The Jewish colonization of the Nablus region – perspectives and continuing developments. *GeoJournal* 21; 349–62

Rowley, G. and S.A. El-Hamdan (1977) Once a year in Mecca. *Geographical Magazine* 49; 753–9

Rowley, G. and S.A. El-Hamdan (1978) The pilgrimage to Mecca; an explanatory and predictive model. *Environment and Planning* A 10; 1053–71

Rowley, J. (ed.) (1979) The Muslim world. *People* 6(4); 3–27

Roy, B.K. (1987) Census count of religions of India (1901–1981) and contemporary issues. *The National Geographical Journal of India* 33; 239–52

Russell, G.W. (1975) The view of religions from religious and non-religious perspectives. *Journal for the Scientific Study of Religion* 14; 129–38

Rutter, E. (1929) The Muslim pilgrimage. *Geographical Journal* 74; 271–3

Sack, R.D. (1976) Magic and space. *Annals of the Association of American Geographers* 66; 309–22

Sack, R.D. (1986) *Human Territoriality: its Theory and History*. London, CUP

Safak, A. (1980) Urbanism and family residence in Islamic Law. *Ekistics* 47 (280); 21–5

Saharan, R. (1989) Customs and rituals of the Malays of Sarawak. *Sarawak Museum Journal, Special Issue* 40, Part 2; 45–61 (in Malaysian)

Saleh, H.A.K. (1990) Jewish settlement and its economic impact on the West Bank, 1967–1987. *GeoJournal* 21; 337–48

Saltini, A. (1978) The countryside in Europe. *Town and Country Planning* 46(7–8); 366–70

Sanger, C. (1991) The ascent of the Aztecs. *Geographical Magazine* 63; 20–3

Sauer, C.O. (1963) *Land and Life*. Berkeley, Calif., University of California Press

Scheffel, D. (1983) Modernization, mortality and Christianity in northern Labrador. *Current Anthropology* 24; 523–24

Schnell, I. (1990) The Israeli Arabs: the dilemma of social integration in development. *Geographische Zeitschrift* 78; 78–92

Schoen, R. and B. Thomas (1990) Religious intermarriage in Switzerland, 1969–72 and 1979–82. *European Journal of Population* 6; 359–76

Schwartzberg, J.E. (1965) The distribution of selected castes in the North Indian Plain. *Geographical Review* 55; 477–96

Seeman, A.L. (1938) Communities in the Salt Lake Basin. *Economic Geography* 14; 300–8

Semple, E.C. (1911) *Influences of Geographical Environment*. New York, Henry Holt

Semyonov, M. (1988) Bi-ethnic labor markets, mono-ethnic labor markets, and socio-economic inequality. *American Sociological Review* 53; 256–66

Sernett, M.C. (1981) *Geographic Considerations in Afro-American Religious History: Past Performance, Present Problems and Future Hopes*. Syracuse University, Department of Geography, Discussion Paper 69, 29 pp.

Shair, I. (1979) Review of the English literature on spatial pilgrim circulation. *Journal – College of Arts, University of Riyadh* 6: 1–5

Shair, I.M. and P.P Karan (1979) Geography of the Islamic pilgrimage. *GeoJournal* 3; 599–608

Shilhav, Y. (1983) Principles for the location of synagogues; symbolism and functionalism in a spatial context. *Professional Geographer* 35; 324–9

Shortridge, J.R. (1976) Patterns of religion in the United States. *Geographical Review* 66; 420–34

Shortridge, J.R. (1977) A new regionalisation of American religion. *Journal for the Scientific Study of Religion* 16; 143–53

Shortridge, J.R. (1978) The pattern of American Catholicism, 1971. *Journal of Geography* 77; 56–60

Shortridge, J.R. (1982) Religion, ch. 8 in J.F. Rooney, W. Zelinsky and D.R. Louder (eds) *This Remarkable Continent* College Station, Tex., Texas A&M University Press

Siddiqi, A.H. (1987) Al-Muqaddasi's treatment of regional geography. *International Journal of Islamic and Arabic Studies* 4; 1–13

Sievers, A. (1987) The significance of pilgrimage tourism in Sri Lanka (Ceylon). *National Geographical Journal of India* 33; 430–47

Sigelman, L. (1977) Multi-nation surveys of religious beliefs. *Journal for the Scientific Study of Religion* 16; 289–94

Silverman, M. (1978) Class, kinship and ethnicity; patterns of Jewish upward mobility in Pittsburgh, Pennsylvania. *Urban Anthropology* 7; 25–44

Simard, J. (1984) The next world as territory. *Cahiers de Géographie du Québec* 28: 303–10

Simoons, F.J. (1961) *Eat not this Flesh; Food Avoidances in the Old World.* Madison, University of Wisconsin Press

Simpson, J. (1986) God's visible judgements: the Christian dimensions of landscape legends. *Landscape History* 8, 53–8

Simpson-Housley, P. (1978) Hutterian religious ideology, environmental perception, and attitudes toward agriculture. *Journal of Geography* 77; 145–8

Singh, R.P.B. (1987a) Emergence of the Geography of Belief Systems (GBS), and a search for identity in India. *National Geographical Journal of India* 33; 184–204

Singh, R.P.B. (1987b) The pilgrimage manada of Varanasi (Kasi); a study in sacred geography. *National Geographical Journal of India* 33; 493–524

Sitwell, O.F.G. (1981) Elements of the cultural landscape as figures of speech. *Canadian Geographer* 25; 167–80

Sitwell, O.F.G. and O.S.E. Bilash (1986) Analysing the cultural landscape as a means of probing the non-material dimensions of reality. *Canadian Geographer* 30: 132–45

Sitwell, O.F.G. and G.R. Latham (1979) Behavioural geography and the cultural landscape. *Geografiska Annaler* 61B; 51–63

Sivignon, M. (1981) Concerning cultural geography. *Espace Géographique* 10, 270–4 (in French)

Smart, N. (1989) *The World's Religions: Old Traditions and Modern Transformations.* London, Cambridge University Press

Smith, C.G. (1968) Israel after the June war. *Geography* 53; 315–19

Social and Community Planning Research (1992) *British Social Atitudes 1992.* Dartmouth Publishing

Soderberg, S. (1991) The Moslems on Malmo. *Nord Revy* 2; 12–15 (in Swedish)

Sopher, D.E. (1964) Landscape and seasons. *Landscape* 13; 14–19

Sopher, D.E. (1967) *Geography of Religions.* New York, Prentice-Hall

Sopher, D.E. (1968) Pilgrim circulation in Gujarat. *Geographical Review* 58; 392–425

Sopher, D.E. (1981) Geography and religion. *Progress in Human Geography* 5; 510–24

Sopher, D.E. (1987) The message of place in Hindu pilgrimage. *National Geographical Journal of India* 33; 353–69

Spate, O.H.K. (1943) Geographical aspects of the Pakistan scheme. *Geographical Journal* 111; 125–36

Spate. O.H.K. (1948a) The partition of India and the prospects of Pakistan. *Geographical Review* 38; 5–29

Spate, O.H.K. (1948b) The boundary award in the Punjab. *Asiatic Review* XLIV; 1–8

Spilka, B., P. Armatas and J. Nussbaum (1964) The concept of God; a factor analytic study. *Review of Religious Research* 16; 154–65

Stanislawski, D. (1975) Dionysus westward; early religion and the economic geography of wine. *Geographical Review* 65; 427–44

Stark, R. and W.S. Bainbridge (1979) Of churches, sects and cults; preliminary concepts for a theory of religious movements. *Journal for the Scientific Study of Religion* 18; 117–31

Stauffer, R.E. (1975) Bellah's civil religion. *Journal for the Scientific Study of Religion* 14; 390–4

Stinner, D.H. *et al.* (1989) In search of traditional farm wisdom for a more sustainable agriculture: a study of Amish farming and society. *Agriculture, ecosystems & Environment* 27; 77–90

Stoddard, R.H. (1968) Analysis of the distribution of Hindu holy sites. *National Geographical Journal of India* 14; 148–55

Stoddard, R.H. (1987) Pilgrimages along sacred paths. *National Geographical Journal of India* 33; 448–56

Stoddart, D.R. (1966) Darwin's impact on geography. *Annals of the Association of American Geographers* 56; 683–98

Stoddart, D.R. (1987) To reclaim the high ground; geography for the end of the century. *Transactions of the Institute of British Geographers* 12; 327–36

Strahler, A.N. (1983) Toward a broader perspective in the evolution-creationism debate. *Journal of Geological Education* 31; 87–94

Straight, S.M. (1989) Russian Orthodox churches in Alaska. *Geographical Bulletin – Gamma Theta Epsilon* 31; 18–28

Stump, R.W. (1984a) Regional divergence in religious affiliation in the United States. *Sociological Analysis* 45; 283–99

Stump, R.W. (1984b) Regional migration and religious commitment in the United States. *Journal for the Scientific Study of Religion* 23; 292–303

Stump, R.W. (1985) Toward a geography of American civil religion. *Journal of Cultural Geography* 5: 87–95

Stump, R.W. (1986a) The geography of religion – Introduction. *Journal of Cultural Geography* 7; 1–3

Stump, R.W. (1986b) Patterns in the survival of Catholic national parishes, 1940–1980. *Journal of Cultural Geography* 7; 77–97

Stump, R.W. (1987) Regional variations in denominational switching among white Protestants. *Professional Geographer* 39; 438–49

Stump, R.W. (1991) Spatial implications of religious broadcasting; stability and change in patterns of belief, pp. 354–75 in S.D. Brunn and T.R. Leinbacj (eds) *Collapsing Space and Time; Geographic Aspects of Telecommunications and Information*. New York, Harper Collins Academic

Sutton, K. (1990) Algeria's vineyards: an Islamic dilemma and a problem of decolonisation. *Journal of Wine Research* 1; 101–20

Suzuki, H. (1981) *The Transcendent and Environments: a Historico-Geographical Study of World Religions*. Yokohama, Addis Abeba Sha

Swan, B. (1990) Geography and faith; a personal perspective. *Geography Bulletin* (Spring); 273–91

Sylvester, D. (1967) The church and the geographer. in R.W. Steel and R. Lawton (eds) *Liverpool Essays in Geography*. Liverpool, Longman

Tanaka, H. (1977) Geographic expression of Buddhist pilgrim places on Shikoku Island, Japan. *Canadian Geographer* 21; 111–32

Tanaka, H. (1981) The evolution of pilgrimage as a spatial-symbolic system. *Canadian Geographer* 25; 240–51

Tanaka, H. (1984) Landscape expression of the evolution of Buddhism in Japan. *Canadian Geographer* 28; 240–57

Tatum, C.E. and L.M. Sommers (1975) The spread of the Black Christian Methodist Episcopal Church in the United States, 1870 to 1970. *Journal of Geography* 74; 343–57

Taylor, E.G.R. (1930) *Tudor Geography 1483–1583*. London, Methuen

Taylor, E.G.R. (1934) *Late Tudor and Early Stuart Geography 1583–1650*. London, Methuen

Taylor, J.L. (1990) New Buddhist movements in Thailand; an 'individualistic revolution', reform and political dissonance. *Journal of Southeast Asian Studies* 21; 135–54

Tomkinson, M. (1969) Seaside city for Mecca's pilgrims. *Geographical Magazine* 42; 95–104

Toynbee, A. (1972) The religious background to the present environmental crisis – a viewpoint. *International Journal of Environmental Studies* 3; 141–6

Trepanier, C. (1986) The Catholic Church in French Louisiana; an ethnic institution? *Journal of Cultural Geography* 7(1); 59–75

Tuan, Y.F. (1968a) *The Hydrological Cycle and the Wisdom of God; a theme in geoteleology.* University of Toronto, Department of Geography, Research Publication 1

Tuan, Y.F. (1968b) Discrepancies between environmental attitudes and behaviour; examples from Europe and China. *Canadian Geographer* 12; 176–91

Tuan, Y.F. (1974) *Topophilia; a Study of Environmental Perception, Attitudes and Values.* New York, Prentice-Hall

Tuan, Y.F. (1976) Humanistic geography. *Annals of the Association of American Geographers* 66; 266–76

Tuan, Y.F. (1978) Sacred space; exploration of an idea; pp. 84–99 in K.W. Butzer (ed.) *Dimensions of Human Geography.* University of Chicago, Department of Geography, Research Paper 186.

Tuan, Y.F. (1984) In place, out of place. *Geoscience & Man* 24: 3–10

Turner, H.W. (1980) African independent churches and economic development. *World Development* 8(7–8); 523–33

Turner, V. (1973) The center out there; pilgrim's goal. *History of Religions* 12; 191–230

Tweedie, S.W. (1978) Viewing the Bible Belt. *Journal of Popular Culture* 11; 865–76

Tyler, C. (1990) Spreading the word. *Geographical Magazine* 62; 12–18

Unstead, J.F. and I.G. Elloway (1965) *A World Survey from the Human Aspect.* London, University of London Press.

Van der Mehden, F.R. (1980) Religion and development in South-East Asia. *World Development* 8(7–8); 545–53

Van Poppel, F.W.A. (1985) Late fertility decline in the Netherlands: the influence of religious denomination, socio-economic group and region. *European Journal of Population* 1: 347–73

Vecsey, C. (1980) American Indian environmental religions; pp. 1–37 in C. Vecsey and R.W. Venables (eds) *American Indian Environments: Ecological Issues in Native American History.* New York, Syracuse University Press

Veglery, A. (1988) Differential social integration among first generation Greeks in New York. *International Migration Review* 22; 627–57

Vellenga, D.D. (1985) Racial and ethnic conflict in a Christian missionary community: Jamaican and Swiss-German missionaries in the Basel Mission in the Gold Coast in the mid-nineteenth century. *Studies in Third World Societies* 26: 201–45

Vessels, J. (1980) Fatima; beacon for Portugal's faithful. *National Geographic* 158; 833–39

Vining, J.W. (1982) The presentation of world religions in early geography texts. *Geographical Perspectives* 49: 17–26

Vogeler, I. (1976) The Roman Catholic culture region of central Minnesota. *Pioneer America* 8; 71–83

Wallace, D. (1992) Dreaming of yew. *Geographical Magazine* (February); 40–3

Wallace, I. (1978) Towards a humanized conception of economic geography; pp. 91–108 in D. Ley and M. Samuels (eds) *Humanistic Geography; Prospects and Problems.* Chicago, Maaroufa Press

Wallace, I. (1985) Coherence is not a luxury; pp. 23–32 in D.B. Knight (ed.) *Our Geographic Mosaic; Research Essays in Honour of G.C. Merrill.* Ottawa, Carleton University Press

Wallace, I. (1986) A theological perspective on humanist geography; pp. 30–4 in S. Mackenzie (ed.) *Humanism and Geography.* Discussion Paper 3, Department of Geography, Carleton University, Ottawa.

Walsh, J. (1992) The sword of Islam. *Time* (15 June); 28–32

Walter, J.A. (1985) Order and chaos in landscape. *Landscape Research* 10: 2–8

Warkentin, J. (1959) Mennonite agricultural settlements of Southern Manitoba. *Geographical Review* 49; 359

Waterman, S. (1989) *Jews in an Outer London Borough, Barnet*. Research Paper – Queen Mary & Westfield College, University of London, Department of Geography 1

Waterman, S. and B.A. Kosmin (1986) The distribution of Jews in the United Kingdom. *Geography* 71; 60–5

Waterman, S. and B.A. Kosmin (1988) Residential patterns and processes; a study of Jews in three London boroughs. *Transactions of the Institute of British Geographers* 13; 79–95

Watson, F. (1961) Pilgrims of Badrinath. *Geographical Magazine* 24; 421–8

Weber, M. (1930) *The Protestant Ethic and the Spirit of Capitalism*. London, Allen & Unwin

Weeks, J.R. (1988) The demography of Islamic nations. *Population Bulletin* 43; 54 pp.

Weissbrod, L. (1983) Religion as national identity in a secular society. *Review of Religious Research* 24; 188–205

Welch, M.R. and Baltzell, J. (1984) Geographical mobility, social integration and church attendance. *Journal for the Scientific Study of Religion* 23; 75–91

Wells-Thorpe, J. (1988) The emerging aesthetic – accident or design? *Landscape Research* 13; 19–22

Whelan, K. (1983) The Catholic parish, the Catholic chapel and village development in Ireland. *Irish Geography* 16; 1–15

Whelan, K. (1988) The regional impact of Irish Catholicism 1700–1850; pp. 253–77 in W.J. Smyth and K. Whelan (eds) *Common Ground; Essays on the Historical Geography of Ireland*. Cork, Cork University Press

White, A.G. (1984) *Religious architecture – Islamic: a Selected Bibliography*. Vance Bibliographies, Architecture Series: Bibliography A–1220, 17 pp

White, G.F. (1985) Geographers in a perilously changing world. *Annals of the Association of American Geographers* 75; 10–16

White, L. (1967) The historical roots of our ecological crisis. *Science* (10 March); 1203–7

Whittle, A. (1990) A pre-enclosure burial at Windmill Hill, Wiltshire. *Oxford Journal of Archaeology* 9; 25–8

Wiebe, D. (1980) Places of worship in Afghanistan and their value in foreign trade. *Afghanistan Journal* 7(3); 97–108 (in German)

Wilber, C.K. and K.P. Jameson (1980) Religious values and social limits to development. *World Development* 8(7–8); 467–79

Wilhelmy, H. (1990) *Bhutan; land of monasteries*. Munich, Beck'sche, (in German)

Williams, L.B. and B.G. Zimmer (1990) The changing influence of religion on US fertility; evidence from Rhode Island. *Demography* 27; 475–81

Williams, R.B. (1987) Negotiating the tradition: religious organisations and Gujarati identity in the United States. *Studies in Third World Societies* 39: 25–38

Wilson, R. (1989) The Islamic Development Bank's role as an aid agency for Moslem countries. *Journal of International Development* 1; 444–66

Wilson, R. (ed.) (1990) *Islamic Financial Markets*. London, Routledge.

Wolf, E.R. (1951) The social organisation of Mecca and the origins of Islam. *Southwestern Journal of Anthropology* 7; 329–53

Wright, G.A. (1990) On the interior attached ditch enclosures of the Middle and Upper Ohio Valley. *Ethnis* 55; 92–107

Wright, J.K. (1947) *Terrae Incognitae*; the place of the imagination in geography. *Annals of the Association of American Geographers* 37; 1–15

Wuthnow, R. (1976) A longitudinal, cross-national indicator of societal religious commitment. *Journal for the Scientific Study of Religion* 16; 87–99

Wynne-Hammond, C. (1979) *Elements of Human Geography*. London, George Allen & Unwin

Yates, W.N. (1983) 'Bells and smells': London, Brighton and south coast religion reconsidered. *Southern History* 5: 122–53

Yiftachel, O. (1991) Industrial development and Arab–Jewish economic gaps in the Galilee region, Israel. *Professional Geographer* 3; 163–79

Yinger, J.M. (1969) A structural examination of religion. *Journal for the Scientific Study of Religion* 8; 88–99

Yinger, J.M. (1977) A comparative study of the substructures of religion. *Journal for the Scientific Study of Religion* 16; 67–86

Yoon, H-K. (1976) Geomantic relationships between culture and nature in Korea. *Dissertation Abstracts International A*, 77–4661; 247 pp.

Young, F.W. (1960) Graveyards and social structure. *Rural Sociology* 23; 446–50

Zelinsky, W. (1961) An approach to the religious geography of the United States; patterns of church membership in 1952. *Annals of the Association of American Geographers* 51; 139–93

Zelinsky, W. (1973) *The Cultural Geography of the United States*. New York, Prentice-Hall

INDEX